Mechanik, Werkstoffe und Konstruktion im Bauwesen

Volume 73

Reihe herausgegeben von

Ulrich Knaack, Darmstadt, Germany

Jens Schneider, Darmstadt, Germany

Johann-Dietrich Wörner, Darmstadt, Germany

Stefan Kolling, Gießen, Germany

Institutsreihe zu Fortschritten bei Mechanik, Werkstoffen, Konstruktionen, Gebäudehüllen und Tragwerken. Das Institut für Statik und Konstruktion der TU Darmstadt sowie das Institut für Mechanik und Materialforschung der TH Mittelhessen in Gießen bündeln die Forschungs- und Lehraktivitäten in den Bereichen Mechanik, Werkstoffe im Bauwesen, Statik und Dynamik, Glasbau und Fassadentechnik, um einheitliche Grundlagen für werkstoffgerechtes Entwerfen und Konstruieren zu erreichen. Die Institute sind national und international sehr gut vernetzt und kooperieren bei grundlegenden theoretischen Arbeiten und angewandten Forschungsprojekten mit Partnern aus Wissenschaft, Industrie und Verwaltung. Die Forschungsaktivitäten finden sich im gesamten Ingenieurbereich wieder. Sie umfassen die Modellierung von Tragstrukturen zur Erfassung des statischen und dynamischen Verhaltens, die mechanische Modellierung und Computersimulation des Deformations-, Schädigungs- und Versagensverhaltens von Werkstoffen, Bauteilen und Tragstrukturen, die Entwicklung neuer Materialien, Produktionsverfahren und Gebäudetechnologien sowie deren Anwendung im Bauwesen unter Berücksichtigung sicherheitstheoretischer Überlegungen und der Energieeffizienz, konstruktive Aspekte des Umweltschutzes sowie numerische Simulationen von komplexen Stoßvorgängen und Kontaktproblemen in Statik und Dynamik.

Dunia Abdullah Agha

3D Paper Printing for the Built Environment

Optimization of the Material Behavior & Production Process to Reach Quality Integration and Dimensional Accuracy

Springer Vieweg

Dunia Abdullah Agha
Institut für Statik und Konstruktion
TU Darmstadt
Darmstadt, Germany

Vom Fachbereich 13 – Bau- und Umweltingenieurwissenschaften (Institut für Statik und Konstruktion) der Technischen Universität Darmstadt zur Erlangung des akademischen Grades eines Doktor-Ingenieurs (Dr.-Ing.) genehmigte Dissertation von Dunia Abdullah Agha aus Babel, Irak.
1. Gutachter: Prof. Dr.-Ing. Ulrich Knaack
2. Gutachter: Prof. Dr.-Ing. habil. Stefan Kolling (Technische Hochschule Mittelhessen)
Tag der Einreichung: 31. August 2023
Tag der mündlichen Prüfung: 29. November 2023

ISSN 2512-3238 ISSN 2512-3246 (electronic)
Mechanik, Werkstoffe und Konstruktion im Bauwesen
ISBN 978-3-658-45075-5 ISBN 978-3-658-45076-2 (eBook)
https://doi.org/10.1007/978-3-658-45076-2

This Springer Vieweg imprint is published by the registered company Springer Fachmedien Wiesbaden GmbH, part of Springer Nature.
The registered company address is: Abraham-Lincoln-Str. 46, 65189 Wiesbaden, Germany

Acknowledgement

To all those who have assisted me in the completion of this thesis, I would like to express my deepest gratitude and appreciation.

First and foremost, I am indebted to my supervisor, **Prof. Dr. -Ing. Ulrich Knaack**, for allowing me to be one of his group researchers and giving me the chance to live abroad where I discover my new path and challenge myself to bring what best of me, furthermore thanks for his tenacious guidance, expertise, and encouragement. His valuable insights, constructive feedback, and commitment to my academic growth have been instrumental in shaping the direction and quality of this research.

I extend sincere gratitude to my co. promoter /supervisor **Prof. Dr.-Ing. habil. Stefan Kolling**, for his indispensable support and meticulous guidance throughout the intricate process of structuring and crafting my dissertation. His expertise has been instrumental in shaping and structuring my study outcomes.

I would like to express my sincere gratitude to the **DAAD - Deutscher Akademischer Austauschdienst** / Organisationseinheit ST33 (Kennedyallee 50/D-53175 BONN) for providing me with the opportunity to conduct my research in Germany through their precious financial support. Their investment has made my research journey possible, allowing me to pursue my academic aspirations and make a significant contribution to my field. Special thanks to **Ms. Dagmar Hosseini-Razi** (first responsible sponsor), **Mr. Mostafa Fathy Mohamed,** and **Ms. Elina Kunkel**.

I expand my heartfelt thanks to the members of my examination committee **(Prof. Dr.-Ing. Uwe Rüppel, Prof. rer. nat. Liselotte Schebek, and Prof. Dr.-Ing. Andreas Eichhorn**) for their valuable time, expertise, and insightful comments on my work. Their input and constructive criticism significantly contributed to the refinement and overall strength of this thesis.

I would also like to express my sincere appreciation to the research group and staff at ISMD, TU Darmstadt (**Silke Wallner, Markus Welzel, Robert Akerboom, Nadja Bishara, Philipp Rosendahl, Marcel Hörbert, Miriam Schuster, Frederic Kreplin, and others**). The support and collaboration I received has been invaluable. Their intellectual contributions, technical assistance, and engaging discussions have greatly enriched my research experience.

To my family, I owe an immense debt of gratitude. Their unwavering support, understanding, and encouragement throughout this journey have been a constant source of strength. Their belief in my abilities and their unconditional love have kept me motivated even during challenging times.

Thank you.

Darmstadt, den April 2024 Dunia Abdullah Agha

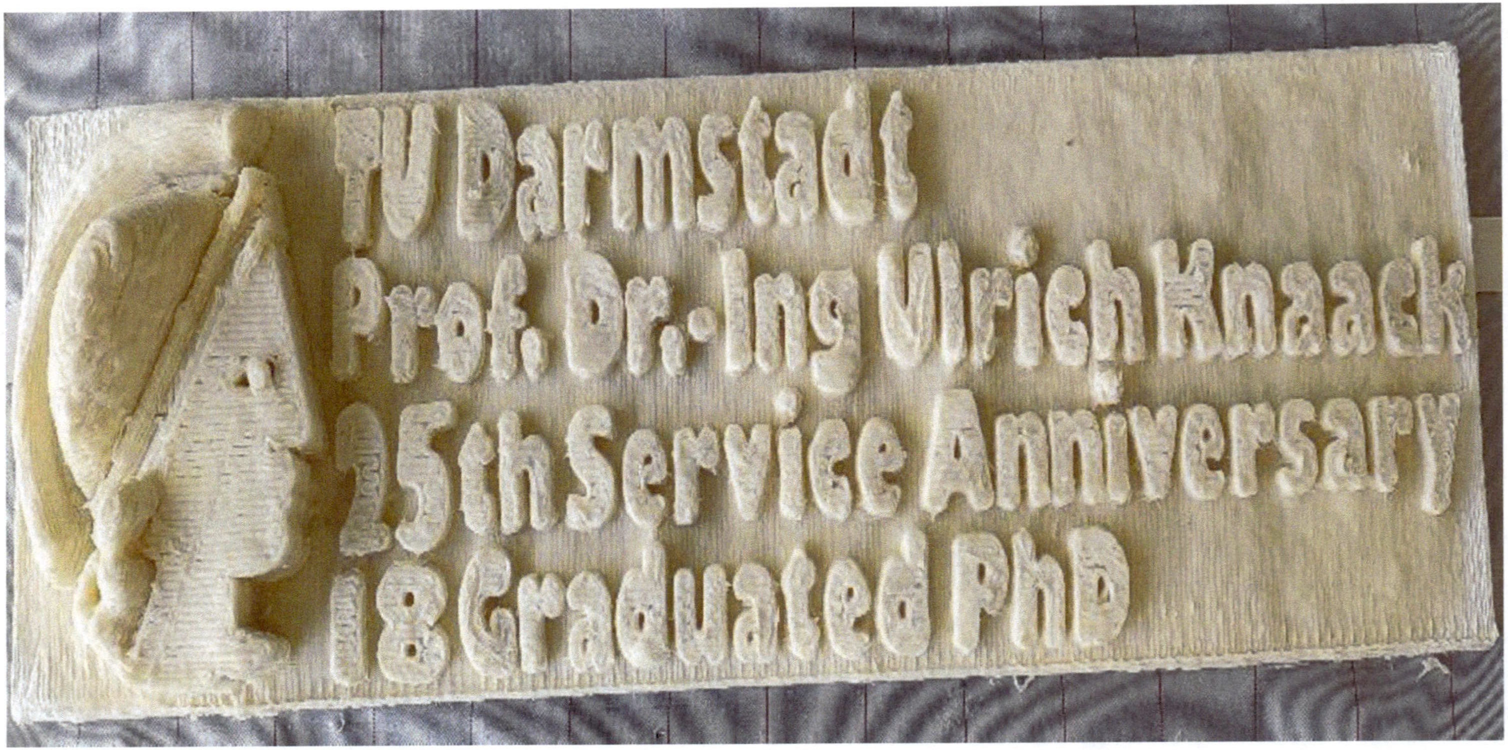

3D printed paper model to congratulate my professor for his 25th service anniversary in 2022.

Dedication

Dedicated to my darlings - Ali and Fahad

Abstract

This research aims to produce and develop products from a new material formulation as bio-based paper material by using additive manufacturing (AM), in precise 3D printing. The approach of producing and developing these products is by optimizing material behavior and production process to reach a high level of quality and accuracy. Potential applications for the built environment of facade engineering are presented which fits the material properties and benefits the most from the complexity provided by the additive manufacturing technology.

The formulation consists of four components with water of ranges (79-68.2%) such as: cellulose of used range (12-16%), carboxymethylcellulose (CMC) of (2.4-6.2%), lecithin of (2.4-4.7%) and a filler of (2.23-15.6%) (such as chalk or undissolved starch or mgso4). Each ingredient has a specific function in the mix. Cellulose is the main structure for the mix. CMC is a hydrogel that provides the adhesive properties of the mix. Lecithin is a fatty substance that improves the extrudability of the compound. Filler increases the number of interparticle contacts and stabilizes the hydrogel, which means improved buildability. The used cellulose is native cellulose fiber (bleached pulp) from different sources (NBHK/NBSK) and with fiber length from 0.2 to 2.1 mm.

The 3D printer used is LUTUM -VormVrij® 3D clay printer version 2.1. The overall printing parameters are air pressure up to (0.3 - 0.85) Mpa, resolution (nozzle diameter of 0.6mm -1.2mm and 1.6mm), and printing speed in the range of (15 – 20) mm/s.

The methodology of the research is to try many mixture formulations to investigate the possible selection of homogeneous pastes with strong potential for AM by extrusion and to achieve buildability (higher height possible while maintaining shape retention and stability). The material would be able to support its own weight and have limited deformation of the printed part during room environmental drying. This was done by focusing on the adjustment and optimization controlled by rheological characterization and printing parameters settings to ensure optimal shape accuracy of 3D printed parts. A printing adjustment guideline and design constraints adapted to the developed paste were proposed.

For nozzle 1.6mm, the best formulation is Cellulose 15%, Aspen type, Water 69.2%, CMC 7.5%, Lecithin 2.8%, and the filler used is Starch of 5.6%.

For nozzle 1.2 mm, the best formulation is using cellulose 15.5%, aspen type, water 71.5%, CMC of 5.8%, lecithin 2.9%, filler used is starch of 4.3%. The nozzle 0.6mm, the best formulation is using cellulose 13.4%, aspen, water 78%, CMC 3.6%, lecithin 2.6%, and filler used is chalk of 2.3%.

The summery results lead to 3D paper printing of products with complex geometries with mechanical properties and a range of capabilities, such as

1. Safety aspect due to use of non-hazardous material,
2. 100% recyclable.
3. Large complex component.
4. Complex part design.
5. Self-supporting material that can carry and hold its own weight while maintaining shape and stability.
6. The buildability of height reached approximately 200 mm without buckling problems or collapsing during the drying process.
7. The wet density has a range of (7.27 - 8.08) kN/m3 and the dry density is (4.02- 5.67) kN/m3.
8. Printing parameters: Air pressure up to 0.3 – 0.85 Mpa
9. Tensile strength up to 4.5 Mpa.
10. Adhesive strength reaches up to 20% of tensile strength.
11. Young's modulus reaches about 6 GPa.
12. Economic aspect of the material and the 1 kg of the material in the range of 10 €/kg.
13. Optimization of the material to achieve the lowest possible shrinkage of the extruded filament, it reached 17.4%.
14. Dimensional accuracy of a 3D printed part after drying achieved 14.8%.

To achieve this goal, this work focused on

- Paste formulations based on short and long cellulose fibers, CMC, lecithin, and filler (indissoluble potato starch, chalk, or Mgso4).
- Room drying.
- Use of 3D printer (LUTUM -VormVrij® 3D clay printer version 2.1).

Keywords: 3D printing, paper, cellulose fibers, bio-based material, 3D paper structure.

Zusammenfassung

Das Ziel dieser Forschung ist, Produkte aus einer neuen Materialformulierung als biobasiertes Papiermaterial mittels additiver Fertigung (AM) im präzisen 3D-Druck herzustellen und zu entwickeln. Der Herstellungs- und Entwicklungsprozess dieser Produkte erfolgt durch Optimierung des Materialverhaltens und des Produktionsprozesses, um eine hohe Qualität und Genauigkeit zu erreichen. Mögliche Anwendungen für die gebaute Umwelt im Bereich der Fassadentechnik werden vorgestellt. Diese sind auf die Materialeigenschaften abgestimmt und profitieren am meisten von der Komplexität, die die additive Fertigungstechnologie bietet.

Die Formulierung besteht aus vier Komponenten mit Wasser wie Cellulose, Carboxymethylcellulose (CMC), Lecithin und einem Füllstoff (wie Kreide oder ungelöste Stärke oder mgso4). Jeder Inhaltsstoff hat eine bestimmte Funktion in der Mischung. Cellulose ist die Hauptstruktur der Mischung. CMC ist ein Hydrogel, das für die Klebeeigenschaften der Mischung sorgt. Lecithin ist ein Fettstoff, der die Extrudierbarkeit der Mischung verbessert. Füllstoff erhöht die Anzahl der Kontakte zwischen den Partikeln und stabilisiert das Hydrogel, was zu einer besseren Verarbeitbarkeit führt. Bei der verwendeten Zellulose handelt es sich um native Zellulosefasern (gebleichter Zellstoff) aus verschiedenen Quellen (NBHK/NBSK) und mit einer Faserlänge von 0,2 bis 2,1 mm.

Als 3D-Drucker wird der LUTUM -VormVrij® 3D-Clay-Drucker Version 2.1 verwendet, wobei die allgemeinen Druckparameter Luftdruck bis zu (0,3 - 0,85) Mpa, Auflösung (Düsendurchmesser von 0,6 mm -1,2 mm und 1,6 mm) und Druckgeschwindigkeit im Bereich von (15 - 20) mm/s sind.

Ziel der Untersuchung ist es, verschiedene Mischungsformulierungen zu testen, um eine mögliche Auswahl homogener Pasten mit hohem Potenzial für AM durch Extrusion zu ermitteln und eine gute Verarbeitbarkeit (größere Höhe möglich bei gleichzeitiger Formstabilität) zu erreichen. Das Material sollte in der Lage sein, sein Eigengewicht zu tragen und eine begrenzte Verformung des gedruckten Teils während des Trocknens bei Raumtemperatur aufweisen. Dies wurde erreicht, indem man sich auf die Anpassung und Optimierung konzentrierte, die durch die rheologische Charakterisierung und die Einstellung der Druckparameter kontrolliert wurde, um eine optimale Formgenauigkeit der 3D-gedruckten Teile zu gewährleisten. Es wurde ein

Leitfaden für die Druckanpassung und Designbeschränkungen vorgeschlagen, die auf die entwickelte Paste abgestimmt sind.
Für eine Düse mit einem Durchmesser von 1,6 mm besteht die beste Formulierung aus 15 % Cellulose, 69,2 % Aspen, 69,2 % Wasser, 7,5 % CMC, 2,8 % Lecithin und als Füllstoff wird Stärke mit einem Anteil von 5,6 % verwendet.

Für die Düsengröße 1,2 mm ist die beste Formulierung: Zellulose 15,5 %, Aspen, Wasser 71,5 %, CMC 5,8 %, Lezithin 2,9 %, Füllmittel Stärke 4,3 %. Düse 0,6 mm, beste Formulierung mit Zellulose 13,4%, Aspen, Wasser 78%, CMC 3,6%, Lecithin 2,6%, Füllstoff Kreide 2,3%.

Die zusammengefassten Ergebnisse führen zum 3D-Papierdruck von Produkten mit komplexen Geometrien, mechanischen Eigenschaften und einer Reihe von Möglichkeiten, wie z.B.
1. sicher durch die Verwendung ungefährlicher Materialien
2. 100% recycelbar
3. große komplexe Teile.
4. komplexes Bauteildesign
5. Selbsttragendes Material, das sein eigenes Gewicht tragen und halten kann und dabei seine Form und Stabilität beibehält.
6. Die Bauhöhe erreicht ca. 200 mm ohne Knickprobleme oder Kollaps während des Trocknungsprozesses.
Das Feuchtgewicht beträgt (7,27-8,08) kN/m3 und das Trockengewicht (4,02-5,67) kN/m3.
8. Druckparameter: Luftdruck bis zu 0,3 - 0,85 Mpa
9. Zugfestigkeit bis zu 4,5 Mpa.
10. Haftfestigkeit bis zu 20 % der Zugfestigkeit.
11. Elastizitätsmodul erreicht ca. 6 GPa.
12. Ökonomischer Aspekt des Materials und 1 kg des Materials in der Nähe von 10 €/kg.
13. Das Material lässt sich optimieren, um die geringste Schrumpfung des extrudierten Filaments zu erreichen, die 17,4 % beträgt.
14. Die Maßhaltigkeit eines 3D-gedruckten Teils nach dem Trocknen erreichte 14,8%.

Zur Erreichung dieses Ziels konzentrierte sich diese Arbeit auf die Entwicklung von
- Pastenformulierungen auf der Basis von kurzen und langen Cellulosefasern, CMC, Lecithin und Füllstoffen (unlösliche Kartoffelstärke, Kreide oder Mgso4).
- Trocknung im Weltraum.
- Verwendung eines 3D-Druckers (LUTUM -VormVrij® 3D Clay Printer Version 2.1).

Contents

List of Symbols

3D	Three-dimensional
AM	Additive Manufacturing
ASTM	American standards for testing material
B.R.	Beating Revolutions
CAD	Computer-Aided Design
CFR	Code of Federal Regulations
CMC	Carboxymethyl cellulose
CNC	Cellulose Nano Crystals
CNF	Cellulose Nanofibrils
DIC	Digital Image Correlation
DIN	Deutsches Institut für Normung
E	Elastic modulus
EC	Ethyl cellulose
ECF	Elemental Chlorine Free
FDM	Fuse Deposition Modeling - FDM Technology
FFF	Fused filament fabrication.
HPC	Hydroxypropylcellulose
HPMC	Hydroxypropylmethylcellulose
ISO	International Organization for Standardization
IGU	Insulated Glass Unit
kN	Kilo Newton- force unit
Mc	Methyl cellulose
mm	Millimeter
MPa	Megapascal
NBHK	Northern Bleached Hardwood Kraft
NBSK	Northern Bleached Softwood Kraft
PLA	Polylactic acid (one of the most used plastics in the AM)
RPM	Revolutions per minute
Rq	Surface roughness
STL	File Format of Stereolithography CAD Software
TCF	Total Chlorine Free
VIC	Visual Image Correlation (Non-Contact, Full-Field Strain & Deformation Measurement System)
W.f	Wood fibers
WRV	Wasserrückhaltevermögen (Water Retention Capacity)

3D paper printed object showing edge details for interlocking.

1 Introduction

Additive manufacturing can be a valuable tool for engineering paper materials for architectural and structural applications. As it deals with these possibilities:

2 In the concept, "AM represents a paradigm shift in design and product construction compared to traditional technology [Paul, C.P., and Jinoop, A.N. 2023], especially for paper material construction.

For the development of the construction with paper material, additive manufacturing is the solution. This is due to several factors:

1. 3D paper printing allows for more freedom and complexity in design, enhancing construction with operational flexibility. With 3D printing, intricate shapes, patterns, and custom designs can be easily created, allowing for greater architectural and engineering possibilities. On the other hand, paper-cardboard sections are limited in terms of design options due to the limited shapes of the sections from pressed paper or corrugated paper, such as layered sheets, tubes, etc. and limited techniques such as sheet layering by additive and subtractive manufacturing, milling, and folding. These sections and techniques do not provide a wide range of complexity. The inherent stiffness and bendability of the material can also limit design options, making it difficult to achieve certain desired shapes or functionalities. Therefore, compared to additive manufacturing technology, it's laborious and unproductive to produce such complex geometries [H. Bruck, Y. Chen, and S. K. Gupta, 2021].
2. Paper-cardboard elements generally have limited structural strength and durability compared to 3D printed paper parts. 3D printing allows the creation of intricate and complex geometries that can be reinforced with internal structures or by designing the print orientation to control the failure mechanism and support mechanisms, increasing their load-bearing capacity and overall durability. In addition, paper-cardboard sections have limited performance, requiring many joints and the need for bonding between layers as an adhesive bond, which localizes failure and increases the possibility of debonding, compared to 3D paper-printed sections which have full bonding between layers. This makes 3D paper printed parts more suitable for applications requiring higher structural integrity. [Tan, X., et. al., 2021], [Chen, Y., & Wang, X., 2019] and [Paoletti, P., et. al., 2020].
3. With its ability to accurately reproduce complex geometries with better aesthetics, 3D paper printing is ideal for precise and efficient manufacturing. These characteristics lead to higher production efficiency and limited material wastage compared to manual paperboard mounting.

D. Abdullah Agha, *3D Paper Printing for the Built Environment*, Mechanik, Werkstoffe und Konstruktion im Bauwesen 73,
https://doi.org/10.1007/978-3-658-45076-2_1

Furthermore, 3D printing allows to produce consistent and uniform profiles, ensuring a more acceptable quality control [Paoletti, P., et. al, 2020] and [Kotnik, T., & Dolšak, B., 2019].

3 The concept of "its properties help to perform the functionalities."

Paper material offers various functionalities that make it fit for engineering purposes. 3D printed paper material is highly flexible and can be easily molded or formed into different shapes and geometries. It is capable of complex geometries, making it useful for the creation of intricate structures and architectural models. Paper material has natural insulating and sound absorbing properties, both thermal and acoustic. It can be used as an insulating element in buildings, where the reduction of heat transfer and the transmission of noise are important. For acoustic purposes, according to measurements of the sound absorption coefficient, both loose and mixed with the glue that builds the structure [A. Trematerra1 and I. Lombardi, 2017], cellulose-based paper material is considered a good system for absorbing sound in a simple, economic, and environmentally friendly way. In addition, when used as an internal architectural part for acoustic purposes, the material also helps as a thermal insulator, could be in contact with humans, because it is not harmful to health, recyclable, renewable, widely available, low cost, and light.

4 The concept " waste and recyclability"

Paper is a renewable and recyclable material that contributes to environmental sustainability by being an environmentally friendly choice for engineering applications. It can contribute to sustainable design practices because of its smaller carbon footprint than many other materials. Conversely, plastics used in 3D printing, especially non-biodegradable plastics, can contribute to environmental problems such as waste and pollution. Increasing bans on plastic products will increase paper production due to waste issues, even though paper is less durable and stiff than plastic. Paper is considered to have the advantage of decomposing much faster than plastic, so it is less likely to be a source of litter or a threat to wildlife. In other words, it's more widely recyclable and environmentally friendly [Lamberti, F.M., et al 2020], and [Hopewell J, et al 2009]

1.1 Research Variables

The factors and characteristics that have been measured, manipulated, or controlled in a research study are the following:

- Paste formulation.

This describes the process of developing a mixture consisting of solid particles dispersed in a liquid medium. Manipulating the paste formulation and the proportions of the ingredients provides valuable insight into the effect of each component within the binder formulation in providing adhesion characteristics for 3D-printed paper materials.

It helps provide a vision for optimizing these factors to achieve the required paper 3D printing aspects.

The specific components and proportions of the paste formulation were varied according to the intended operation, such as viscosity and rheological properties, and application, such as providing sufficient strength as a self-supporting material to print large slender elements. This was done to ensure extrudability, buildability, and adequate adhesion. The variables investigated were:

- Selection of raw materials and ingredients

The necessary ingredients were identified to give the paste the desired properties.

- Proportions of components in the mix formulation

The appropriate proportions of each ingredient were determined based on the desired functionality, compatibility and performance required, depending on viscosity, rheological properties, and printability.

This was done by adding thickeners, adhesion components and adjusting the concentration of the liquid medium. Finally, the formulated paste is tested to ensure that it meets the desired specifications, such as viscosity, extrudability, buildability, stability, and other relevant parameters. It's then essential to re-design and make the necessary adjustments if the results are not within the required range.

- ➢ Nozzle size

To investigate different resolutions, the extrusion process was studied for different nozzle resolutions (1.6 mm, 1.2 mm, 0.6 mm). Extrudability and buildability were considered, and the compound formulation was improved to gain better performance.

- ➢ Printing parameters

These refer to the settings and parameters that are adjusted to control the printing process and achieve the desired results, such as print speed, layer height, infill setting, flow rate and retraction settings. These parameters were varied and redesigned according to the results of the structural integration to control the dimensional accuracy of the 3D paper printed elements.

- ➢ 3D models design

The design of complex 3D models and the printing results were studied to examine the efficiency of printing and to discover the challenges of 3D printing with paper material that could be encountered for the purpose of redesigning and modifying the model and to see the printability and limitations. The dimensional accuracy and structural integration for each 3D model is examined.

These variables are explored within the chapters, as shown in Fig. 1.2.

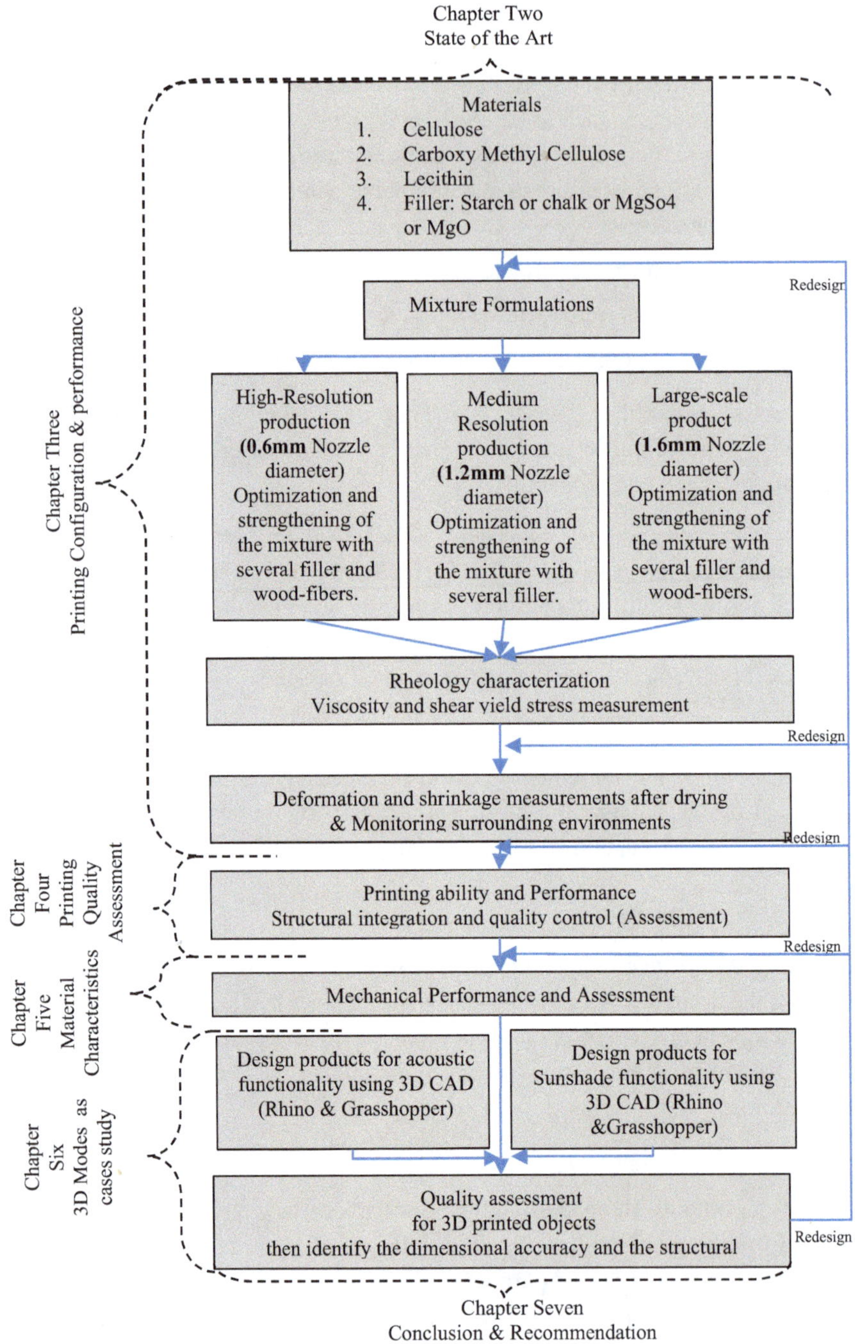

Figure 1.2 Scheme of research variables studied within chapters.

1.2 Research Objectives

The research objectives include:

1. To consider the implementation of advanced manufacturing technology for the production of elements as an innovative formulation for the paper material from a cellulose-based material. In order to be able to produce a complex geometry without self-supporting material.
2. To find paper pastes that fit the nozzles (1.6 mm, 1.2 mm, 0.6 mm) and achieve extrudability, buildability, and adequate adhesion.
3. To optimize the production process to improve product quality, production process efficiency, shrinkage control, and dimensional requirements.

These objectives were achieved by:

- Trial and error to find the best compound formulations and measure the rheological characterizations, both viscosity and shear yield stress.
- By trial and error to find the best printing parameters.
- By measurement of deformation and shrinkage after drying
- By measuring the mechanical performance.
- By creating 3D models and using them to specify the details and challenges of the printing performance and then evaluate the results.

1.3 Research Questions

The formulation of the paper paste was designed, then, the geometries are printed using a 3D printer with the designed paper paste. From this point, the investigation is carried out to examine the readiness of this design paste to follow the requirements of the production process in terms of printability, buildability, structural integration, and fulfill the manufacturing of a complex geometry.

Redesign of the paper paste is required and printing parameters in case of not achieving these challenges. The quality of these printed elements is investigated in terms of the printing outcomes and challenges to introducing the design into a 3D paper-printed model in terms of structure integration and production efficiency, and dimensional accuracy.

The main question that the whole dissertation will cover is:

How to produce 3D paper printed models with better quality and dimensional accuracy?

The quality here indicates by the nice outcomes and referred to uniform lines with smooth extruded outlines and have acceptable aesthetics with geomatical integration and considerable accuracy. The research carried out successive procedures that answer this main question and clarified by each chapter until reaching the goal in the final chapter, which producing a 3D paper functionalized model.

Each process has been considered as a millstone that filters the successive steps and separates the paths according to the product requirements. Therefore, the following sub-questions will guide the principles for these millstones.

- Chapter one:

Why is AM needed to produce paper structures in the field of engineering for the built environment?

What are the main objectives achieved through the research?

- Chapter two:

What are the overall parameters that influence the quality of the 3D printed objects?

What are the previous studies that have investigated the use of cellulose as multiple forms within AM?

- Chapter 3:

What is the formulation and mix design of the pastes/mixtures in terms of (changing the type of cellulose, type of filler, proportions of the components and using the wood fibers)?

What are the optimal mixtures suitable for the nozzles in terms of rheological properties to meet extrudability and buildability?

What are the solutions used in this study to minimize shrinkage?

How can the quality of 3D paper printed objects be improved and optimized in terms of the production process?

- Chapter 4:

What are the results of the evaluation of the printing performance and the assessment of the printing quality?

What are the defects of the 3D paper printing results?

Chapter 5

What is the mechanical behavior of the material from different aspects such as filament orientation, adhesive strength, nozzle size and strengthening the paste with wood fibers?

What are the proper print parameters that give better mechanical performance for solid objects?

- Chapter 6:

What are the potential applications for 3D paper models?

What are the results of evaluating the quality and dimensional accuracy of the 3D models?

- Chapter 7:

What are the most valued results of this research?

What are the design limitations for 3D printed paper models from this research?

What are the avenues opened by the endpoints of this research?

1.4 Thesis Structure

The thesis is divided into six chapters:

Chapter One presents the main ideas of the research, including explicit statements of significance specific to the topic studied, and illustrates the main problems and questions of the research.

Chapter Two reviews the literature in the context of the research in parts:

Part I presents a thorough examination of cellulose resources, covering types, production processes, resulting products and their impact in the field. In this section, the specific type of cellulose used in the research is identified and the different types are distinguished according to the production process. Part II focuses on the use of cellulose as a paper-based material within additive manufacturing (AM) technology. It provides a detailed overview of its various applications and discusses primary strategies to overcome the challenges involved. It also explores conceptual frameworks aimed at unlocking the potential of paper-based materials to achieve specific goals within AM processes. The motivation behind the research and its main objectives and contributions are summarised.

The **third chapter** describes in detail the approaches followed for all the stages of the experimental work, from the material used, the mix design of the compounds, the mixing procedure, to the manufacturing by 3D printing. The rheological behavior and shrinkage measurements for all the mixes used are also provided. In addition, it offers solutions and optimizations to the problems of printing performance, geometric accuracy and structural integration of 3D paper printed objects.

Chapter four shows an evaluation that has been carried out to provide guidelines to produce 3D paper printed objects.

Chapter five presents the mechanical properties and material response for 3D paper printing in terms of stress-strain behavior, Young's modulus and failure mode for the experimental work conducted in chapter three.

Chapter six presents the details of the evaluation of the 3D paper printed models. Identification for the printing efficiency and the effect factors of different printing results for these 3D models were also presented in terms of structural integration, and efficiency of printed geometry and dimensional accuracy. The 3D models were designed to satisfy the functionality of 3D printed paper models as potential applications for shading elements by optimizing the passive solar radiation energy, and for acoustic purposes for interior partitions or cladding acoustic panels with aesthetic and complex geometries.

Chapter seven summarizes the conclusions drawn from the present study and the design considerations in the main of the evaluations and limitations for: the printing performance and mechanical performance for 3D paper printed objects for the building environment. The chapter will then give prospective recommendations for future work.

1.5 Methodology

The goal of this research focuses on producing 3D paper printed structures that achieve functionality for the built environment. The production approach in the process must be developed and optimized to reach these customized products with better quality and desired accuracy. The method for presenting this approach signifies following several steps till reach the goal. These steps will be representing sectors such as: optimizing a mixture according to the extrusion ability and buildability and controlling the deformation and shrinkage after drying. Furthermore, for dealing with buckling issues, large-scale printing, and for manufacturing objects with better material properties.
The investigation of the different sectors involved will be based on engineering principles for manufacturing a product within additive manufacturing. These engineering principles will be represented with the engineering process that matches the objectives of this dissertation. The objectives clarify an essential planned system that covers the essence of all the contents of this research. There are several parts of the mentioned system that are not regular for the common additive manufacturing since this research investigates a new input material for AM technology, which are the mixtures that have been studied from mainly cellulose. This is considered something novel and modern in the fields of research and industry. Therefore, some steps and additional major sectors are being linked to the followed engineering process that fit the interest of objectives of the dissertation.
The engineering process has been followed described in Fig. 1.1 for design, optimization, and manufacturing with the quality control that would achieve the structural integration and performing the purposes and functionality that it has been designed for. Redesign is necessary to make a loop in the engineering process for optimizing and finding the most optimal solution that performs an outrun of the issue. The redesign represents in the current investigation by moderating the values for the same variable or other input parameters that are the basis for the design case for each sector.
A conceptual framework for the variables related to the research was illustrated by a visual representation for the successive parameters that impact the followed approach, as shown in Fig. 1.3.

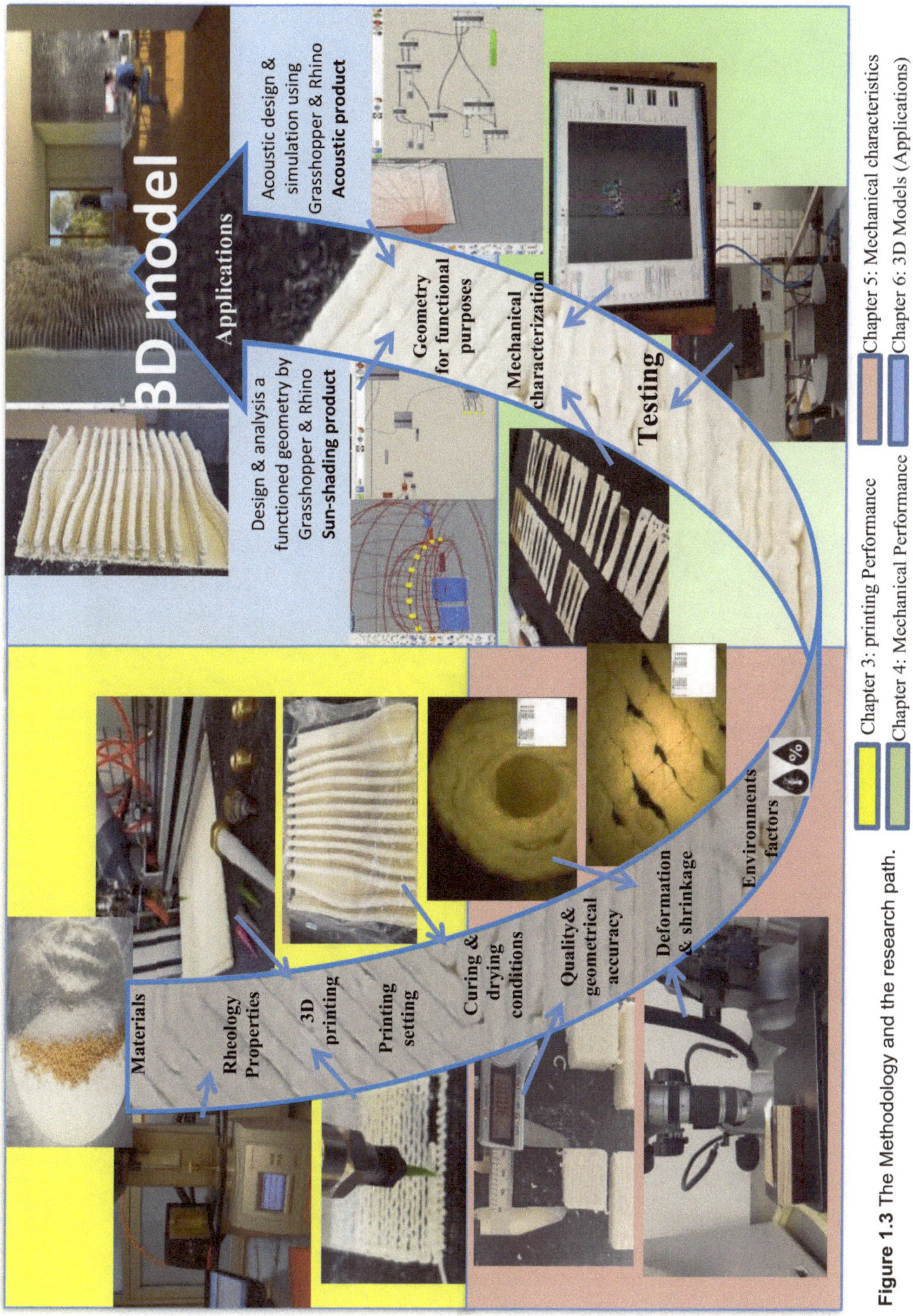

Figure 1.3 The Methodology and the research path.

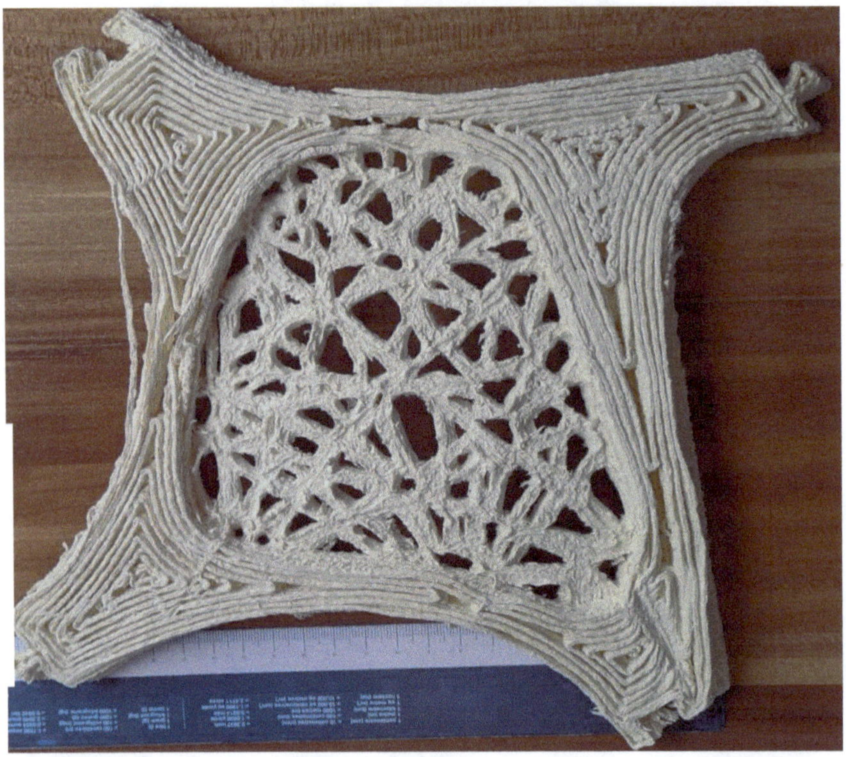

3D String shade wall (a), and 3D paper printed model for the string wall part.
This 3D model is presented at BAMP exhibition – Biennale in Venedig 30. Juni 2021.

2 State of the Art

Additive manufacturing has indeed brought about a significant evolution in product development and manufacturing processes. This evolution is continuously expanding as new technologies, materials, and applications are developed [Redwood, B. 2017]. This is qualifying AM at high industrial levels using different materials. The number of materials used by AM is constantly increasing, giving the potential to implement more adopted materials that AM can handle and treat according to their properties and functionality. [Mahmood A, 2022]. Cellulose, a natural polysaccharide found in the cell walls of plants, has gained significant interest in 3D printing applications due to its abundant availability, renewable nature, and biodegradability. The incorporation of cellulose-based materials into 3D printing processes offers several advantages and opens new possibilities in various fields. Cellulose-based materials used in AM can include cellulose acetate, cellulose nanocrystals (CNC), cellulose derivatives and cellulose-based composites. These materials can be processed into filaments, powders, or liquids suitable for various 3D printing techniques [Xiaoyu Bi, Runzhou Huang, 2022].
This chapter reviews and explores the state of the art for various aspects of the production of AM paper products from cellulosic materials using different additive manufacturing technologies, including design and manufacturing processes, material properties and material formulation development.

2.1 Cellulose

The cellulose forms that are adequately available in the market are lignocellulose, bleached pulp, dissolving pulp. However, to understand the uses and implementation of cellulose within AM, first must review its structure, types and its functioning, and its available forms. Cellulose is a versatile biopolymer with a variety of formulations. The most common types of cellulose are:

1. Native cellulose (cellulose fiber)

It refers to the natural form of cellulose found in plants. Cellulose fibers are the main load-bearing component in trees and plants due to the high modulus of its crystalline part, which can reach 140 GPa [Daniele R., 2019]. The type of cellulose used in the study is the pulp cellulose fibers from the first purification stage, as indicated in Fig. 2.1.
Cellulose fibers are used for industries such as papermaking and textiles. It's used also in food additives, pharmaceuticals and cosmetics that is typically found in younger plant tissue [Brethauer, S. et. al, 2020] and [Gauss Ch. et al, 2021].

D. Abdullah Agha, *3D Paper Printing for the Built Environment*, Mechanik, Werkstoffe und Konstruktion im Bauwesen 73,
https://doi.org/10.1007/978-3-658-45076-2_2

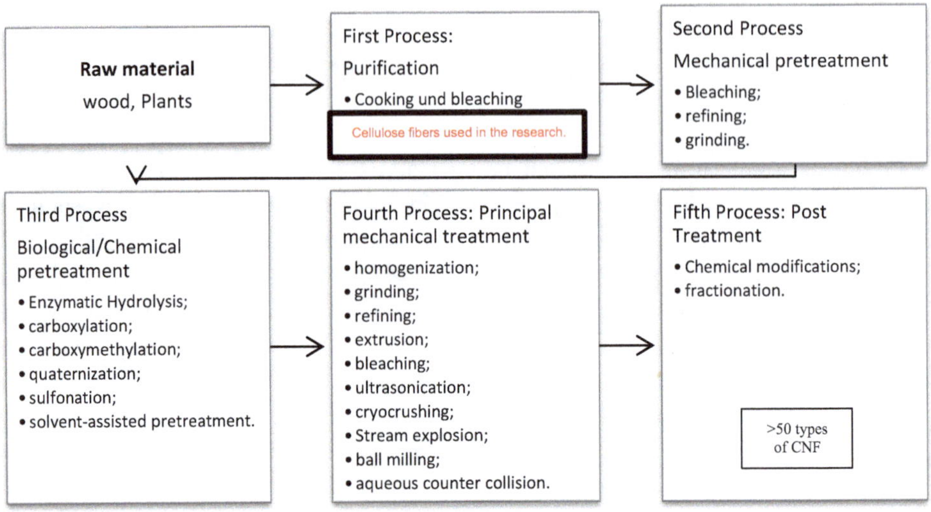

Figure 2.1 Wood fibers treatments and up to five stages to produce CNF [Thibaut, C. 2020].

2. Nanocellulose

- Nanocrystals (CNC): are produced by breaking down cellulose fibers into smaller crystalline particles. It has unique characteristics, including high strength, high aspect ratios and excellent optical clarity. CNC is used in nanotechnology, biomaterials, coatings, and polymer reinforcement.
- Cellulose Nanofibrils (CNF): are long and slender structures derived from cellulose fibers. They have high strength. CNFs are used in a wide range of applications including films, coatings, gels and as reinforcement in nanocomposites.
- Bacterial cellulose (BC) is a gel-like substance composed of interconnected cellulose nanofibers produced by certain types of bacteria [Chen et al., 2011; Moon et al., 2011; Torres et al., 2012].

3. Cellulose Derivatives

Cellulose Microfibril can be chemically modified to produce cellulose derivatives with specific properties. Industries such as pharmaceuticals, food additives, adhesives and personal care products use these derivatives. The types of available and widely used cellulose derivatives are shown in Fig. 2.2.

- Carboxymethylcellulose (CMC)

It is a water-soluble polymer derived from cellulose fibers by chemical modification.

The production of CMC generally involves: 1) cellulose extraction through pulping. 2) Alkali treatment with NaOH. 3) Carboxymethylation by monochloroacetic acid (-CH2COOH) or its sodium salt with a catalyst. 3) Neutralization and washing: 4) Drying and Finishing. [S. Slotte, 2021] and [Gauss et al, 2021].

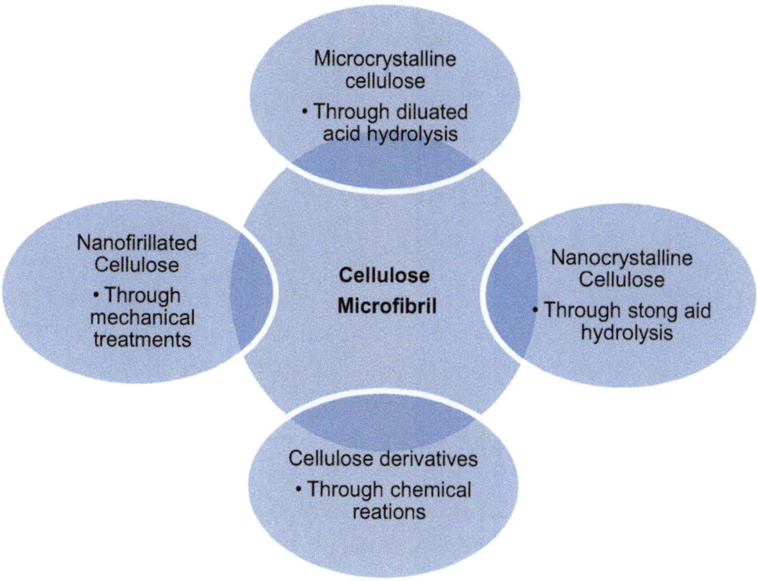

a

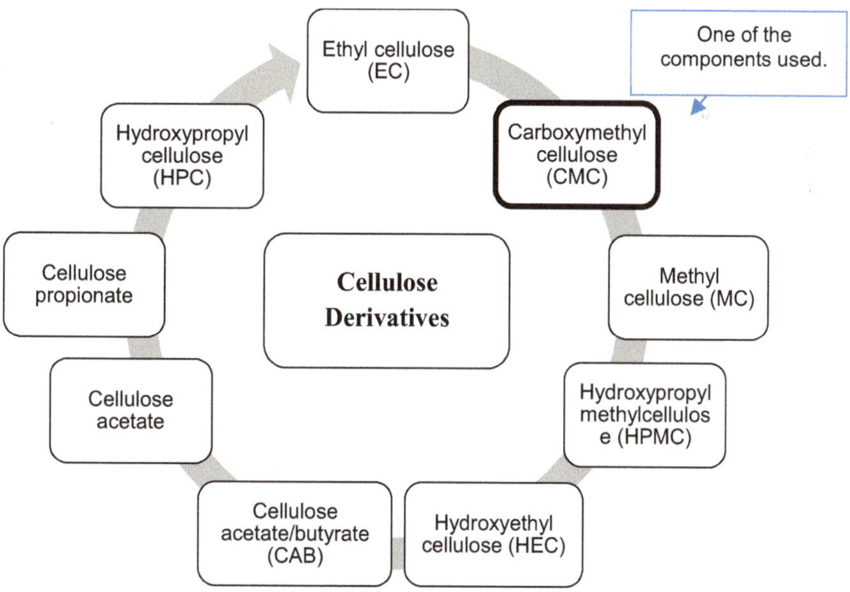

b

Figure 2.2 The microstructure of the cellulose according to different preparation methods [Kunkun Tu, 2022] (a), and cellulose derivatives [J.K. Pandey et al., 2015] (b).

4. Regenerated Cellulose

It is produced by the dissolution of cellulose in a solvent and then regeneration into a solid or film form. Viscose rayon and cellulose films are common examples. Regenerated cellulose has applications in textiles, films, and membranes [Shamsuri, A. et al, 2021].

2.2 The Use of Cellulose in AM

Today, many forms of cellulose derivatives are used in additive manufacturing. Thermoplastic filaments are the first application using cellulose based material as cellulose-based composites in AM, according to the review by Gauss et al. 2021.
Wang et al. 2018, Dai et al. 2019 and Gauss et al. 2021 report on the use of cellulose in additive manufacturing and present data according to the nature of the contribution of cellulose in materials developed for AM. According to these data, the development of composites for biomedical applications such as wound dressings, prostheses, and tissue engineering has been the main use of cellulose in additive manufacturing.
The data from Gauss et al. 2021 shows that the use of cellulose-based materials with 3D printed materials has encouraging and promising results for shape changing behavior and as a rheology modifier or reinforcing agent in field of materials science, engineering, and chemistry. According to this review, the number of publications within cellulose-based composites is 477 in between of 48,458 documents containing only the terms '3D printing' or 'additive manufacturing' in the period (2006-2021).

2.2.1 Thermoplastic Cellulosic Composites in AM

Thermoplastic cellulose-based composites are filaments treated with temperature of materials that combine cellulose fibers or particles with a thermoplastic polymer matrix. Research and development of thermoplastic cellulose composites continues to improve their properties, optimize manufacturing processes, and explore new applications. Some Properties are listed below, along with approximate ranges, according to reviews by Wang et al. 2018, Dai et al. 2019, Gauss et al. 2021, and Utz, J et al. 2022:

- Thermoplastic cellulosic combination:
  - Cellulose Reinforcement: can range from 10%wt to 50%wt.
  - Thermoplastic Polymer Matrix: ranges from 20%wt to 90%wt.
  - Polymer can vary widely, including polyethylene (PE), polypropylene (PP), polylactic acid (PLA), polyethylene terephthalate (PET), and others.
  - Additives and Fillers: To enhance specific properties or processability of the composite, typically in the range of 1% to 10% by weight, may be added, such as plasticizers, coupling agents, flame retardants, or processing aids.

- Mechanical Properties:
  - Tensile Strength: In general ranges from 10 MPa to 718 MPa [Yu Liu, 2019].
  - Young's Modulus: Typically ranges from 1 GPa to 19 GPa.
- Thermal Properties:
  - Melting Temperature: Typically ranges from 100°C to 250°C.
- Physical Properties:
  - Density: Usually ranges from 0.8 kg/m^3 to 1.5 kg/m^3.
- Printing parameters
  - Nozzle Diameter: From 0.2 mm to 0.6 mm.
  - Extrusion Temperature: Normally ranges from 180°C to 250°C.
  - Print Speed: 20 mm/s to 100 mm/s.
  - Layer Height: 100 μm to 300 μm.

Tab. 2.1 shows a summary for different researchers investigating thermoplastic cellulosic composites.

2.2.2 Cellulose Use in Inks, Gels, and Suspension Concentrations

Dissolved cellulose, nanocrystal-cellulose or derivatives are mostly used in this field and dispersed in a liquid medium to create inks with specific viscosities, rheological properties and solids content tailored for different AM techniques. These cellulose-based formulations offer unique properties and functionalities that make them suitable for various additive manufacturing processes according to specified application. Tab 2.2 shows some formulation that used cellulose with different activity. Inks formulated with cellulose-based materials enable precise and controlled deposition of cellulose-based materials, allowing the fabrication of complex structures layer-by-layer. Tab. 2.3 summarizes some characteristics that qualified cellulose for functioning within the formulation adopted for 3D paper printing.

Table 2.1 Thermoplastic cellulose-based composites in additive manufacturing [Wang, Q. et. al, 2018] and [Gauss et al, 2021].

Materials	Filament diameter mm	FDM printer	Nozzle size mm	Printing temperature	Potential application	Reference
nanocrystal Cellulose / poly(e-caprolactone) (0–10%)	3	Thing-O-Matic 3D printer	0.3	185	Bone tissue engineering	Hong (2015)
ColorFabb recycled. wood flour/PLA. (15%)	2.85	Prusa i3 rework 3D printerwas	0.3	210	4D printing actuation functionality	Le Duigou et al. (2016)
Wood flour/PLA (%)	1.75	Self-assembled FDM 3D printer	0.4	210	Functional, loadbearing applications	Extrusion Tao et al. (2017)
Micro-nanocellulose/ PLA (0–30%)	1.75	Z603S, Jgaurora	0.4	190	Structure Materials extruder	Wang et al. (2017)
Cocoa shell waste/ poly(e-caprolactone) (0–50%)	1.75	Prusa i3 Hephestos 3D printer	0.3	120	Household / biomedical application	Tran et al. (2017)
Cellulose fiber/PLA (0–20%)	2.85	Lulzbot TAZ 5 3D printer		210	automotive industry	Kearns (2017)
Cellulose nanocrystals polyvinyl alcohol (2–10%)	1.7	Sharebot Next Generation 3D printer	0.35		improve the performance of PVOH	Rigotti et al. (2017)
Hydroxypropyl cellulose/Domperidone (80–90%)	1.75	Replicator 2X	0.2	210	Drug delivery	Chai et al. (2017)
Nanocellulose (NC) (0.5–3%)	-	Next Generation desktop-3D printer	0.35	210 °C for PLA and 180-200 °C for PHBH	enhancement properties of PVOH	Rigotti , et al. (2018)
NFC (10-40%) PLA	1.68	Commercial desktop FDM unit (Solidoodle 3)	0.4 and 0.2	180-215	Material development	Tekinalp et al. (2020)
BC – bacterial cellulose (1–2.5%)	PLA/ Pickering emulsion	Funmat HT 3D printer	0.4	180-195	Material development	Li et al 2020
NFC (1-5%) PLA Me	1.75 ± 0.05	M3036FDM desktop 3D printer	0.4	210	Material development	Wang et al. (2020)
CNC (5–10%)	1.75	Vellemen K8200 3D printer	0.6 and 0.4	170	Material development	Giubilini et al. (2020)
Commercial cellulose fiber containing reference material. 20%	-	fused granular fabrication (FGF) by Brinter®	0.8	185-224	Researching	Immonen K, et. al, (2021)

Table 2.2 The main functions that cellulose is used for [Gauss et al, 2021].

Cellulose Activity in the Mix	Characteristics that qualify cellulose for functionality	Area of application
Rheology modifier	Hydrogen bonds, which increase viscosity and restrict water movement, are formed by cellulose chains. During processing, this allows for easier flow. Due to high shear rates, the hydrogen bonds re-form, and the cellulose structure is restored.	Cellulose has suitable rheological properties for 3D inks. It flows smoothly at high shear rates and solidifies quickly after printing.
Binder	Some of the cellulose-based material has shear thinning behavior qualifying it to be a choice of binder for 3D printing powders	It is essential to understand these properties to be able to use cellulose in various fields, such as paints and glues
Excipient	Cellulosic hydrophilicity allows for controlled swelling upon contact with water.	This property of cellulose can be also exploited in applications such as biomedicine, drug delivery and 4D printing
Matrix	When dissolved in solvents, high molecular weight cellulose is highly viscous and shear thinning. This means that as the shear rate is increased, the viscosity decreases, allowing it to flow more easily during processing	For 3D printing methods such as the material extrusion process, high viscosity solubilized cellulose is suitable for producing dimensionally stable prints.
Anisotropic reinforcement Or solid particles	The inherent strength of cellulose fibers provides exceptional mechanical properties. These include high strength, high stiffness, and low density.	In emerging fields such as bio-based materials and nanotechnology, cellulose's unique properties make it desirable. That's why it's often used as a filament or used in formulating bio inks.

2.3 Synthesis of the state of art

From review, formulation based on cellulosic material serves several key functions when employed in various applications. These functions are dependent on the purpose and the properties that the mixture formulation provides, including its role as a structural component. Each researcher has his invented kind of ideas and purposes that control his results. These results provide strength and rigidity that are required from the materials to employ the application design for.

However, despite the advancements in technology, such as additive manufacturing, achieving high-quality and dimensionally accurate results remains a challenge across various materials used in AM. Furthermore, in engineering applications, the ability to scale up production for large-scale products, particularly in building components, is crucial. The starting point in this context involves identifying and utilizing suitable formulations and materials that meet these specific requirements. The selection of an appropriate paste formulation is vital to ensure desired performance and compatibility with the intended use.

Table 2.3 Cellulose-based composites inks, gels, and suspension concentrations in AM.

Materials & formulation	Cellulose role mm	Printing Method	Main outcomes	Potential application	Reference
Methylcellulose/ alginate/H_2O (0–9%)/(0–3%)		Biofactory Bioprinter, nozzle diameter 0.21–0.25	Solidification is Crosslinking.	Tissue engineering	Li et al. (2017)
0.4 wt. % CNF, 1.1 wt. % MMT, 2 wt. % MC, and various amounts of glyoxal and TA	Methylcellulose as table foam	Direct Ink Writer, Bioplotter, nozzle diameter = 0.84 mm, pressure= 0.5 bar	Foams displayed a pronounced shear thinning behavior with the viscosity dropping from 10^4 to 4Pa·s	Renewable nanomaterials	Voisin et al. 2018
cellulose acetate at 25-30 wt.% in acetone	Dissolved cellulose	1.0 MPa extrusion pressure, 20 mm s−1 print speed, and a layer height of 0.18 mm	The quasi-immediate evaporation of the acetone (~1 min) during the printing accelerates hardening and the shrinkage compensation of every printed layer, maintain dimensional stability of the part.		Pattinson et Hart, 2017
Quaternized hemicelluloses, QH, 50 wt.% CMC 50 wt.%	Modified by alkaline extraction method	Not specified	The blend film (QH and CMC) of (1:2 m/m) had a tensile strength of 65.2 MPa. The CMC addition has strong electrostatic interactions and enhances hydrogen bonding with QH.	Coating and packaging application-blend Films	Gao et al. 2018
Purified cellulose (fiber <200µm) 20% or wood four (grain <500µm) Sawdust-Chitosan 2.4%	Matrix Reinforcing	Nozzle 7mm, 12 bar air pressure. Intense postprocessing (coated and polished)	This material has excellent bonding properties with wood (hold 33.5±7.3 kg.cm-2). low density of 0.4 g.cm-3, tensile strength of 11 MPa and a Young's modulus of 240 MPa.	Internal part of a wind turbine blade.	Sanandiya et al, 2018
Micro cellulose fiber (15 - 45) wt.%, CMC (5-20) wt.%	cellulose fibers (Carl Roth) ≤ 200 µm	Modified Cartesian 3D printer/extrusion system. Nozzle 0.71mm, using ethanol drying	Conductivity 102-103 S.m-1 for PLA/carbon composite. Young's modulus is about 2-3 GPa. Best formulation (cellulose 0.3: CMC 12.5: water 57.5) wt.% give 36.5% shrinkage. Using ethanol bath reduces shrinkage to 26.4%	Not specified	Thibaut, 2020
Starch (10, 9, 70) %, water (76, 70, or 25)%, Micronized cellulose (14, 12, 5)%, sawdust 9%	Matrix Reinforcing, cellulose optimum ratio 10 - 20 %.	robotic printer – Lutum 2.0, 2 bar air pressure 0.3 mm nozzle diameter	The mixing procedure is applying heat to activate the gelation of starch with water. Adding sawdust helped to reduce the fragility and improve quality. The retraction index was between (24.12-27.65) %.	Architectural Buildings components	T. Campos et al 2022

Additionally, cellulose offers a high surface area for interactions, making it useful for adsorption or filtration processes. Its hydrophilic nature allows for water retention and swelling properties. Furthermore, cellulose acts as a renewable and biodegradable material, making it environmentally friendly. These functions make cellulose a versatile and valuable resource in numerous industries. The behavior and response of every compound is different due to a variety of factors including ingredient selection, ratio, process conditions, and desired results. Because of this complexity, it is not possible to generalize specific cases across the full range of compound formulations. To understand its unique behavior and optimize its performance, each formulation requires careful consideration and analysis. This combination highlights how important it is to have a tailored approach and specific considerations for compounding in different applications.

2.4 Motivation of the Research

The inspiration of the research is stated in Chapter One. The strategy of the research is to develop a mixture formulation with cellulose fiber (short fiber length of 0.75mm and long fiber cellulose of 2.1mm) for paper material and a functional one compatible with 3D printing technology to produce complex geometries.

The characteristics and the range of capabilities of this production are as follows:

- Large scale complex building component and Complex part design,
- Resolution depends on nozzle diameter of 0.6mm -1.2mm and 1.6mm,
- Printing speed in the range of 15 - 20 mm/s,
- Air pressure up to 0.3 - 0.85 Mpa,
- Buildability reached approximately 200 mm in height without any buckling problems or collapse during the drying process.
- Self-supporting material that can maintain and hold its own weight.
- Tensile strength reaches up to 4.5 Mpa,
- Adhesive strength reaches up to 20% of tensile strength,
- Young's modulus reaches about 6 GPa,
- Safety aspect by using non-hazardous material,
- Economic aspect of the material:1Kg of the material in the range of 10 €/kg,
- Achieving minimum shrinkage possible, it reached to 17.4%,
- Recyclability 100%,
- Dimensional accuracy was reached up to 14.8%.

To achieve this goal, this work focused on:

- Paste formulations based on short and long cellulose fibers, CMC, lecithin, filler (indissoluble potato starch, chalk or Mgso4).
- Use of 3D printer (LUTUM -VormVrij® 3D clay printer version 2.1)
- Room environmental drying.

Printed panel (400x400x10) mm for shading purposes.
One line wall thickness of 10 layers with 10 mm height without buckling effect.

3 Printing Configuration & Performance

This chapter presents how to produce and improve 3D printed paper elements from pulped cellulose to achieve a better degree of structural integration, accuracy, and quality. The considerations made to achieve this production are: i) the ability and ease of extrusion of the mixture, ii) the optimization of the mixtures to fit the nozzles for minimum shrinkage and maximum buildability, as well as iii) the control of the printing parameters. Thus, the production would be controlled with a better printing performance in terms of retention and stability as a static structure after printing from fresh state to get dried. This the printing configuration are carried out starting from the selectivity of the materials used, the mixture formation and portion design, up to the manufacturing process.

3.1 3D Printing Configuration and its Quality

In the case of additive manufacturing technology, there are some essential aspects to consider producing a 3D part with the desired quality and structural integration. More specifically, the parameters shown in Fig. 3.1 are critical to achieving a perfect balance between part production and quality.

An important determinant of quality is dimensional accuracy. It's evaluated as the dimensional convergence with the required and regularity of the geometry to ensure a limited tolerance, when compared with the designed model. However, improving dimensional accuracy is one of the major scientific challenges in improving product quality using 3D printing.

In fact, there are many challenges in developing integrated 3D paper printing production with controlled quality and high efficiency. Among them, the biggest problem is "shrinkage", which plays a very important role in geometric accuracy and especially aesthetic appearance. This chapter introduces the technique followed to improve the production according to the approach mentioned in Fig. 3.1.

3.2 Input Parameters

The input parameters such as raw material selection, compound design and manufacturing technology are essential to define the final production results as it is stated in Fig. 3.1.

D. Abdullah Agha, *3D Paper Printing for the Built Environment*, Mechanik, Werkstoffe und Konstruktion im Bauwesen 73,
https://doi.org/10.1007/978-3-658-45076-2_3

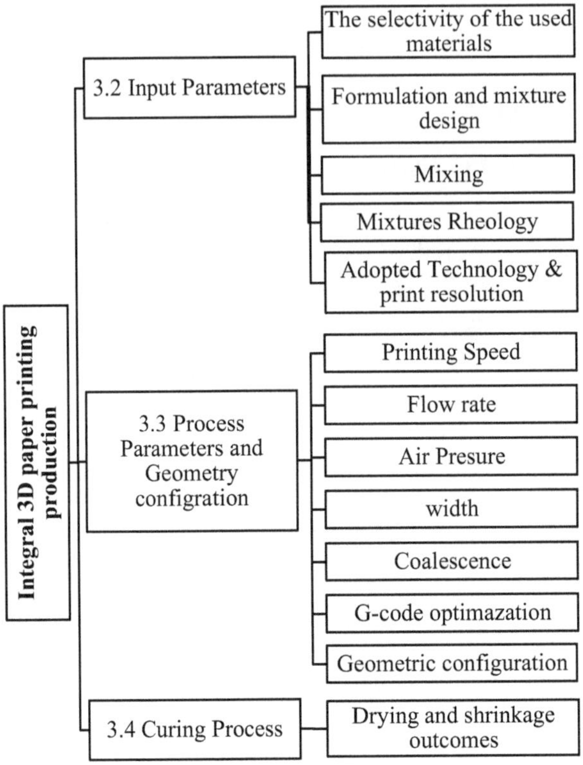

Figure 3.1 The Key parameters affecting the quality of 3D paper printing.

3.2.1 Raw Material and its Selectivity

One of the most important factors in defining product characteristics and ensuring their quality is the selectivity of the materials used. The materials used are the backbone for the creation of a product. The raw materials selected must meet the required specifications and be compatible with the intended production process.

The appropriate materials used are based on their activity and functionality in the formulation of the product. Thus, the final functionality and use potential of the product is basically defined by its resource and configuration, which would have wide areas, such as providing specified characteristics of the finished product according to the customer's desire. Therefore, several considerations are considered when selecting the materials used, including company evaluation, purity of the material within the supplier's certificate of analysis, potency, and composition. [21 CFR 111.70]. The data sheets for the materials used are given in Appendix A.

3.3.1.1 Cellulose

Cellulose is a natural polysaccharide which is the major structural component of plant cell walls. It is the most abundant organic polymer on earth, found in various plant materials including wood, cotton and agricultural residues. Cellulose consists of repeating glucose units linked by beta-1,4-glycosidic linkages [Siqueira, G., Bras, J., & Dufresne, A., 2010]. Several types from different regions and companies in the paper industry have been used, as follows:

1. Nanocrystals, cellulose powder, Sigma-Aldrich/Merck in Germany.
2. Short cellulose fiber, Aspen PHK, DMI Canada.
3. Long cellulose fiber, NBSK_ECF, Kiefer STENDAL Germany.
4. Long cellulose fiber - SÖDRA Golden Sweden.
5. Short cellulose fiber SÖDRA- Blue Sweden.

The cellulose properties were measured with the Valmet Fiber Image Analyzer "Valmet FS5" (Fig. 3.2) and carried out at the Institute of Paper Technology and Mechanical Process Engineering/TU Darmstadt. The device is used by pulp and paper mills to test the actual value prospects of their raw material for optimum product quality. The properties examined by the Valmet FS5 are shown in Fig. 3.2. In
Tab. 3.1 shows the cellulose properties taken from the companies' technical data sheets.

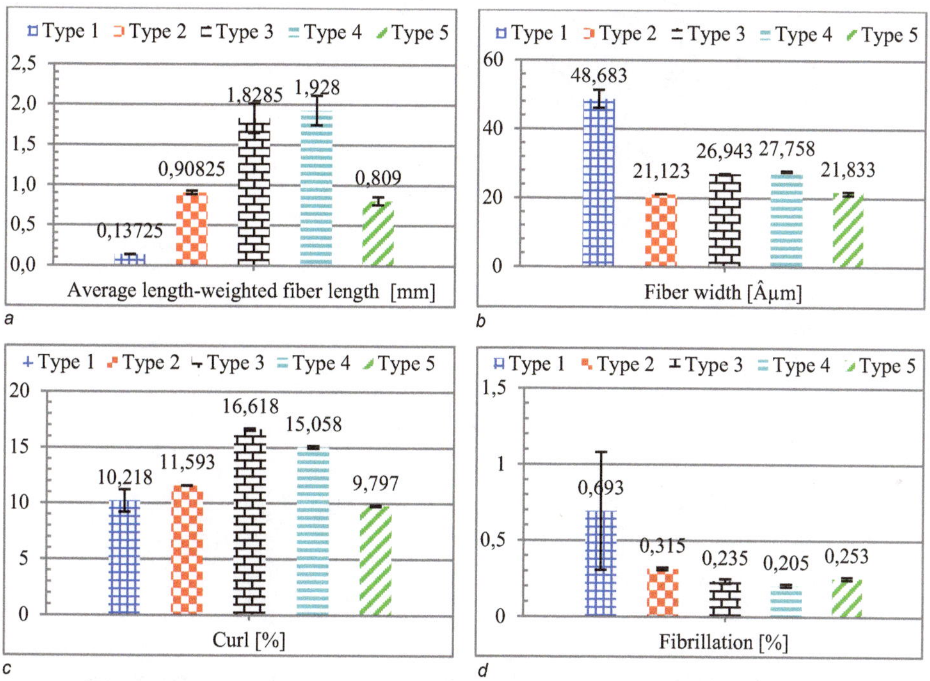

Figure 3.2 Properties comparison of the cellulose types for: (a) Fiber length; (b) Fiber width; (c) Curl, and (d) Fibrillation.

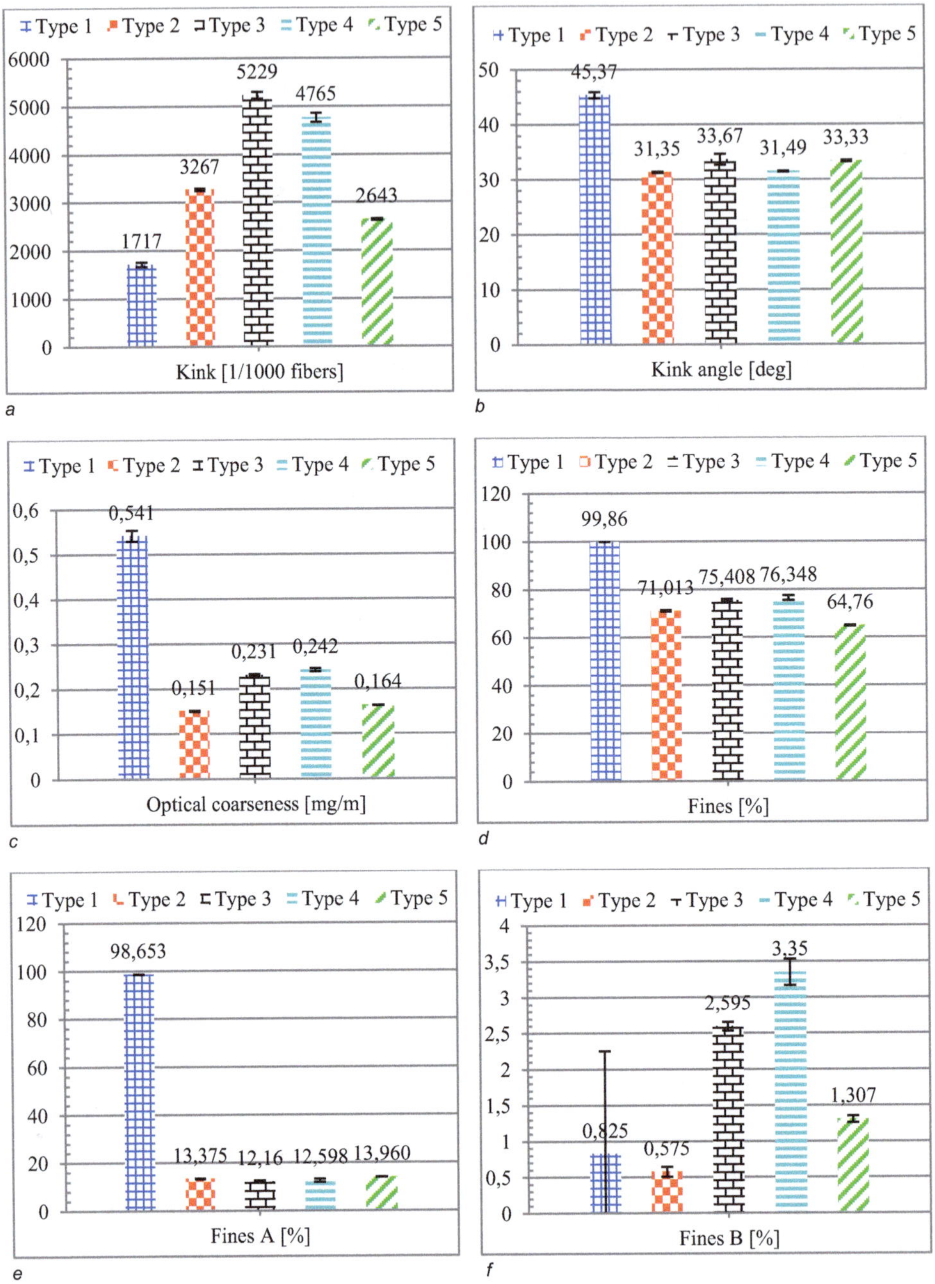

Figure 3.3 Comparison of the cellulose types for: (a) kink, (b) kink angle, (c) optical coarseness, (d) fines; (e) fines A, and (f) fines B.

Table 3.1 Properties of the cellulose from the company data sheet.

Index	Type1	Type 2	Type 3	Type 4	Type 5
Product identifier	Cellulosepulver	Peace River Pulp	Pulp Stendal ECF	blue F	Gold birch Z
Product company	Sigma-Aldrich/Merck in Germany	Daishowamarubeni International Canada	Mercer STENDAL, Germany	SÖDRA-Sweden	SÖDRA-Sweden
Species	Cotton linters	NBHK, Trembling Aspen 85%, Balsam Poplar 15%	NBSK, Pine 40-70%, Spruce 30-60%, ECF	bleached Kiefer sulphate pulp, TCF	NBHK made from birch
Fiber length, mm	0.2	0.75	2.10	2.10	0.9
Fiber width, μm	-	27.4	28.2	30.5	21.5
Coarseness, mg/m	-	0.1	unavailable	0.135	0.08
Tensile index, by 2000 B.R, kNm/kg	-	42*	72.5	74	92
Tear index, by 2000 B.R, N.m^2/kg	-	4.1*	17.6	13.8	unavailable
Water retention capacity, WRC**, %	-	58.1	57.27	54.21	56.71

* By 1500 Beating revolutions, B.R.
** Calculation mentioned in Appendix C.

3.3.1.2 Carboxymethyl Cellulose (CMC)

A versatile and widely used polymer derived from cellulose, a natural polysaccharide found in plant cell walls. CMC is produced by chemically modifying cellulose through carboxymethylation, introducing carboxymethyl groups (-CH2-COOH) into the cellulose backbone. The desirable properties of CMC make it suitable for a wide range of industrial applications:

1. Thickening and stabilizing ingredients: These include food, personal care, and pharma products. It adds viscosity and improves texture, stability, and suspension properties.
2. Water attraction: The high moisture retention properties of CMCs make them useful in applications like waterborne coatings, adhesives, and drilling fluids. It retains moisture and prevents drying, by improving the overall performance of these products.
3. Film forming property: Once dissolved in water, CMC forms a clear, flexible and protective film. These films provide barrier, adhesion, and surface protection in applications such as coatings, paper, textiles, and packaging [Laith, A. and Al-Hashimi, A., 2019].

4. Binding agent: Across various industries such as ceramics, cleaning products and building materials, they act as binders, improving the handling properties of these materials through increased cohesion and strength.
5. In pharmaceutics: It can facilitate the manufacturing of tablets by acting as a binder, separating agent or controlled releasing agent [Rajabi, H., & Fathi, M., 2017].

The characteristics of the type used are presented on Tabs. 3.2 and 3.3. Shrinkage calculations (Tab. 3.5) were performed using the microscope (Section 3.4.1) as shown in Fig. 3.5. Viscosity measurements were performed as described in Section 3.2.5.

Table 3.2 CMC properties from the Company's Data Sheet.

Type	Product identifier	Product company	Appearance	Water Solubility	viscosity cps*	Grain size mesh**
Type 1	Methylcellulose	Würzteufel GmbH, Germany	creamy white, powder	Soluble	75 000	100-130

* 1cps= 1 mPa-s.
** 100 mesh= 0.150mm.

Table 3.3 CMC measured properties within hot water of 45C° before and after drying.

Index	Shrinkage %	Viscosity* [mPa·s]	Observed behavior
5% CMC & 95% water	-	1.45E11 ±3E10	As shown in Fig. 3.5, it was found that as the ratio of CMC to water content increased, the adhesive properties of the CMC gelation decreased, and it behaved like a solid material with lower adhesive properties.
10% CMC & 90% water	48.5± 2	1.8E12 ±2E11	
20% CMC& 80% water	32.7± 3.2	1.1E13 ±2E12	
30% CMC & 70% water	22 ± 4.1	1.00E14 ±1.8E13	

* Measured according to Fig. 3.4 and Fig. 3.4, using the microscope, labeled in paragraph 3.4.1.

3.3.1.3 Soy-Lecithin

This is a naturally occurring emulsifier composed of phospholipids, which contributes to the stabilization of a dispersion and enhances the dispersibility of fat and water in a wide range of food and industrial applications. It also improves texture, shelf life and wetting properties during the preparation of doughs [G.R. List, 2015] and [Qingyi Xu, et.al, 2011]. The properties of the lecithin granules used are listed in the data sheet in the Tab. 3.4. Further details can be found in appendix A.

Table 3.4 Soy-Lecithin properties from the Company's Data Sheet.

Product identifier	Source	Appearance	Water Solubility	Purity	Grain size mm	Relative Density
Lecithin-Granulat	Mivolis, dm in Germany	yellow to amber granular solid	soluble	Pure Soja-Lecithin	≤ 2*	1.0302

* Measured by the microscope (Section 3.3.1).

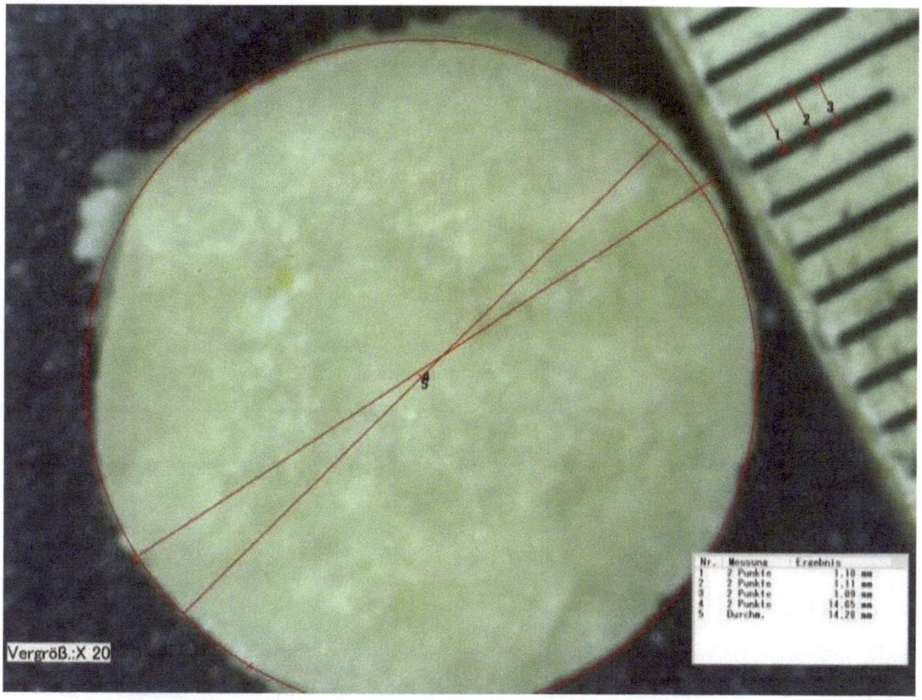

Figure 3.4 20% CMC specimen under microscope for shrinkage measurement.

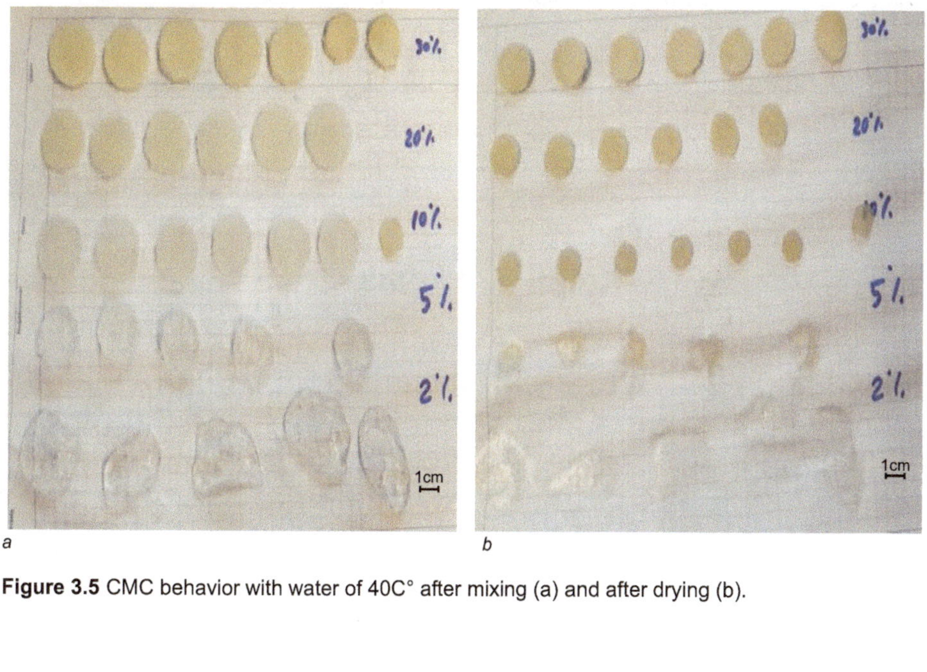

a *b*

Figure 3.5 CMC behavior with water of 40C° after mixing (a) and after drying (b).

3.3.1.4 Starch

In the food industry, it's widely used to thicken, stabilise and gel food products, giving them viscosity, structure and stability. In the paper and textile industries, it is used as a sizing agent to give paper strength and smoothness, and to improve yarn strength and smoothness in textiles [Baranwal, J et. al, 2022].
The properties are displayed in Tab. 3.5.

Table 3.5 Starch properties from the Company's Data Sheet.

Product identifier	Source	Appearance	Water Solubility	Average granule size µm	Relative Density	Source
03967 Stärke	Monolith Mitte GmbH/Germany	white solid powder	Gelatinized* at 60C°	≤20	1.5	From potato

* The gelatinisation of starch with hot water occurs by breaking the intermolecular hydrogen bond and disrupting the order of the micelle skeleton in the starch granules [Xu Fan et. al, 2021].

3.3.1.5 Chalks

Chalk is a soft, permeable sedimentary rock consisting mainly of limestone, formed mainly from calcium carbonate ($CaCO3$) [Fabricius, I. L. 2007]. It is included in the mix because it is a common material in the building environment. Tab. 3.6 shows the properties of the chalk used according to the material's technical data sheet.

Table 3.6 Properties of chalk from the data sheet of the company.

Index	Product company	Appearance	Water Solubility	Purity	Grain size mm
Calcium carbonate	Wolfgang Kenter Kalk-Laden	a white color	Insoluble	>98% CaCO3	≤ 20 µ (0.02 mm)

3.3.1.6 Magnesium Sulphate MgSo4

The physical properties are illustrated in Tab. 3.7.

Table 3.7 MgSo4 properties from the company datasheet.

Product identifier	Brand	Appearance	Density g/cm³	Water Solubility	Decomposition temperature °C
Magnesium sulfate Product Number: M7506	SIGALD from company: Sigma-Aldrich Chemie GmbH	Form: solid Color: white	2,66 at 20°C	300 g/l at 20 °C - completely soluble	1.124

3.3.1.7 Wood Fibers

The properties are shown in Tab. 3.8 and Fig. 3.6. The sieve analysis is shown in Fig. 3.7.

Table 3.8 Wood fibers properties from the company's data sheet.

Product identifier	Brand	Appearance	Density g/cm^3	Purity	Grain size mm	Raw material
Holzfasern als Substratausgangsstoff (RAL-GZ 250/5-1)	Gütegemeinschaft Substrate für Pflanzen e.V, Germany	yellow solid fibers	0.55 at 20°C	No Foreign matter (e.g., plastics, metals, Glass)	According to the used sieve, Fig. 3.7.	Wood chips made from chemically untreated softwood

Figure 3.6 Type of woof fiber used.

Figure 3.7 Sieve analysis for the wood fibers conducted at TU Darmstadt, graded from top to bottom as: 1.0, 0.5, 0.4, 0.25, 0.125mm, then pan.

3.3.1.8 Water

Taped water is used for the mixtures.

3.2.2 Formulation and Mixture Design

Several trials were carried out to find a suitable mixture in terms of extrusion capability depending on nozzle size, buildability, shrinkage, adhesion and strength. Furthermore, the size of the die has a great influence on the choice of the type of cellulose to avoid clogging, especially for the fine die; where for a 0.6 mm die, powder cellulose was usually used. Tab. 3.9, Tab. 3.10, and Tab. 3.11 document these trial mixes for 1.2mm, 1.6mm and 0.6mm nozzles. Other challenges have also been encountered in relation to mixed design when printing a large object, this must be given a crucial priority as printing here is for structural design and engineering-built environment and not for small sizes. Therefore, buckling and collapsing issues must be considered, so the mix should be designed to be able to support its own weight in the wet phase within the drying process until it is completely dry. The drying time is at least a week, depending on the mass of the object. This can be achieved by using the lowest possible water content, increasing the cellulose content and/or using fibrous cellulose, fillers, and the addition of another fiber's percentage for reinforcement.

The printable mix consists of four components with water. Each component of the mix contributes to a specific purpose to achieve a compatible paste with all the above considerations.

Cellulose is the main structure of the mix, acting as a skeleton for the paste. The cellulose particles or fibers are held together by an adhesive material, which in this case is carboxymethyl cellulose "CMC". CMC is a hydrogel material which when mixed with hot water would form a gel; the activity of CMC within the mix is as a thickener/rheology modifier and provides the adhesive property between all the components of the mix.

To stabilize the gel formation of CMC with water, fillers are used to increase the number of interparticle contacts, which are effective in improving the buildability factor. On the other hand, to improve the flowability and to help the paste to deform by extrusion through the die, lecithin is applied within the mixture formulation for this pursuit, where lecithin is an organic material of fatty substances, its oily nature helps the cellulose with other components to be able to deform and extrude outside the nozzle. [Abdullah Agha and Knaack, 2022].

Table 3.9 Trial mixture formulations and the proportions for 1.2-, 1.6-, 0.6- mm diameter nozzle.

Paste	Variable	Cellulose* % by wt.	Water % by wt.	CMC* % by wt.	Lecithin % by wt.	Filler* % by wt.	Wood-fiber % by wt.
1 or Rh.1	Changing cellulose	13 Type 5	79.6	2.5	2.5	2.5 chalk	0
2 or Rh.2		13 Type 4	79.6	2.5	2.5	2.5 chalk	0
3 or Rh.3	Increasing Lecithin content	12.8 Type 5	78.7	2.4	3.6	2.4 chalk	0
4 or Rh.4	Increasing CMC and Lecithin content	12.7 Type 5	77.8	3.5	3.5	2.4 chalk	0
5 or Rh.5	Increasing CMC content	12.8 Type 5	78.7	3.6	2.4	2.4 chalk	0
6 or Rh.6	Without filler or fatty material**	37.5 Type 5	50	12.5	0	0	0
7 or Rh.7	Paste 6 with adding fatty material	35.5 Type 5	47.2	11.8	5.6	0	0
8	Change proportions	14.5 Type 4	77.4	3.34	2.6	2.23 chalk	0
9		15 Type 4	74.6	3.8	3.8	2.3 chalk	0
10 or Rh.8	Changing cellulose and filler type	15.5 Type 1	74.6	3.8	3.8	2.3 starch	0
11 or Rh.9		15.5 Type 3	74.6	3.8	3.8	2.3 starch	0
12 or Rh.10		15.5 Type 2	74.6	3.8	3.8	2.3 starch	0
13		15.5 Type 2	74.6	3.8	3.8	2.3 chalk	0
14 or Rh.11		15.5, Type 3	74.6	3.8	3.8	2.3 Mgso4	0
15 or Rh.12	Change proportions	15.5 Type 3	74.6	3.05	2.7	3.4, starch	0
16 or Rh.13	Tested after 6 days of mixing	15.5 Type 3	74.6	3.8	3.8	2.3, starch	0
17 or Rh.14		15.5 Type 3	74.6	4.2	3	2.7, starch	0
18 or Rh.15	Rotting paste tested after 1 month of mixing	15.5 Type 3	74.6	3.8	3.8	2.3 chalk	0

* Material type according to section 3.2.1.1 Raw material and its Selectivity.
** Mixture formulation from C. Thibaut, 2020.

In this research, the smallest three nozzles available with the printing machine (LUTUM-vormvrij version 2.1) are investigated, which are 0.6mm the pink one, the grey colour 1.2mm and the green one 1.6mm. The test mixes are started with the 1.2 mm medium nozzle to find the right mix fit without blocking. Then these amounts for the components in Table 3.9 would be approximate starting values to estimate the amount of mixture components for the other nozzles such as 1.6mm and 0.6mm. A detailed demonstration of the print quality of these pastes would be given in chapter four.

Table 3.10 Additional trial mixture formulations and the proportions for 1.6 mm diameter nozzle.

Paste	Variable	Cellulose* % by wt.	Water % by wt.	CMC* % by wt.	Lecithin % by wt.	Filler* % by wt.	Wood-fiber % by wt.
19 or Rh.16	Adding wood fibers (passing from 1mm sieve – remaining on 0.5mm)	15.5 Type 2	74.6	3.8	3.8	2.3 chalk	+2
20 orRh.17		15.5 Type 2	74.6	3.8	3.8	2.3 chalk	+3
21 or Rh.18		15.5 Type 2	74.6	3.8	3.8	2.3 chalk	+5
22 or Rh.19	Adding wood fibers (passed from 0.5mm sieve – remained on 0.4mm)	15.5 Type 2	74.6	3.8	3.8	2.3 chalk	+2
23 or Rh.20		15.5 Type 2	74.6	3.8	3.8	2.3 chalk	+3
24 or Rh.21		15.5 Type 2	74.6	3.8	3.8	2.3 chalk	+5
25 or Rh.22	Changing filler type and increasing its percentage	16 Type 2	74	4	3	3 chalks	0
26 or Rh.23		16 Type 3	74	4	3	3 Mgso4	0
27 or Rh24		16 Type 2	74	4	3	3 Starch	0
28 or Rh.25		13.5 Type 2	64.4	3.5	2.6	15.6 Starch	0
29 or Rh.26		14.8 Type 2	69	3.7	2.8	9.7 Starch	0
30 or Rh.27	Change proportions	15, Type 2	69.2	7.5	2.8	5.6, Starch	0
31 or Rh.28		15.5, Type 2	71.5	5.8	2.9	4.3, Starch	0
32 or Rh.29		13.4, Type 2	78	3.6	2.6	2.3, chalk	0

The principle followed in designing the mix is to find a formulation with less water content without affecting the consistency of the mix, so manipulating with the other components by increasing until a printable paste is obtained for each specific nozzle. The basic objectives of stiffening the paste are to reach a certain level of buildability and have a better shape retention after printing and during drying. The challenge is to keep the sample stable under its own wet weight after printing until drying and to have good shape retention to avoid collapse. Other aims are to reduce shrinkage and deformation after drying and to maintain dimensional accuracy and print quality.

For the set (pastes 1 to 9 in Table 3.9), the variation here is to investigate the performance of using the adhesive element (CMC), the oily substance (lecithin) and the filler effect within the compound formulation.

Table 3.11 Mixture formulations and the Proportions of the Mix design for 0.6 mm diameter nozzle.

Paste N1.6	**Variable**	**Cellulose*** % by wt.	**Water** % by wt.	**CMC*** % by wt.	**Lecithin** % by wt.	**Filler*** % by t	**Wood-fiber** % by wt.
33 or Rh.30		15.8 Type 1	68.7	6.2	4.7	4.7 Mgso4	0
33 or Rh.31	Adding wood fibers (Passing from 0.25mm sieve – remaining on 0.125mm)	15.8 Type 1	68.7	6.2	4.7	4.7 Mgso4	+0.4%
35 or Rh.32		15.8 Type 1	68.7	6.2	4.7	4.7 Mgso4	+0.8%
36 or Rh.33		15.8 Type 1	68.7	6.2	4.7	4.7 Mgso4	+1.2%
37 or Rh.34		15.8 Type 1	68.7	6.2	4.7	4.7 Mgso4	+1.6%
38 or Rh.35		12 Type 5	82	2.4	2.4	2.4 Chalk	0
39 or Rh.36	Adding wood fibers (passing 0.4mm sieve – remaining on 0.25mm)	12 Type 5	82	2.4	2.4	2.4 Chalk	+0.32
40 or Rh.37		12 Type 5	82	2.4	2.4	2.4 Chalk	+0.96

For the group (pastes 9 to 14), the proportions of the components were varied to find out the behavior and the best ratios in terms of viscosity and yield shear by changing the variables.

In the group (pastes 17 to 23), the efficiency of the paste was checked over time. Group (pastes 17 to 21) the effect of the addition of wood fibers was studied in particular when using the same percentage but different size ranges.

Group (pastes 24 to 31) illustrates the importance of reducing water and using alternatives to compensate for flowability in terms of changing filler type, indicating the appropriate one and specifying the appropriate ratios.

Pastes 32 to 36 are specific to the 0.6 mm nozzle and show the percentages of wood fibers that could be used and extrudable.

Pastes 37 to 39 are specific to the 0.6 mm nozzle and indicate the percentages of wood fibers studied that could be extruded.

3.2.3 Mixing Process Procedure

The performance of the mixture is represented by the microstructure of its components and how the components are bonded together to form a network in which cellulose fibers are bonded together and adhere to each other by interlocking with the bonding material and other components to form a mat. This interlocked fibrous network is achieved by an effective mixing method to achieve homogeneity and even distribution of the components within the mat, particularly for those with limited water content. The correct mixing procedure has been determined through trial and error and research into the appropriate process to achieve the most efficient method of achieving homogeneity. To achieve homogeneity, the subsequent steps are followed, as shown in Fig. 3.8:

1. The cellulose was placed in an oven for 24 hours at 40°C to remove any water content in the cellulose so that it wouldn't affect the weight. The dried cellulose of any given mixture is then weighed immediately after being taken out of the oven and then cut into square pieces of about 20x20 mm. The pieces are then immersed in a well-sealed container of water for (2-3) weeks. Before weighing the cellulose, it is mixed for 2 minutes with a planetary mixer using a 2-wire stainless steel whisk with a shaft (1) speed: 105 rpm, to ensure complete disintegration.
2. The water soaking the cellulose is replaced with hot water of 45°C through a filter sieve / 400mesh and should be at least ¾ of the total amount of water in the mixture. The cellulose and hot water are weighed together.
3. Dissolve the lecithin with ¼ of the total amount of water in the mixture until a creamy consistency is obtained, using an electric hand mixer. The time required is approximately 5-10 minutes, depending on the amount to be liquefied.
4. Moisten the planetary mixer before use.
5. The dissolved lecithin is added to the mixture (cellulose & water) and then everything is mixed together with a planetary mixer at a speed of 408 rpm for 10 minutes. The fat is added first to facilitate the mixing process, which gives the ingredients additional fluidity. In this way, mixing is easy.
6. Gradually add the CMC and mix for 15-20 minutes.
7. Finally, the filler (chalk or starch or magnesium sulphate) is added, and the components are mixed for 15-30 minutes, depending on the consistency of the mixture. The pastes 6 and 7 in Tab. 3.9 are mixed for 60 minutes.
8. If wood fibers are used for the mixtures with wood fibers, the wood fibers are distributed during mixing in step 7 and the mixing time is increased by 10 minutes.

Figure 3.8 Disintegration for cellulose by cutting into 20mm square (a), mixing with water(b), and the final consistency of the mixing.

3.2.4 Rheology

A comprehensive rheological characterization of the pastes is essential to understand how rheology applies to the properties of flowable materials due to their viscoelastic behavior. It is important to target extrudability optimization within stiffness ranges and to meet the different requirements during the manufacturing process and application demands. Rheology studies the deformation behavior and flow of matter. This characterization is mainly defined by the consistency behavior influenced by viscosity and the yield stress caused by shear stress [Glatthor & Schweizer, 1994]. The high shear yield stress correlates with higher buildability and better stability during production until drying. Conversely, viscosity is associated with ease of extrusion. Viscosity is the resistance provided by the material to the movement of the system. Successful extrudability of pastes requires control of process and rheological parameters. These parameters are mainly chemical and physical. They have a strong influence on the rheology of a compound, such as: water/solid content, bulk density, type and surface area of the raw materials, the final geometric configuration, and the surrounding environment [Kuder & Shah, 2010].

3.2.4.1 Rheometer Preparation and Setup

The experimental measurements were carried out using the rheometer ©Anton Paar/ MCR 302/SN81834655 in the laboratory of the TU Darmstadt, Germany, shown in Fig. 3.9. More pictures are shown in Appendix B. The tests were conducted with the shear-controlled measurement setting, rotational shear experiments. The measurements were performed in stepwise controlled shear rate mode between 0.01 sec^{-1} - 1000 sec^{-1} in 40 steps; and the measurement time is up to 1200 sec [Yüce, C., and Willen bacher 2017]. Each experiment is performed with a rotating upper plate and a fixed lower plate. The upper plate has a diameter of 20 mm and a smooth surface. The parallel lower plate has a diameter of 25 mm, and its geometry has a roughness (Rq = 2 - 4 µm). The room temperature is air-conditioned to 20°C±2. The gap between the upper and lower plates was studied to be optimized to choose the adequate behavior to the reality and to adapt the fiber length. The principle of using a gap larger than the fiber length is to avoid pressing the fibers or controlling their movement during the test. So that the result is not affected if some fibers are oriented in the longitudinal direction of the gap distance. This allows the fibers in the tested paste to be randomly oriented and to move freely during the rotation of the rheometer without being compressed by a gap smaller than their length. The 3 mm gap should be filled with a volume of 5.89 mL of paper paste mixed homogeneously.

However, for the powder-cellulose paste, the effect of changing the gap distance between 1 mm and 5 mm was studied, as shown in Tab. 3.12. For most other tests, the gap is controlled at 3 mm.

Table 3.12 The gap-distance investigation for the set-up of the rheometer.

Paste	Variable	Gap space	Cellulose	Water	CMC	Lecithin	Filler
			% by weight	% by weight	%by weight	% by weight	% by weight
Rh.8	Changing gab space for Rheometer setups	3mm	15.5 powder	74.6	3.8	3.8	2.3, starch
Rh.8		1mm	15.5 powder	74.6	3.8	3.8	2.3, starch
Rh. 8		5mm	15.5, powder	74.6	3.8	3.8	2.3, starch

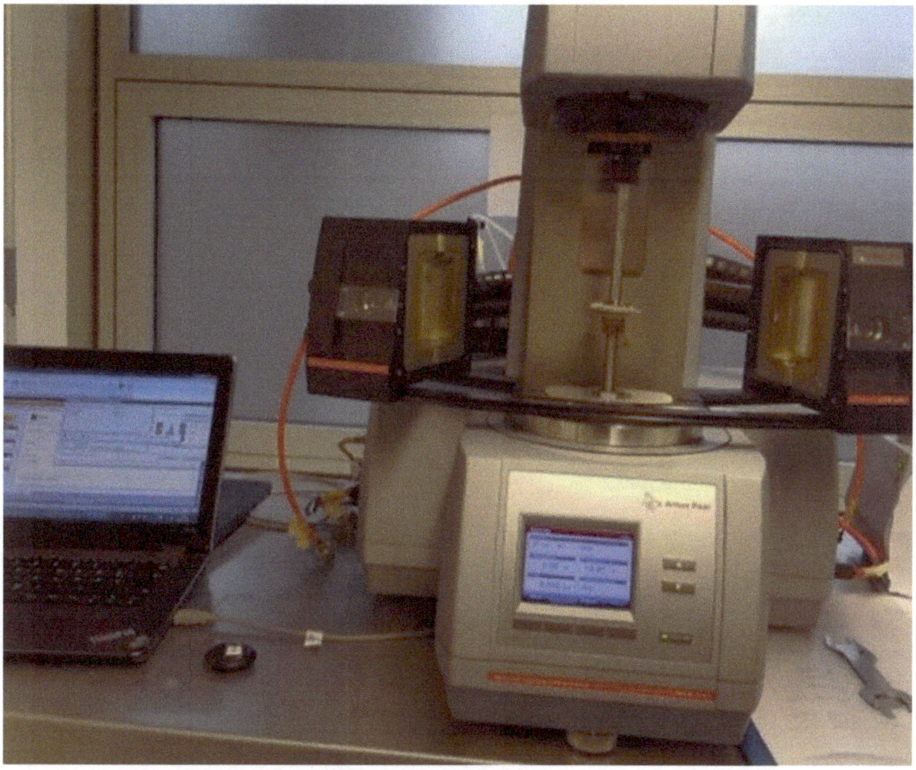

Figure 3.9 Used rheometer ©Anton Paar/ MCR 502.

3.2.4.2 Measurement Procedure

- The rheometer setting is adjusted as described in Section 3.2.4.2 Rheometer Preparation and Setup. The motion profile is set as pasty.
- The mixture must be homogeneously mixed before use.
- A small amount of the mixed paste is weighed so as to fit approximately the volume of the gap in the plates and according to the specific density. It is then applied to the lower plate of the parallel-plate rheometer and the upper plate brought into a measuring position. This amount should be about 5.89 ml and not much more, so that the paste doesn't compress too much, the result then would be affected by this compression. Nor should it be less than this height, as this would not fill the gap well. By taking in consideration, the top surface of the paste should be flat and even.
- Due to the pressing, the normal force on the paste must be reduced to near zero, so waiting time about 5-10 minutes is placed before starting the measurement.
- The apparent viscosity or apparent shear deformation is plotted logarithmically against the apparent shear rate.

It's important to note that the calculated stresses and shear rate values should be treated as apparent or nominal, since the actual deformation and stresses inside the gap are not known.

3.2.4.3 Rheology Behaviour

From the curves plotted (logarithmic shear deformation vs logarithmic shear stress) it can usually be seen that the rheological behavior obtained can be divided into three phases. The first phase is an almost linear relationship in which the shear rates increase continuously until the yield shear is reached. The adhesion between the components and the components themselves have a significant effect here in stiffening the viscoelastic behavior until it starts to yield, in which case the resistance to wall slip dominates.
In the second phase, the deformation of the paste increases monotonically and rapidly. This phase is curvilinear, short and almost indistinct, except for the tangents of the first and third phases. This means that at a critical shear rate or specific angular velocity, the paste creeps out of the gap while the apparent viscosity and shear stress drop sharply due to the collapse of the sample, as shown in Fig. 3.10 (a). Accordingly, this region of the curves shows a characteristic kink. The third phase begins at the critical stress, called the apparent yield stress, τy, where the elastic reversible deformation clearly transitions to irreversible flow, so that the measurement points are widely spaced and have larger values.

This yield stress is a key factor in paste formulation, especially when designing the mix for additive manufacturing. It helps to identify the paste to be classified according to the level of buildability. A high yield stress is desirable to ensure shape retention after printing and stability by increasing the buildability so that the lower layers can take more load from more layers above without collapsing. Typically, this stress is distinguished by the kink in the deformation versus shear stress curve. The intersection of two tangents is an example of the value of this yield stress, as shown in Fig. 3.10 (b).

Tab. 3.13 and Fig. 3.11 show all the yield shear stress and apparent viscosity results calculated from the shear deformation-stress curve and the shear viscosity logarithmic relationship for all the mixtures. The yield shear stresses in Tab. 3.13 are determined as in Fig. 3.10. The apparent viscosity was determined as an average of at least three samples from the average of the apparent viscosity values of the first phase of the paste behavior.

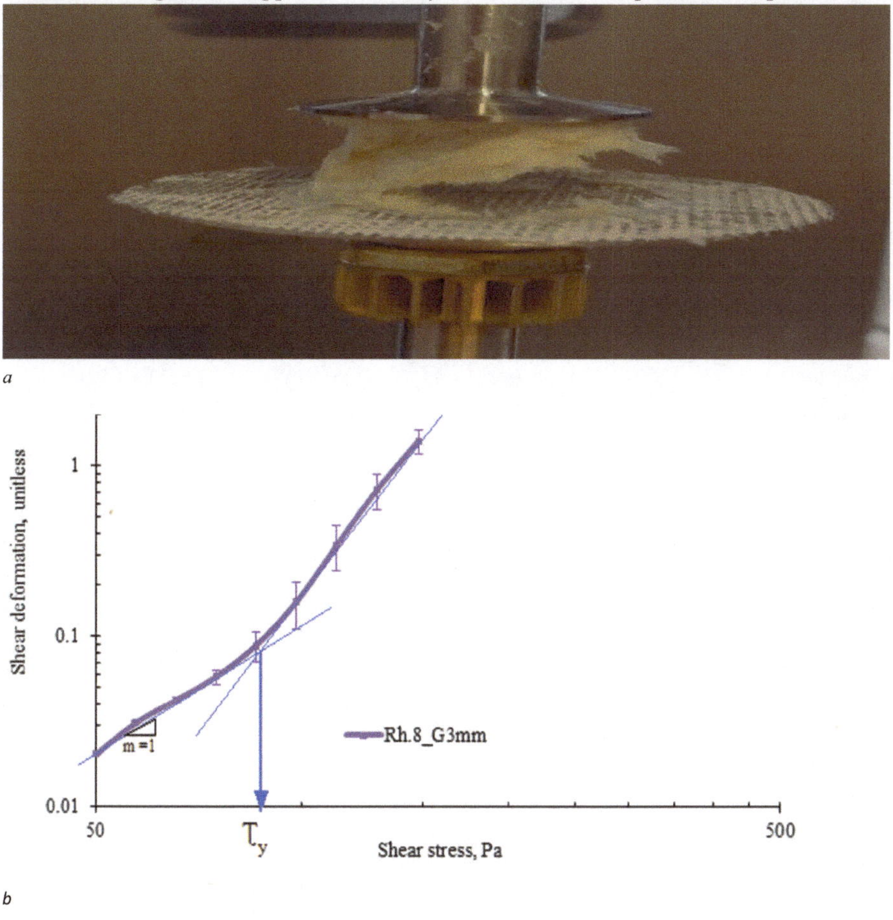

b

Figure 3.10 The paste at failure (a) and laying the intersection of the tangent for yield stress locating.

Table 3.13 The values of yield stress and apparent viscosity for the pastes.

Paste from Tab.3.9	Shear Yield stress, Ty Pa	Apparent Viscosity mPa·s
Rh.1 (paste 1)	381 ± 48	1.26 E13 ± 0.6E12
Rh.2 (paste 2)	450 ± 90	1.96E13 ± 0.8E13
Rh.3 (paste 3)	290 ± 97	1.2 ± 0.5E13
Rh.4 (paste 4)	436 ± 12	1.35 E13 ± 0.7E13
Rh.5 (paste 5)	405 ± 43	1.65E13 ± 0.8 E13
Rh.6 (paste 6)	450 ± 65	31.8E13 ± 3 E13
Rh.7 (paste 7)	270 ± 20	7.11E13 ±1 E13
Rh.8 (paste 10)	98.4 ± 10	0.166E1 ± 0.05 E13
Rh.8_ Gap 5mm	96.5 ± 7	0.173E13 ±0.07E13
Rh.8_ Gap 1mm	88.2 ± 8	0.264E13 ± 0.7 E13
Rh.9 (paste 11)	430 ± 55	1.25 E13 ± 0.55E13
Rh.10 (paste 12)	147.5 ± 18	0.775E13 ± 0.23 E13
Rh.11 (paste 14)	290 ±36	0.962E13 ± 0.42E13
Rh.12 (paste 15)	436.6 ±55	1.92E13 ± 0.2E13
Rh.13 (paste 16)	220 ± 10	1E13 ± 0.22E13
Rh.14 (paste 17)	169 ± 22	0.96E13 ± 0.2E13
Rh.15 (paste 18)	129 ± 20	0.38E13 ± 0.15E13
Rh.16 (paste 19)	655.5 ± 50	3E13 ± 1E13
Rh.17 (paste 20)	333 ± 40	2.8E13 ± 1.3E13
Rh.18 (paste 21)	254 ± 35	2.5E13 ± 0.55E13
Rh.19 (paste 22)	290.5 ± 40	1.98E13 ±0.4E13
Rh.20 (paste 23)	150 ± 22	2.4E13 ± 0.66E13
Rh.21 (paste 24)	147 ± 45	2.5E13 ±0.9E13
Rh.22 (paste 25)	381 ± 20	1.84E13 ± 0.45E13
Rh.23 (paste 26)	984 ± 63	1.48E13 ± 0.58E13
Rh.24 (paste 27)	436 ± 25	1.55E13 ± 0.73E13
Rh.25 (paste 28)	572.5 ± 28	3.5E13 ± 1.2E13
Rh.26 (paste 29)	720 ± 95	2.67E13 ± 0.92E13
Rh.27 (paste 30)	655.5 ± 40	6.14E13 ± 1E13
Rh.28 (paste 31)	572.5 ± 72	2.83E13 ± 0.6E13
Rh.29 (paste 33)	193 ± 7	1.35E13 ± 0.5E13
Rh.30 (paste 34)	140 ± 20	0.84E13 ± 0.4E13
Rh.31(paste 35)	98 ± 5	0.414E13 ± 0.1E13
Rh.32 (paste 36)	75 ± 10	0.328E13 ± 0.1E13
Rh.33 (paste 37)	65 ± 10	0.33E13 ± 0.05E13

3.2.5 Rheology

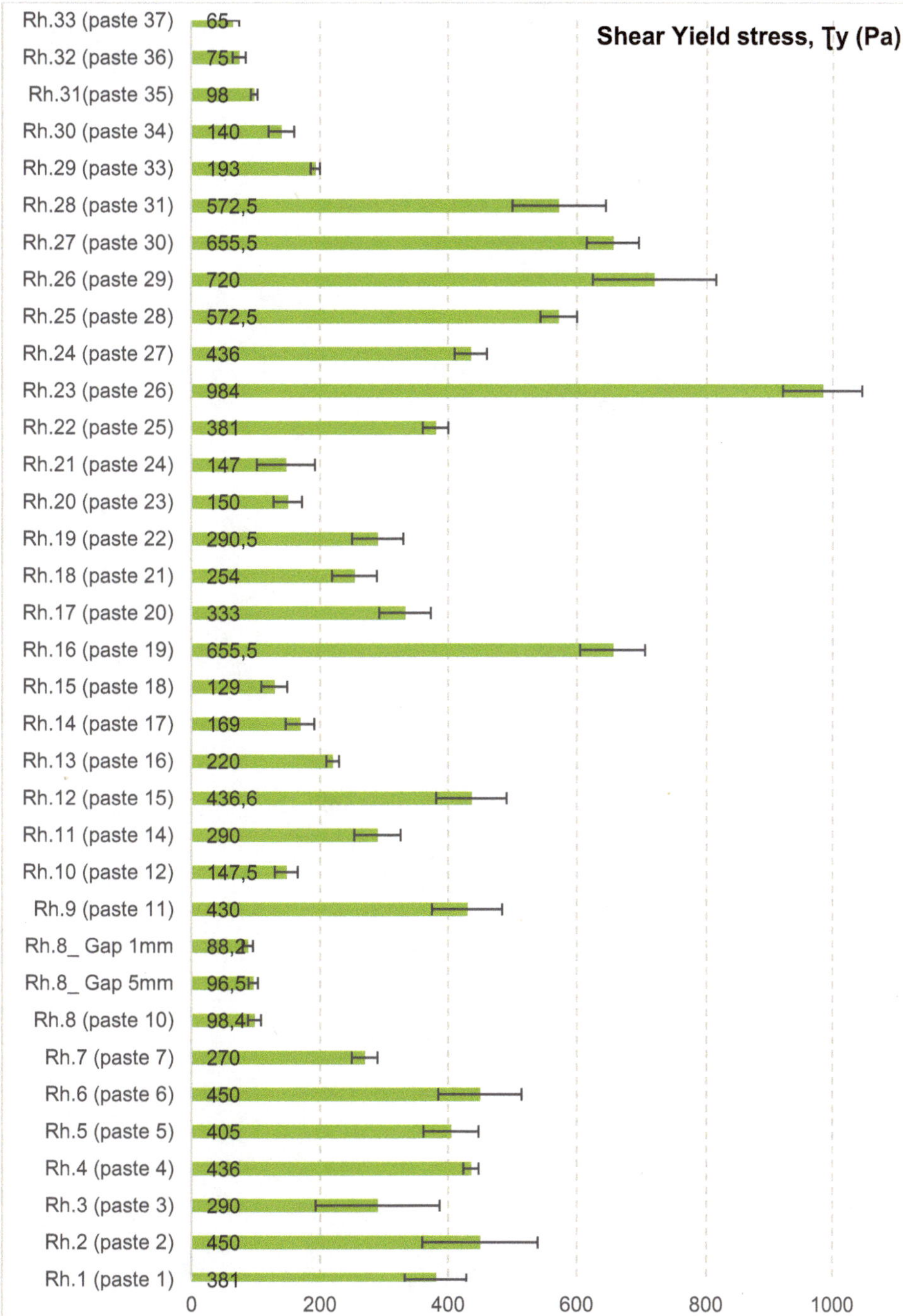

Figure 3.11 The results of the rheology behavior for all pastes.

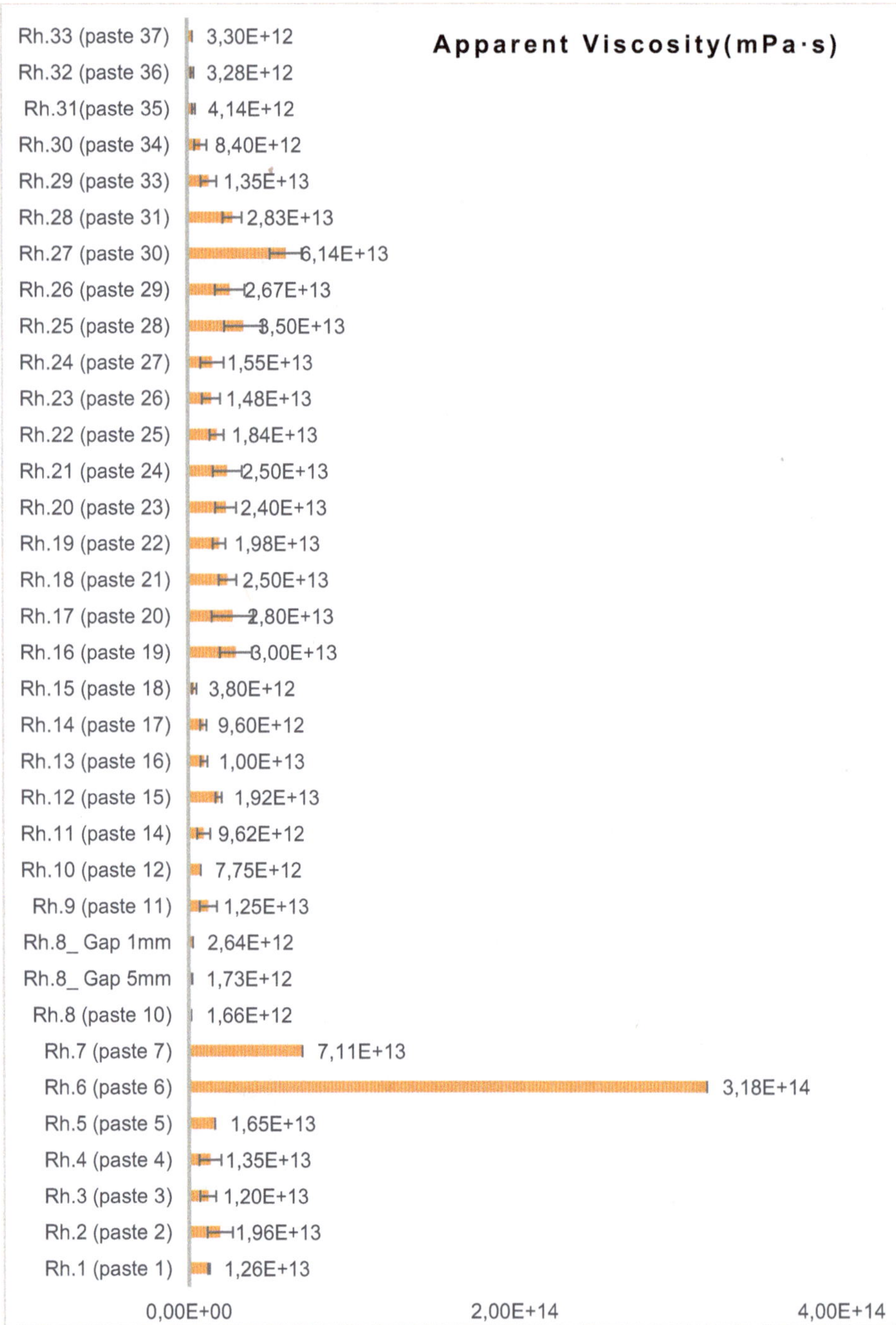

3.2.4.4 Measurement Setting (Gap Distance)

Tab. 2.13 and Fig. 3.12 show that the results of Rh8-3mm gap and Rh8-5mm gap are both closer together than Rh8-1mm, although the paste used for (Rh8-3mm gap, Rh8-5mm and Rh8-1mm) is the same and it's homogeneous, isotropic, and identical as the cellulose is powder. This means that reducing the gap distance reduces the potential for deformation and gives an inexact stiffer behavior due to less material resulting in less flowability and less free movement. Reducing the gap distance also reduces the yield stress because a 1mm gap distance does not have enough material inside to allow adhesion to participate in the resistance. In conclusion, a 3mm gap is more efficient to proceed with the study of the other tests as it is suitable greater than the cellulose fiber length.

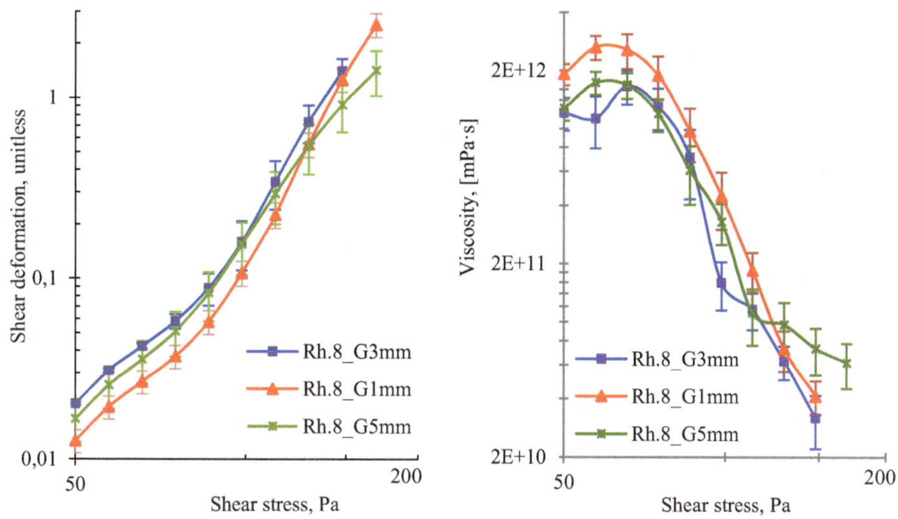

a *b*

Figure 3.12 The shear deformation-stress (a); and (b) shear viscosity logarithmic relationship for changing gap distance of Rh.8.

3.2.4.5 Changing Cellulose Type

Fig. 3.13 shows that changing the type of cellulose to cellulose with longer fibers (Rh.2) (cellulose type 4 instead of type 5, properties in section 3.1.1.1) gives a remarkable effect on the rheology behavior in terms of increasing the yield stress and the apparent viscosity compared to the paste of cellulose with short fibers (Rh.1); by 78.43% and 107.94% respectively, due to the higher shear strength of the cellulose with long fibers.

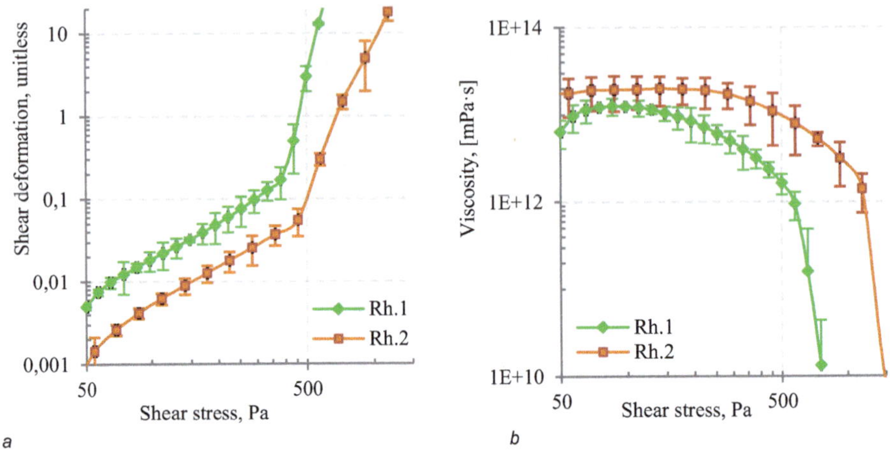

Figure 3.13 (a) The shear deformation-stress; and (b) shear viscosity logarithmic relationship for changing cellulose type of (Rh.1 and Rh.2).

3.2.4.6 Impact & Optimal Ratio of Lecithin & CMC

The importance of the addition of fatty substances (lecithin) and CMC within the formulation is illustrated in Fig. 3.14. Rh.1 is the reference paste. Rh.3 is a paste with increased lecithin ratio. Rh.4 is a paste with increased both CMC and lecithin content. Rh.5 is a paste with increased CMC. When the lecithin content is increased by 46% and the water content is reduced by 1.13% of the total weight (Rh.3 compared to Rh.1), the flowability is significantly improved; the yield stress and apparatus viscosity are reduced by 23.9% and 5% respectively. While the results are better when both the lecithin and CMC are increased by 40% and the water content is reduced by 2.3%. The viscoelastic behavior of Rh.4 is initially closer to Rh.1, but it becomes stiffer consistency after about 220 Pa. Both the yield stress and the apparatus viscosity of Rh.4 increased by 14.4% and 7.1%; respectively compared to Rh.1.

In meanwhile, increasing only the CMC content lowers the flowability and gives stiffer consistency. The yield stress and the apparatus viscosity of Rh.5 increase by 6.3% and 30.9% compared to Rh.1. However, increasing CMC only gives a better rheology behavior than increasing only the lecithin, see Rh.3 compared to Rh.5 in Fig. 3.15. Both the yield stress and the apparatus viscosity of Rh.5 have increased by 14.5% and 3.2% respectively, compared to Rh.3. It can be concluded that it is preferable to increase the percentage of both CMC and lecithin rather than increase only lecithin or CMC alone when designing the subsequent mix.

Finally, Fig. 3.14 confirms that Rh.4 (which increases both CMC and lecithin) has the best behavior in this series. Rh.4 has a better buildability compared to Rh.5, Rh.1 and Rh.3, where the yield stress of Rh.4 increased by 50%, 71.7% and 31% compared to Rh.5, Rh.1 and Rh.3; respectively.

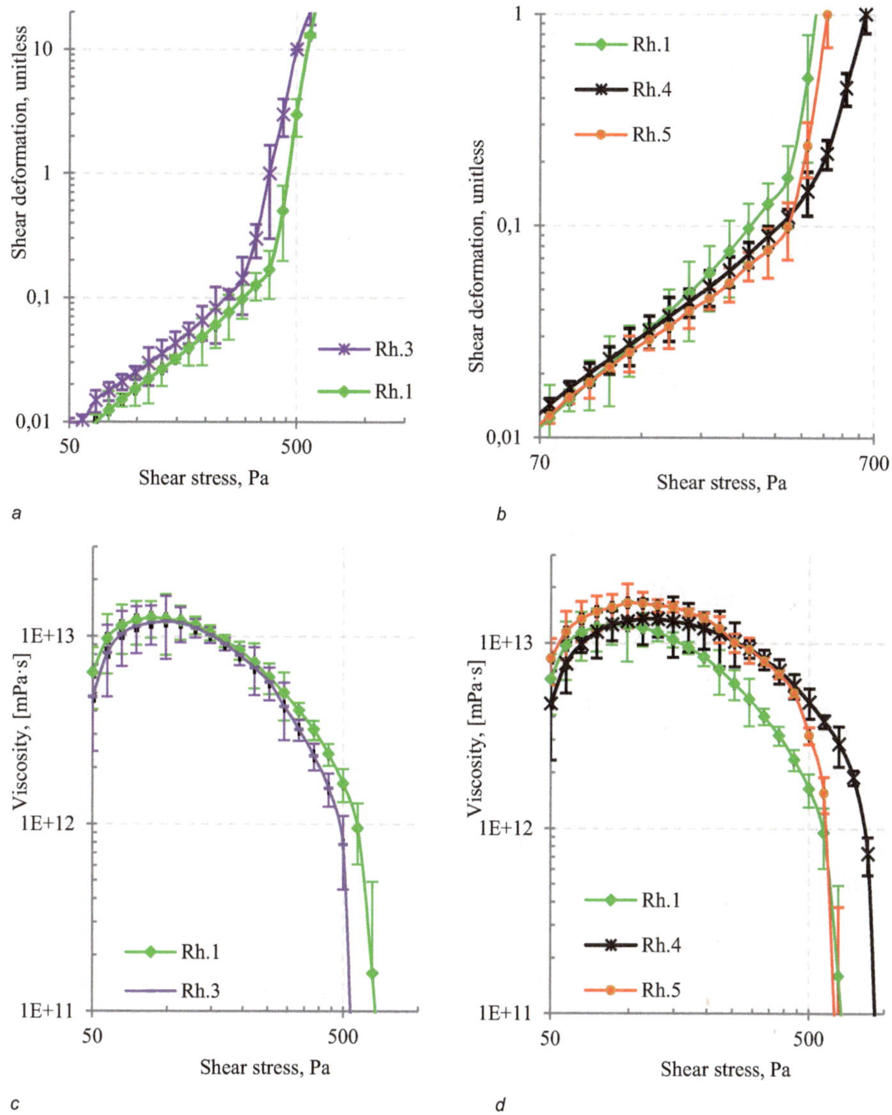

Figure 3.14 The shear deformation-stress for Rh.1 & Rh.3 (a) and Rh.1, Rh.4, and Rh.5 (b); and the shear viscosity logarithmic relationship for Rh.1 & Rh.3 (c) and for Rh.1, Rh.3, Rh.4, and Rh.5(d).

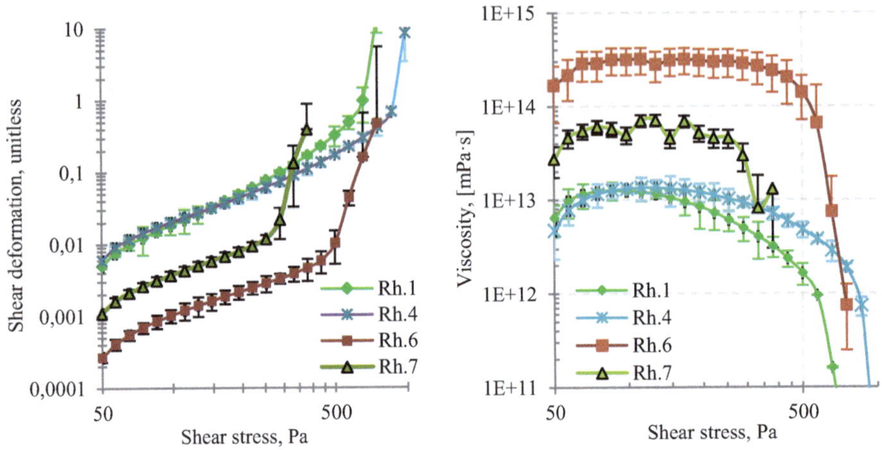

a *b*

Figure 3.15 The shear deformation-stress (a); and shear viscosity logarithmic relationship (b) for (Rh.1, Rh.4, Rh.6, and Rh.7).

Fig. 3.15 confirms the essence of adding a fatty substance and filler content into the paste formulation. Where, Rh.6 (only CMC without lecithin and filler) is an extremely stiff paste, particularly; mixing the components together is not easy to reach homogeneity.
So, adding only lecithin 5.6% by weight (Rh.7 in comparison with Rh.6) increases the flow-ability and shifted the behavior; in which the stiffness was reduced by 312.5% and 77%, in terms of shear deformation at a load equal to 254 Pa and apparent viscosity; respectively, as depicted in Fig. 3.15. Therefore, it confirms that fatty substances have a significant impact to change the rheology and consistency of the paste. Hence, it's concluded from Fig. 3.15 that Rh.4 (using lecithin and filler) has better rheology and consistency in terms of ease in extrudability and more potential to have a buildability characteristic. Even when using more water content by 57.4% and 66.7% (Rh.4 in comparison by Rh.6 and Rh.7; respectively), the yield stress is larger by 50% and 195.5% (Rh.4 in comparison by Rh. and Rh.7; respectively), and apparent viscosity is reduced by 95% and 79% (Rh.4 in comparison with Rh. and Rh.7; respectively).

3.2.4.7 Changing Cellulose with Reduced Water Content

Fig. 3.16 shows the effect of changing the type of cellulose, keeping the other components and proportions the same. Rh.9 (long fiber cellulose, type 3) has a remarkable increase in yield stress of 97.3% and 196.5% and in apparent viscosity of 29.7% and 505.4% compared to the paste of short fiber cellulose, type 2 (Rh.9) and powder cellulose (Rh.8); respectively. The use of short cellulose/type 2 (Rh.10) has a better yield stress of 49.9% and apparent viscosity of 366% in comparison with Rh.8 (paste with powder cellulose).

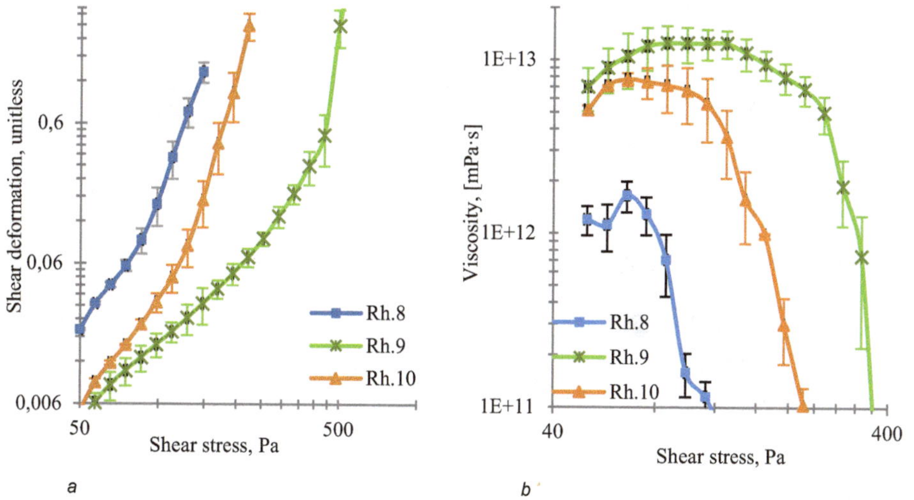

Figure 3.16 The shear deformation-stress (a); and shear viscosity logarithmic relationship (b) for (Rh.8, Rh.9, and Rh.10)

3.2.4.8 Effect of Changing Filler Type and its Proportion

Rh.9 is a paste with lone fiber cellulose. Rh.11 is like Rh.9 with using Mgso4 instead Starch. Rh.12 is a paste like Rh.9 but with reducing CMC and Lecithin by 20% and 29% by weight; respectively. Fig. 3.17 shows that Rh.11 has a behavior closer to that of Rh.9. This leads to the conclusion that the use of magnesium sulphate (Rh.11) has reduced the stiffness by 32.6% in terms of yield shear stress compared to the same paste but using starch (Rh.9). The explanation lies in the hydration of Mgso4 compared to starch, since magnesium sulphate is easily soluble in water. However, magnesium sulfate forms after drying a new chemical form that absorb water and crystallizes [Larysa Okhrimenko et.al, 2017], [Pub©hem], and [Tim Baksa et.al. 2007]. The starch is not able to dissolve and not gelatinized when the water of the mixture has a temperature lower than 45°C, because the starting temperature of starch gelatinization is at least 56°C, as shown in Fig. 3.18 [M. Schirmer et.al., 2015] and [A. Briffaza et.al. 2014].

The paste with magnesium sulphate (Rh.11) has a softer rheological behavior by 35% than the same portion paste but with starch (Rh.9) in terms of apparent viscosity. From the reaction concept of Mgso4 and starch in water, it would be possible to increase the filler content. To investigate the effect of increasing the filler content in the mix, increasing the starch to 46% (Rh.12) increased the yield shear and apparent viscosity by 1.5% and 53.6% respectively, compared to the same mix (Rh.9) but with CMC and Lecithin reduced by 20% and 29% by weight; respectively.

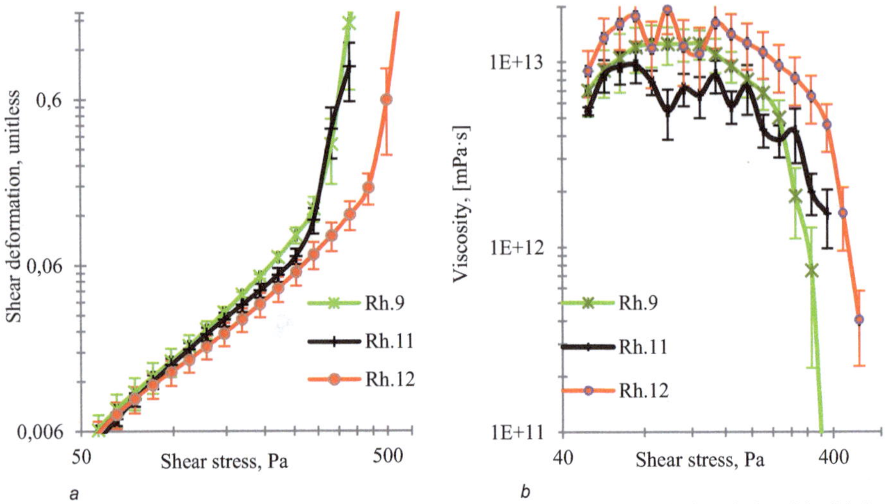

Figure 3.17 The shear deformation-stress (a); and shear viscosity logarithmic relationship (b); for (Rh.9, Rh.11 and Rh.12)

The degree of starch behavior in water depends on the treatment temperature and water content, in which the starch starts to dissolve at a temperature equal to 56°C [A. Briffaza et.al. 2014]. According to Markus Schirmer et.al. 2015, the process of developing gelatinization of the starch is starting from the native stage with a temperature equal to approximately 60°C, and it has low viscosity at the beginning of gelatinization.

3.2.4.9 Ageing Effect of the Fresh Mixture

Fig. 3.18 (a) confirms that the shear yield stress was reduced by 48.8% after 6 days of mixing and by 70% after one month of mixing; comparing Rh.9 with Rh.13 with Rh.15, respectively. With time, the paste is rotten, and the fungus causes black spots that change the yield shear of the material, since the mixture is made from a natural organic substance (CMC; food glue); this makes it susceptible to rotting and softer with time. The paste with more CMC (Rh.14) becomes worse, the shear yield stress decreases by 60.7% and by 23% when compared with a fresh paste (Rh.9) and with the rotted paste after the same aging duration. For this reason, the paste is usually not valid after days of mixing unless it is kept cooled. Figure 3.18 (b) shows that the apparent viscosity of Rh.9, Rh.13 and Rh.14 is approximately the same. In other words, the buildability and the extrudability decreases as the fresh mix ages. The apparent viscosity decreases by 63.8% because the decomposition of the rot produces water in the mix. In other words, the buildability and stability of an old mix decreased over time and that became worse when using higher CMC ratio.

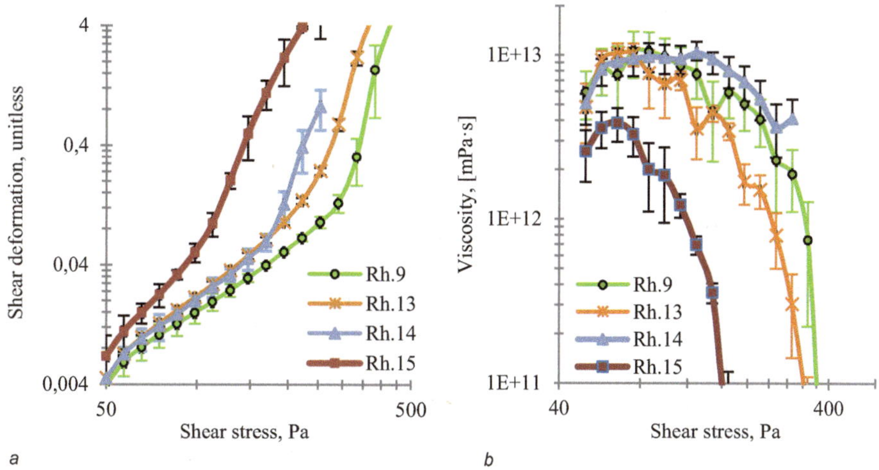

Figure 3.18 The shear deformation-stress (a); and the shear viscosity logarithmic relationship (b) for ageing effect of the fresh mixture (Rh.9, Rh.13, Rh.14 and Rh.15)

3.2.4.10 Change Wood Fibers Ratio and Ranges

Fig. 3.19 shows that the most acceptable wood fiber content is 2% that passes through the (1mm) sieve and remains on the (0.5mm) sieve. In addition, Fig. 3.19 (a) and (b) show that increasing the wood fiber content ratio without increasing the CMC content does not result in a good benefit. The reason for this is that the existing CMC content with the addition of additional wood content ratio would not be enough to cover and manage the adhesion for all the surface areas of all the fibers (cellulose and the wood fibers); where CMC is the basic ingredient to manage the adhesion and bonding between the fibers. This means that the paste with a larger number of fibers has less adhesion bond and less slip resistance.

Using 3% and 5% of (1-0.5) mm wood fibers instead of 2% reduces the yield shear stress by 49.2% and 60.8% respectively when comparing Rh.17 and 18 with Rh.16 as shown in Figure 3.19(a). Correspondingly, the use of 3% or 5% instead of 2% of (0.5-0.4) mm wood fibers reduce the yield shear stress by 48.4% (Rh.20 vs. Rh.19) and 49.4% (Rh.21 vs. Rh.19), respectively, to half the value, as shown in Fig. 3.19(b).

However, the viscoelastic behavior of the paste with (0.5-0.4) mm fiber length, which has a larger percentage, has a stiffer viscoelastic performance. That is, Rh.21 is stiffer than Rh.20 in the first phase of the rheological behavior and Rh.20 is stiffer than Rh.19, as drawn in Fig. 3.20. The explanation is that the deformation due to motion in the viscoelastic region (flow deformation progress) faces more significant obstacles, so the flow path would be longer and take more time to slip [Lei Pu et. al 2021].

Fig. 3.19 (a) confirms that the viscoelastic behavior for Rh.16, Rh.17 and Rh.18 (the pastes with fiber length of (1 - 0.5) mm) remains convergent. The reason is that the number of long fibers is lower than the number of short fibers within the same weight percentage, comparing the pastes with (1- 0.5) mm fiber length with the pastes with (0.5- 0.4) mm fiber length. This means that the number of obstacles is lower, and these percentages (2%, 3% and 5%) do not give a significant behavior clearly within the viscoelastic phase for the pastes with (1- 0.5) mm fiber length, that is why the behaviors of the pastes with long fibers are different from those of the pastes with short wood fibers.
Fig. 3.19 (c), (d) and (e) shows the effect of the wood fibers on the behavior of the shear stress-deformation relationship, which clarifies the difference between the longer fibers (1- 0.5) mm and (0.5-0.4) mm. It's obvious that the rheological behavior of the (1 - 0.5) mm pastes is stiffer than that of the (0.5 - 0.4) mm pastes. However, the value of the yield shear stress is reduced by about half when using a fiber length of (0.5-0.4) mm instead of (1-0.5) mm, since

1. By 55.7% (Rh.19 compared to Rh.16, Fig.3.19 (c)).
2. By 55% (Rh.20 compared to Rh.17, Fig.3.19 (d)).
3. And by 42.1% (Rh.19 compared to Rh.16, Fig.3.19 (e)).

In another expression, Rh. 16 has a higher value in yield shear stress by 96.8% 158%, 125.6% 337% and 345.9% compared to Rh.17, Rh.18, Rh.19, Rh.20 and Rh.21 as shown in Fig. 3.19 (f).
The theoretical explanation is that the bond between the fiber and the matrix surrounding the longer fiber is greater; and the yield shear strength of the matrix is greater. If the fiber is so short; the matrix deforms around the fiber, the friction or bond is small, then slippage or debonding occurs easily without resistance and no stress transfer and enforcement by the fiber would occur [Gibson 2007].
Figure 3.21 shows that there are no significant differences between the values of apparent viscosity for Rh.16, Rh.17, Rh.18, Rh.19, Rh.20 and Rh.21, which have a range of (2E13-3E13) mPa-s.

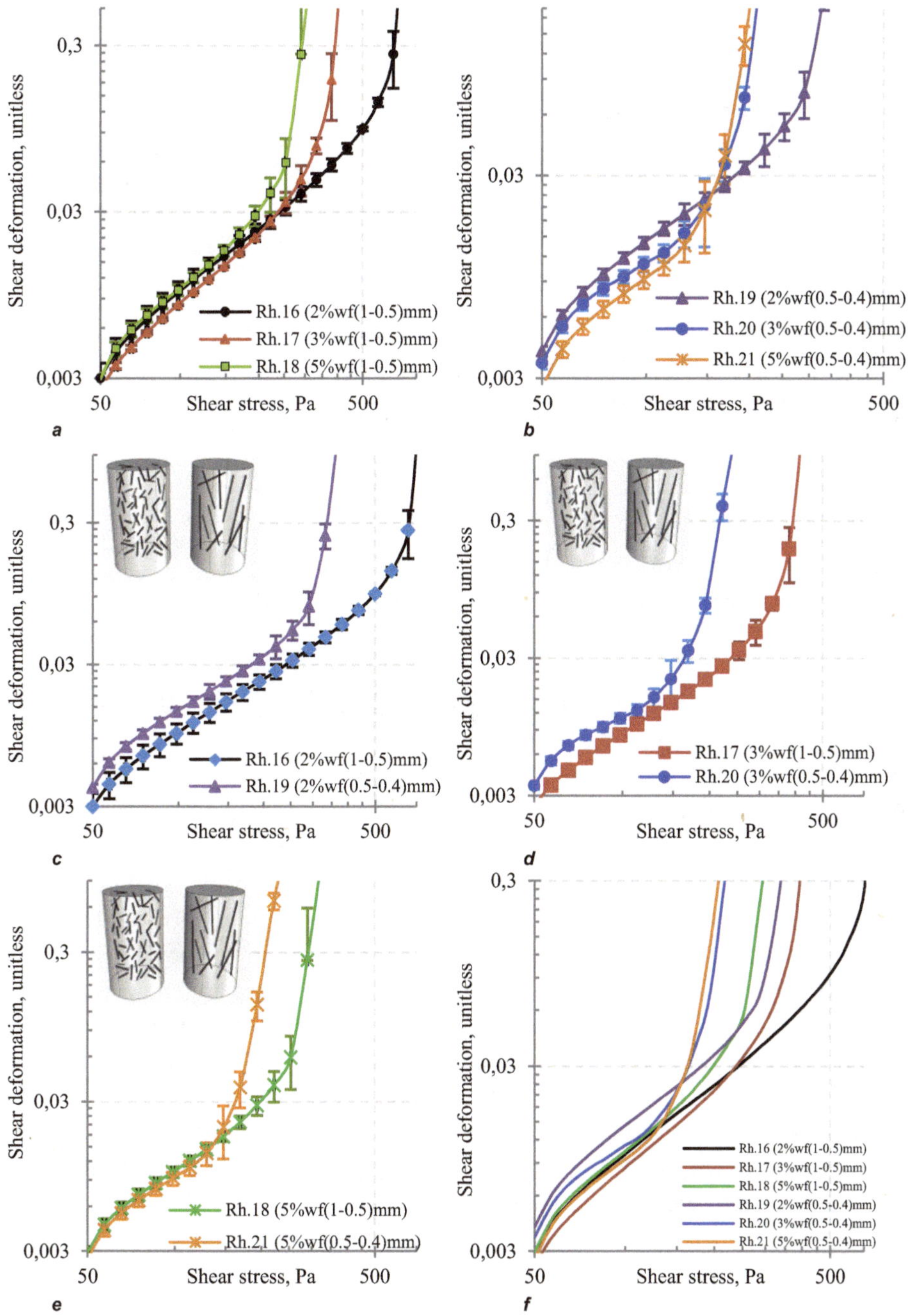

Figure 3.19 The shear deformation-stress relation for Rh.16, Rh.17, and Rh.18 (a); Rh.19, Rh.20 and Rh.21 (b); Rh.16 and Rh.19 (c); Rh.17 and Rh.20 (d); Rh.18 and Rh.21 (e); and Rh.16 to Rh.21 (f).

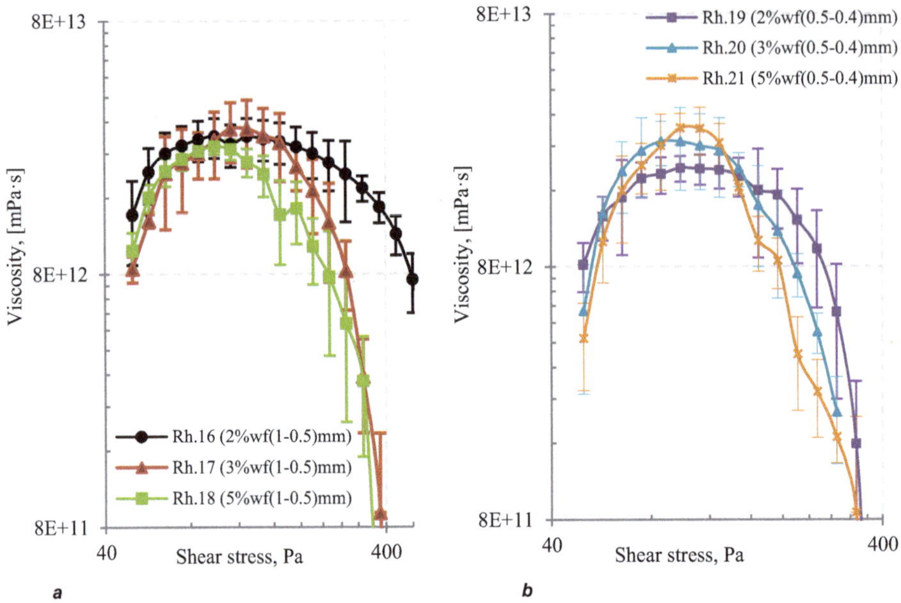

Figure 3.20 The shear- apparent viscosity logarithmic relationship for Rh.16, Rh.17, and Rh.18 (a); and Rh.19, Rh.20 and Rh.21 (b).

3.2.4.11 Changing Filler Type and Ratio with Water Reduction

The wettability behavior of the fillers used with water was studied to understand their effectiveness within the mix. The conclusion from the previous work is that the starch is insoluble in water; this gives the potential to reduce the water content within the mix. Water evaporation causes shrinkage problems. Reducing the amount of unnecessary water is therefore the objective to limit shrinkage and consequently improve quality and workability. Fig. 3.21 shows the result of mixing the filler types used at 70% with the water amount of 30%. Chalk is insoluble in water, but water penetration into the porosity of the chalk forms a colloidal concentration of surfactant. [Manish Thakker, 2015]. Fig. 3.21 pictures the flowability of the filler samples when mixed with water after 6 seconds of turning the pot from standing to lying position. These films are mixed from 40 % filler (starch) or 60% water. The chalk sample flows homogeneously, while the starch sample pours only the free water, and the starch remains separate and slowly flowing.

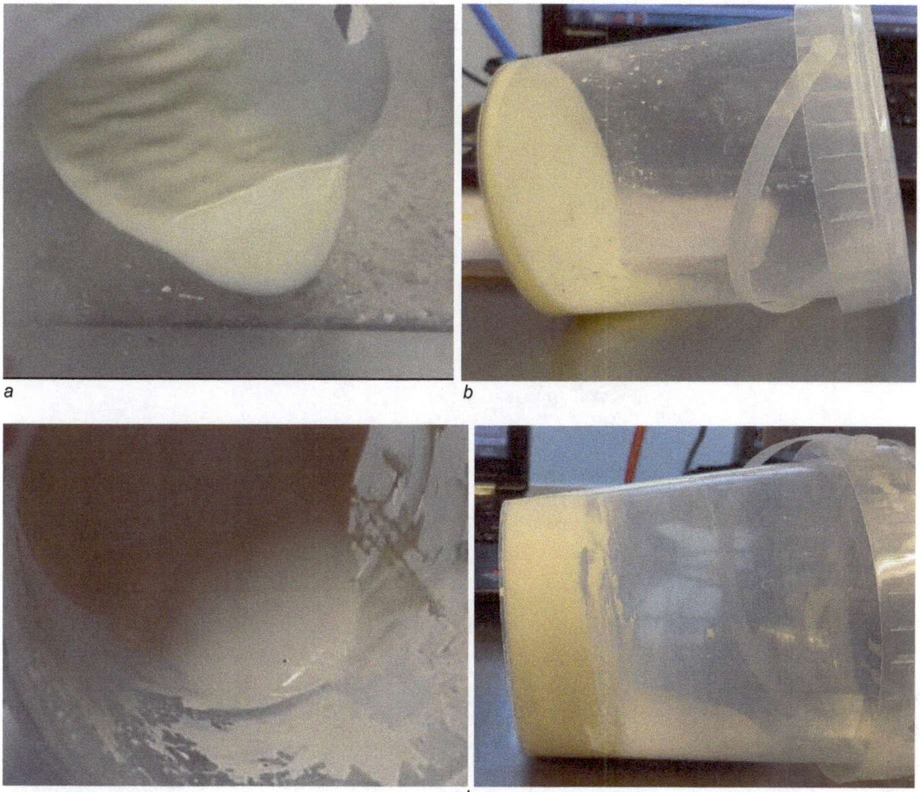

a b c d

Figure 3.21 Flow-ability of the reaction 40% water with 60% starch from the front view (a) from side view (b), and for wetting 60% chalks from front view (c); and from sideview (d).

Therefore, the viscoelastic behavior of the paste with starch Rh.24 is weaker in the early stages than that of the paste with chalk Rh.22, using the same water content in both mixes, due to the wetting behavior of the starch. However, the starch filler paste (Rh.24) is 12.6% higher in shear yield stress than (Rh.22), giving a better buildability because the unsolved starch used here as an interparticle stabilizes the flow. The apparent viscosity of Rh.22 is 15.8% higher than that of Rh.24, as shown in Fig. 3.22.

The drying samples in Fig.3.23 indicate that shrinkage of chalk is bigger than shrinkage of starch, that enhances the starch preference in the choose of the material of mixture formulation too.

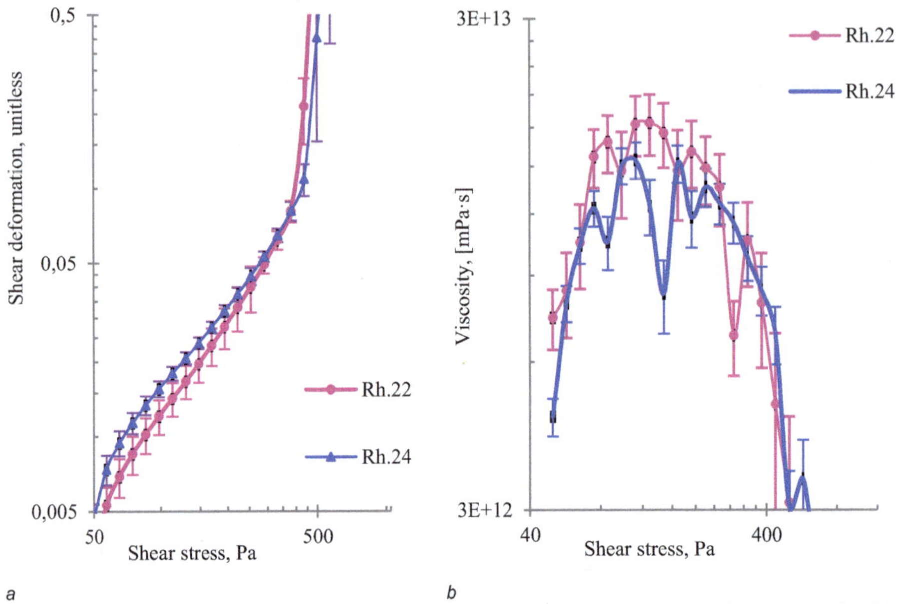

Figure 3.22 The shear deformation-stress (a); and shear viscosity logarithmic relationship (b) for changing filler type (Rh.22, Rh.23, and Rh.24).

Figure 3.23 Dried samples of starch left and chalk right.

Magnesium sulphate exploits the unrestricted water in the mix and crystallizes as hexahydrate magnesium causing a stiff paste [Larysa Okhrimenko et.al, 2017] and [Tim Baksa et.al. 2007]. The magnesium sulphate solidifies and crystallizes when wetting the 90% magnesium sulphate with 10% water forming hexahydrate magnesium, as shown in Fig. 3.24.

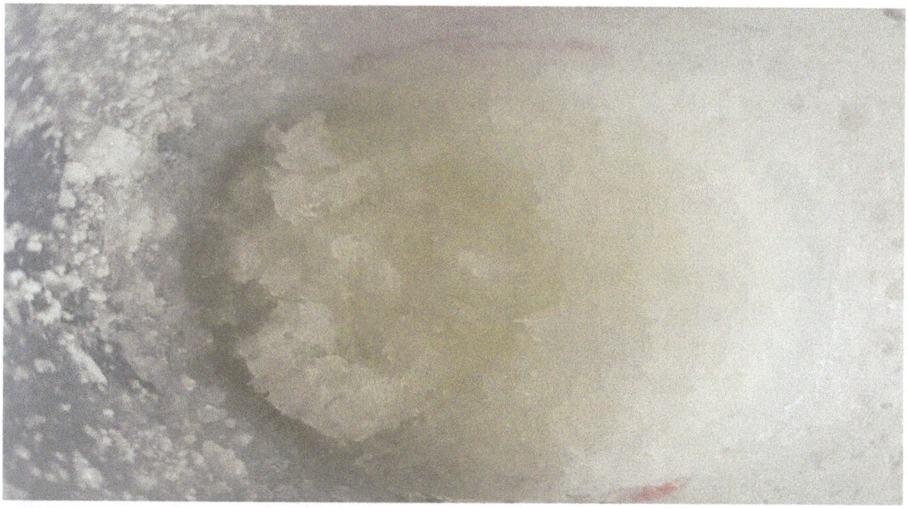

Figure 3.24 Dried sample of magnesium sulphate mixed with water.

Fig.3.25 shows the effect of reducing water with manipulating the component proportion using the magnesium sulphate paste from 74.6% to 74% (Rh.23 compared with Rh.11). Rh.23 has a shear yield stress 125.7% higher than Rh.11. The apparent viscosity of Rh.23 is higher than Rh.11 by 49.6%.

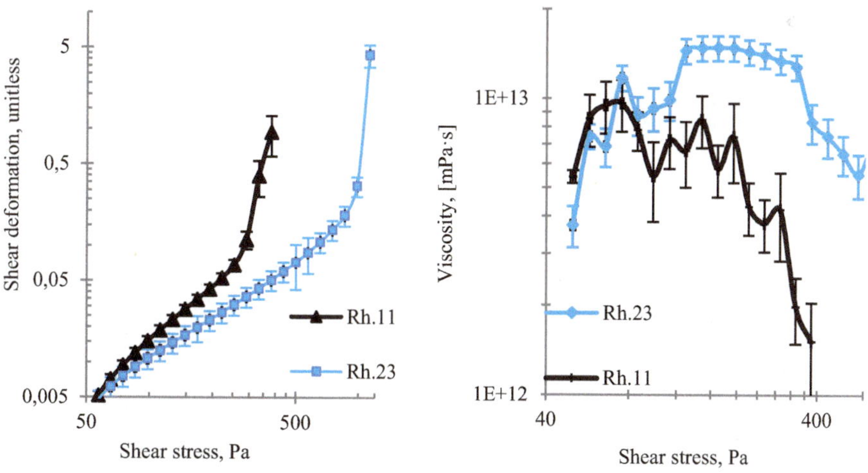

Figure 3.25 The shear deformation-stress (a); and shear viscosity logarithmic relationship (b) for changing filler type (Rh.11, and Rh.23)

Increasing the filler content has a positive effect up to the optimum and then a negative effect on the results. Increasing the filler up to 223% (Rh.26 compared to Rh.24) increases the shear yield stress up to 65%, while increasing this percentage up to 420% (Rh.25 compared to Rh.24) increases the shear yield stress up to 31.3%. This indicates that an increase of 223% is better than 420%. The apparent viscosity for Rh.25 and Rh.26 is convergent by different equal to 31%, and they both increase by 132.7% and 76.8% respectively; than Rh.24, as shown in Fig. 3.26.

In conclusion, a certain amount of increase of the filler with starch has a direct influence on the results. Therefore, three mixtures (Rh.27, Rh.28 and Rh.29) were developed according to the nozzle (1.6 mm, and 1.2 mm); respectively, as shown in Fig.3.27.

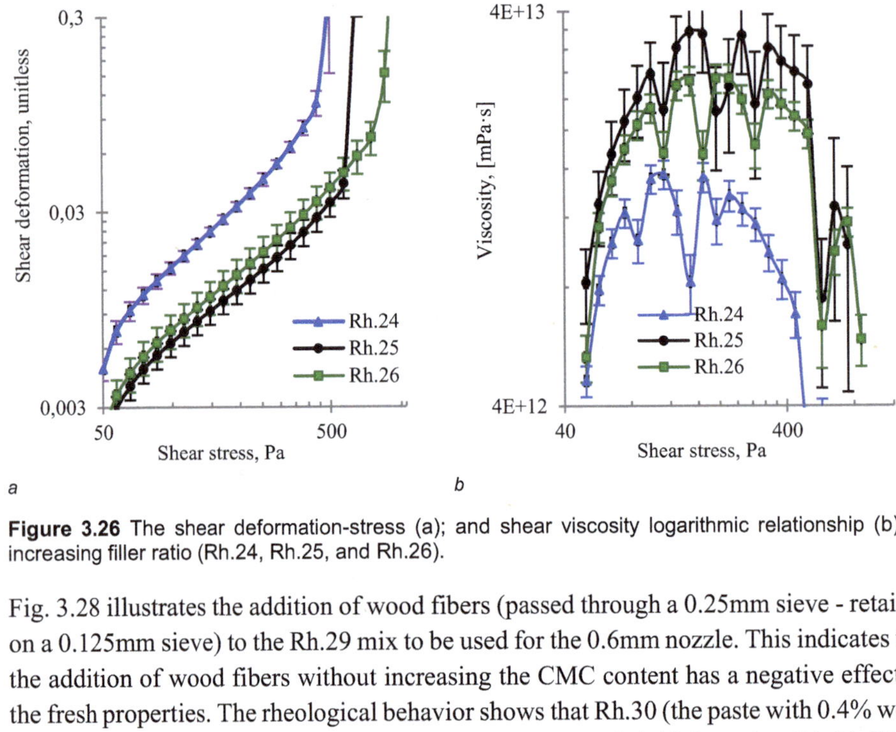

Figure 3.26 The shear deformation-stress (a); and shear viscosity logarithmic relationship (b) for increasing filler ratio (Rh.24, Rh.25, and Rh.26).

Fig. 3.28 illustrates the addition of wood fibers (passed through a 0.25mm sieve - retained on a 0.125mm sieve) to the Rh.29 mix to be used for the 0.6mm nozzle. This indicates that the addition of wood fibers without increasing the CMC content has a negative effect on the fresh properties. The rheological behavior shows that Rh.30 (the paste with 0.4% wood fibers addition) has a better wood fibers content in terms of yield shear than Rh.31, Rh.32 and Rh.33.

Also, Rh.30 is convergent by the apparent viscosity with the reference paste (Rh.30; powder cellulose with MgSO4 as filler), while other percentages have divergent results and are dissimilar to the Rh.30.

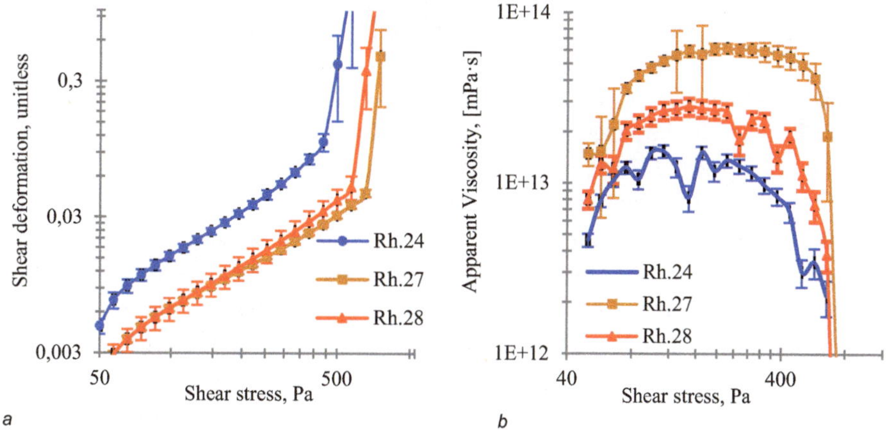

Figure 3.27 The shear deformation-stress (a); and shear viscosity logarithmic relationship (b) for (Rh.24, Rh.27, Rh.28 and Rh.29)

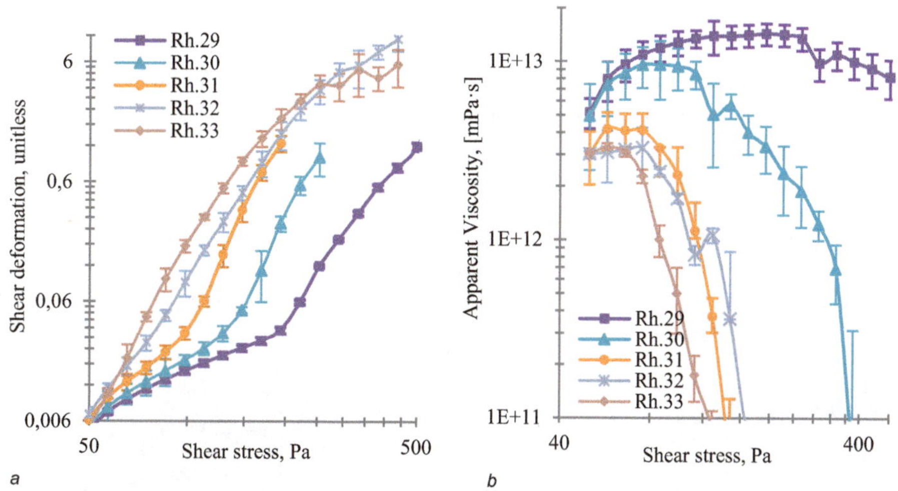

Figure 3.28 The shear deformation-stress (a); and shear viscosity logarithmic relationship (b) for (Rh.29, Rh.30, Rh.31, Rh.32 and Rh.33).

3.2.4.12 Summery for the Rheology Behaviour

From the above parametric factors show that each of the components has an impressive impact on the modification of the properties of the compound. These tests for the trial blends were done to find the suitable blend components and their proportion in terms of rheology behavior of yield shear stress and viscosity to have a reasonable evaluation in values to predict the extrudability, buildability and provide a proper adhesion property depending on the type of ingredients used and its proportion.

3.2.5 Adopted Technology & Print Resolution

The LUTUM -VormVrij® 3D clay printer version 2.1 shown in Fig. 3.29 was used. The printer is adjusted to fit the paper material instead of the clay paste, through manipulating the printing parameters. This printer is specified as material extrusion-AM technology, also known as free-form fabrication or fused filament fabrication (FFF). This technology is based on extruding paths through material that is pumped and extruded through a nozzle, and it is constructed in a layer-by-layer pattern [Redwood, B.et al, 2018]. The material extruded by this printer is a semi-liquid or highly viscous-pasty material that deposits a filament along a defined path. After extrusion, as the material solidifies, it forms a bond between layers depending on the adhesion strength formulated from the material mix [R. Naboni & N. Jakica, 2022]. The material then solidifies and should gain sufficient strength to support the subsequent layers printed on it. However, in certain configurations, such a cantilever requires a particular arrangement to support the print, depending on the properties of the extruded material. In such geometric design cases, treated with a specific slope, others need a supporting material. However, printing was carried out without the use of support materials. The extruded mixture should be homogeneous and have specific characteristics in terms of viscosity, consistency, and adhesion. These characteristics depend on the designed proportioning of the components of the mixture and characterize the paste in terms of extrusion capability and as a self-supporting material after extrusion.

a *b*

Figure 3.29 LUTUM v2.1 3D printer (a), and its modified nozzles (b).

Build Space and Print Bed

It may seem that the printer has a small footprint, but it has a large print bed and is capable of printing large volumes of up to 0.5 m^2 with dimensions reaching 750x750x500 mm. The media feeder influences the size of the printed image. Three sizes are available: 10, 5 and 0.5 liters. Figure 3.29(a) shows the large and small containers. The small container can be refilled while the print is paused, but the dimensions of the nozzle position should at least be reserved. The print bed used during printing was a 12mm thick screen-printing plate/plywood. Printing is done on the rough side of plywood. The principal advantage of this roughness is that it helps the first layer of the print to adhere to the bed. On top of the plywood, a 0.25m^2 "peel ply linen roll" is placed to improve the quality of the base surface and to prevent the print from sticking to the plywood in the case of large surface contact between the print and the plywood. A photo of the used Peel ply fabric is shown in Appendix A.

Extrusion Mechanism

The motion system for the printer in Fig. 3.30 is controlled by two actions associated with the extrusion of the extruder. The first is by air pressure. It is provided by an external resource. However, the printer is designed with air pressure capacity reaching up to 1 bar. The machine is equipped with mini pressure regulator, this manometer can control the inlet air pressure. The amount of air pressure could be set manually at the beginning of printing and should remain constant during printing for quality control. The second is the motor, which is connected to a rotating screw, controlled by values imposed by the G-code. This value is set within the filament setting/extrusion multiplier and is selected depending on the stepper motor, which is proportionally related to the material flow. Fig. 3.30(a) shows the details of the printer screw.

Nozzle Size

The nozzle has been built according to the demands. The available nozzle sizes are 0.6-, 1.2-, 1.6-, 2.5-, 3-, 5- and 7-mm diameter from left to right as shown in Fig. 3.29(b). The shape of the 0.6 or 1.2- or 1.6-mm nozzle is a hollow truncated cone with an apex diameter of 2.5 mm and a length of 25 mm. These nozzles are plastic and require two additional adapters to fit into the printer as shown in Fig. 3.30 (b). The 2.5mm nozzle is a long hollow plastic cone with a length of 50mm and an upper diameter of 3mm. This nozzle is mounted directly without an adapter. The 3mm metal nozzle has a conical shape and is mounted directly on the printer.

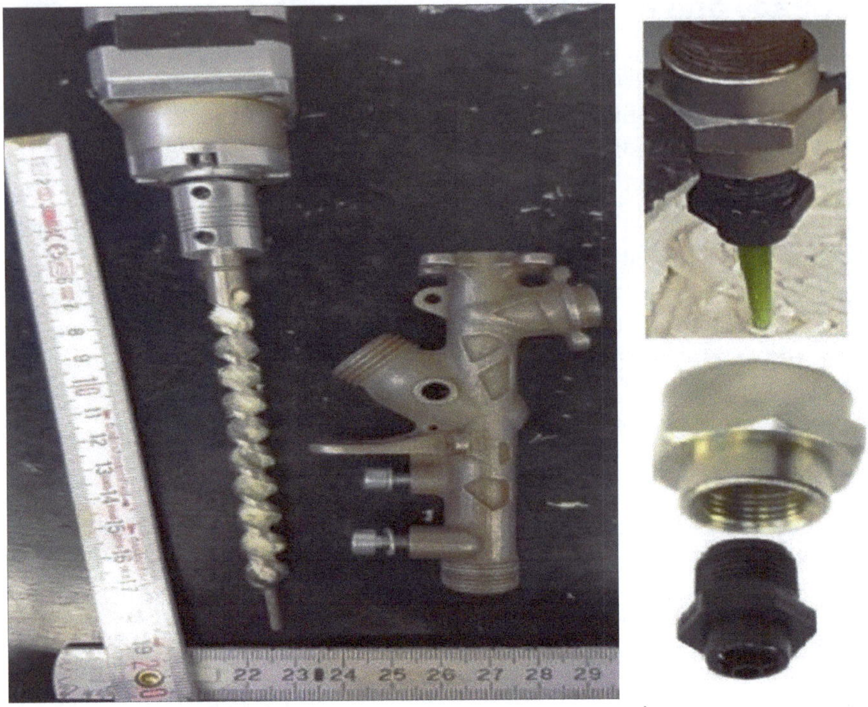

a b

Figure 3.30 the motor component and extrusion assembly unit ©LUTUM-vormvrij version 2.1 (a), and caulking nozzle- and nickel universal of the nozzles (b).

3.3 Process Parameters & Geometry Configuration

One of these effects that dominates the geometric accuracy of 3D printed objects is the slicer setting and printing parameters. The slicer is a program that converts a 3D model into a G-code script that can be decoded by the firmware of any machine. The super powerful and popular slicer used in this study is PrusaSlicer, which uses an expert mode so that all settings are controlled. Repetiers-Host is also used to estimate the print time and keep track of the nozzle movement.

These settings are adjusted according to the 3D printer and the material to be printed, to obtain good print quality and to minimize the divergence between the printed product and the design geometry. Several attempts were made to achieve balanced process parameters and to evaluate changing these values by trial and error: comparing with the desired result for the geometry.

Before starting, some red flags are considered [R. Devicharan and Raghav Garg, 2019]:
- The build platform should be level.
- The nozzle should not start too far or too close to the bed.
- The speed of the first layer should be slower than the sequence over layers.

Slicer settings cover all aspects of achieving the print configurations, such as:

3.3.1 Nozzle Move-Speed

Printing speed is a priority for controlling production speed. Obviously, it is related to the speed of finishing a product manufactured. Because the speed of finishing is related to the cost, and this is related to satisfying the consumer need within a faster delivery time. Therefore, the maximum possible speed is used so that it does not cause any printing errors, and then other process parameters are adjusted accordingly. The default values are used in the first trials; these are specified for plastic printing. The default value of the speed is 60 mm/s; it is a high rate and so fast that it causes irregularities in the movement and undesirable vibrations in the machine; consequently, it loses the correct coordinates when moving; the new position for printing after moving is not in the right place. So, this speed is not suitable for paper paste. Therefore, the value has been gradually reduced until it is suitable. Fig. 3.31 shows an example of a high speed.

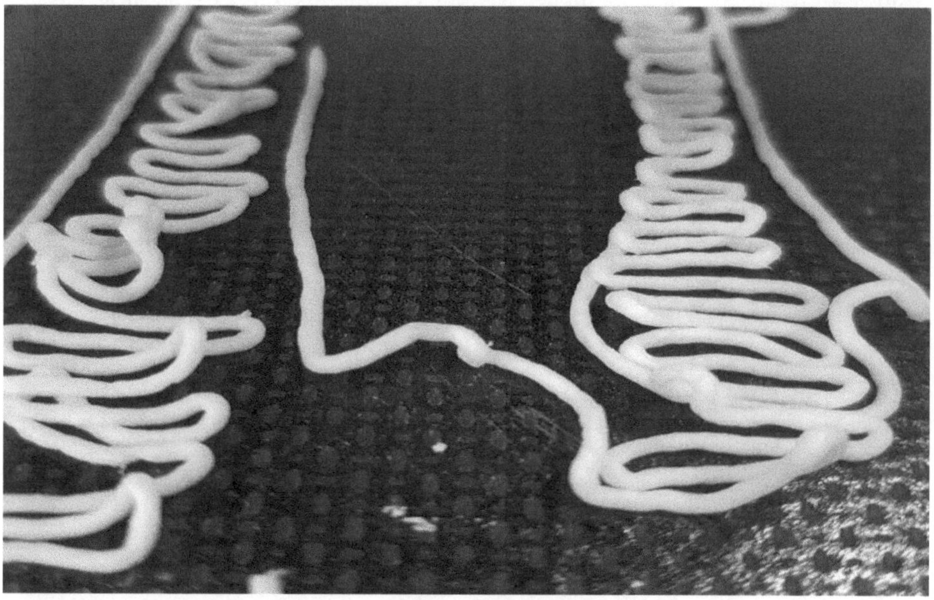

Figure 3.31 High-speed printing.

Travel speed is limited to a value that does not cause the filament to pull when the machine moves to another print position, as shown in Fig. 3.32.

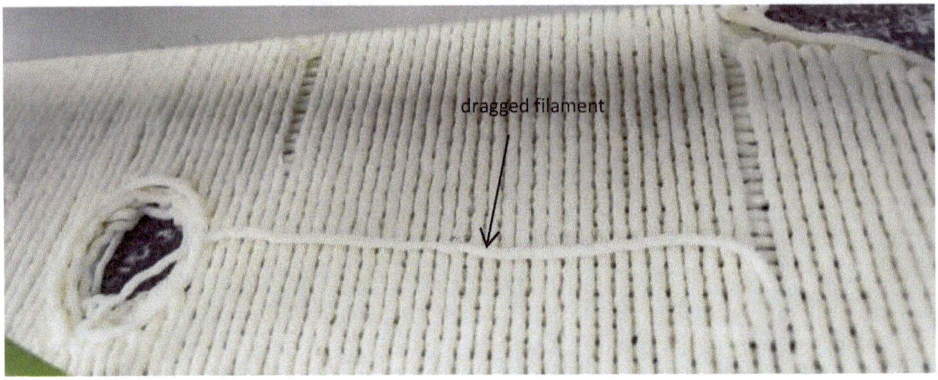

Figure 3.32 A pulled filament during printing as the machine moves to another print position.

The speed of the first layer should be lower than other printing speeds because the paste must stick to the bed; Fig. 3.33 shows this problem. The speed used is high even when the flow rate is increased; it still doesn't give enough time for the extruded filament to stick to the bed. The process parameters used in Fig. 3.36 are: speeds=30mm/s (circumferential speed, infill speed, first layer speed, gap infill speed and max print speed), nozzle diameter=1.6mm, extrusion multiplier=1.4, air pressure=0.2Mpa, 100% linear density. Fig. 3.34 shows the appropriate speed used to balance the process parameters.

Figure 3.33 High first layer print speed prevents filaments from sticking to bed.

Figure 3.34 An example of balancing process parameters, where the speed is 18mm/s.

A theoretical simulation for 3D printing is considered to simplify the trial-and-error process of selecting initial values for the remaining process parameters.

3.3.2 Extrusion Speed and Layer Height

The extrusion process is complex and dynamic, depending on the feedstock and the machine extrusion mechanism. The machine extrusion mechanism used in this study is an accumulation of three technologies, which are pneumatic extrusion (air pressure), screw-based extrusion and syringe-based extrusion. A mathematical model is adopted according to Fuwen Hu et al, 2021 to calculate an approximated values for the process parameters to start with then using try and error to find balance values. This model has relationships linking the process parameters. Some assumptions are made for the operation of this model, such as that the paste is homogeneous, the flow is steady and the flow is smooth. According to these assumptions, the flow rate is equivalent in all sections and independent of time.

The flow rate in the extruder: $V_1 = Q\,dt = \frac{\pi}{2}\,W\,H\,D\,n\cos\theta\;t$ (3.1)

Where V1 is the flow of fluid volume through a surface per unit time t, v is the flow velocity, A is the cross-sectional vector area/surface, W is the channel width of the screw, H is the discharge depth, D is the screw diameter, n is the screw rotation speed, θ is the screw spiral angle, and t is the duration time.

The flow volume rate in the nozzle: $V_2 = \frac{V_e\, t\, \pi\, d^2}{4}$ (3.2)

Where: V_e is the paste extrusion speed, and d is the nozzle diameter.

The flow volume rate in the extruded tips: $V_3 = V_{xy}\left(h\, d + \frac{\pi h^2}{4}\right)$ (3.3)

Here: h is the layer height, d is the diameter of the nozzle, and Vxy is the 3D printing scan speed, it refers to the speed of the print movement.

From the equivalent the equations of V_1 and V_2, we get V_e :

$$V_e = \frac{2\, W\, H\, D \cos\theta}{d^2}\, n \quad (3.4)$$

It can also be shown from the equivalence relation of V2 and V3 that the filament extrusion velocity Ve and the scanning velocity Vxy are given by:

$$\frac{V_{xy}}{V_e} = \frac{\pi d^2}{4\, h\, d + \pi\, h^2} = \frac{1}{\frac{4}{\pi}\frac{h}{d} + \left(\frac{h}{d}\right)^2} \quad (3.5)$$

The reformulation of equation 3.5 is to have a relationship linking the process parameters. So, the equation for the flow rate equilibrium of the extrusion process is:

$$h = \frac{\sqrt{4 + \pi^2 . \frac{Ve}{Vxy}} - 2}{\pi} . d \quad (3.6)$$

Where h is the layer height, Ve is the extrusion velocity, Vxy is the scanning velocity and d is the diameter of the nozzle tips.

Equation. 3.6 hypothetically confirms the relationship between layer height, nozzle diameter, print speed and extrusion flow rate.

In the Prusa Slicer software, the print speeds are set in the Print Settings → Speed menu. The flow rate is defined by the extrusion multiplier. It can be found in Filament Settings → Filament → Extrusion Multiplier. Technically, this value is related to the speed of rotation of the screw drive inside the feed mechanism of the printing unit.

3.3.3 Process Parameters Optimization

The process parameter selection methodology (Fig. 3.35) is presented by specifying empirical values for the die movement speed (scanning speed) and the extrusion multiplier (the flow rate). The layer height is calculated using the empirical values with the current nozzle diameter using equation 3.6. A value for the track width is also bounded by trial and error to fit a suitable coalescence and achievable surface contact. Consequently, it affects the convergent dimensional accuracy between the print results and the designed geometry and the characterization of the product geometry.

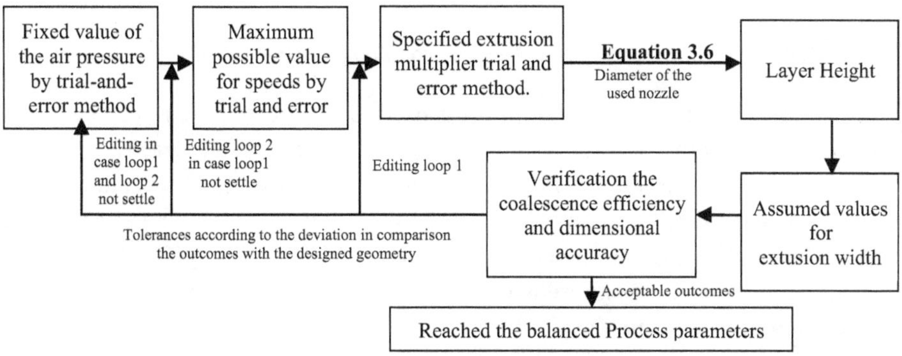

Figure 3.35 Schematic of the optimization methodology for selecting the process parameters used.

3.3.4 Extrusion Width & Coalescence of Layers and Walls

This is also referred to as the line width. The line width is the central distance between two adjacent lines. It affects print quality and print speed. It increases the contact between the adjacent walls, reduces the pores between them and increases the build factor, but within an optimum value. On the contrary, if it is used with a value lower than the optimum and the other process parameters are constant, it would lead to oversizing in the dimensions and corners [Patrich Ferretti et al, 2021] [Osama Habbal, 2021]. The coalescence of the filaments increases some of the physical and mechanical properties, in addition to the bonding or adhesion of the filaments to each other. Accordingly, they solidify together with the previously printed layers to form a solid, cohesive part; this solidification depends on the paste properties and the coalescence configuration.

The coalescence of the printed filaments is mainly controlled by the line width, the layer height and the flow rate. To achieve full coalescence, the filaments are pressed against the previous one; by limiting the layer height up to (50 - 75%) of the nozzle diameter and adopting the extrusion width (between 60 - 200% of the nozzle diameter) to adjust the other parameters in balance with the flow rate.

The pressing force and partial re-wetting of the underlying material promotes the bonding of the new layer to the previously printed layer. This also means that the filament is squeezed from a circular to an elliptical shape (Fig. 3.36).

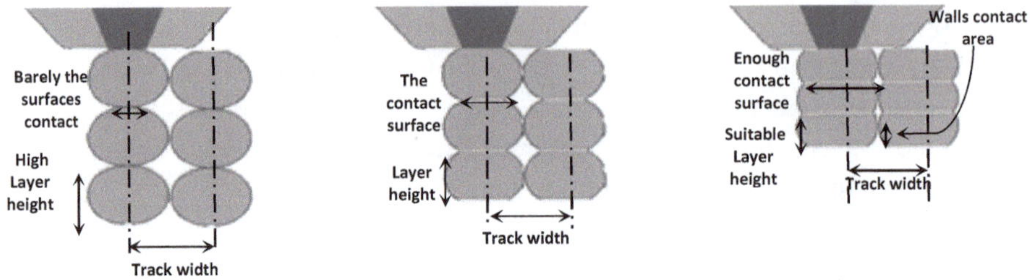

Figure 3.36 Schematic illustration of filament coalescing using the same line width and varying layer height.

Fig. 3.37 shows different settings of the process parameters affecting the coalescence of a 50mm cube with 100% filling. The process parameters are all speeds= 18mm/s (circumferential speed, infill speed, first layer speed, gap infill speed and max print speed), nozzle diameter=1.6mm, layer height=1.2mm, extrusion multiplier=0.8, air pressure=0.2Mpa, 100% linear fill density, extrusion width= 300%; Fig. 3.37(a).

Fig. 3.37(b) shows the printing results of the same previous setting but with Extrusion Multiplier=0.9. In meanwhile Fig. 3.37(c) displays the results of the same setting as in Fig. 3.37(a) but with a layer height of 1mm. The process parameters in Figures 3.41(a), (b) and (c) didn't achieve total coalescence, even though the filling was 100%.

Fig. 3.37 (d) shows a better coalescence than Fig 3.37(a), (b) or(c).

Fig. 3.38 shows the effect of changing only the flow rate on the coalescence of 100% filling with Extrusion Multiplier=0.9, 1 and 1.2.

a *b*

Figure 3.37 Coalescence results of a 50mm cube affected by the width line (a), with the same width line increasing the flow rate (b).

c d

Figure 3.37 continued: Coalescence results of a 50mm cube affected by the width line with the same previous process setting decreasing the layer height (c) and a better Coalescence (d).

Figure 3.38 Coalescence outcomes affected by the flow rate of 100% filling.

3.3.5 Filling

Fill has a direct effect on the physical and mechanical properties of the printed part. The type of fill and the orientation of the filaments lead to the inherent anisotropic behavior or orthotropic characterization. The printed elliptical filaments, when stacked side by side, form joints between them like small valleys. They are weak points that allow the crack to propagate. The layering and filling give 3D printed parts a rough surface finish and a layered appearance, as shown in Fig. 3.39.

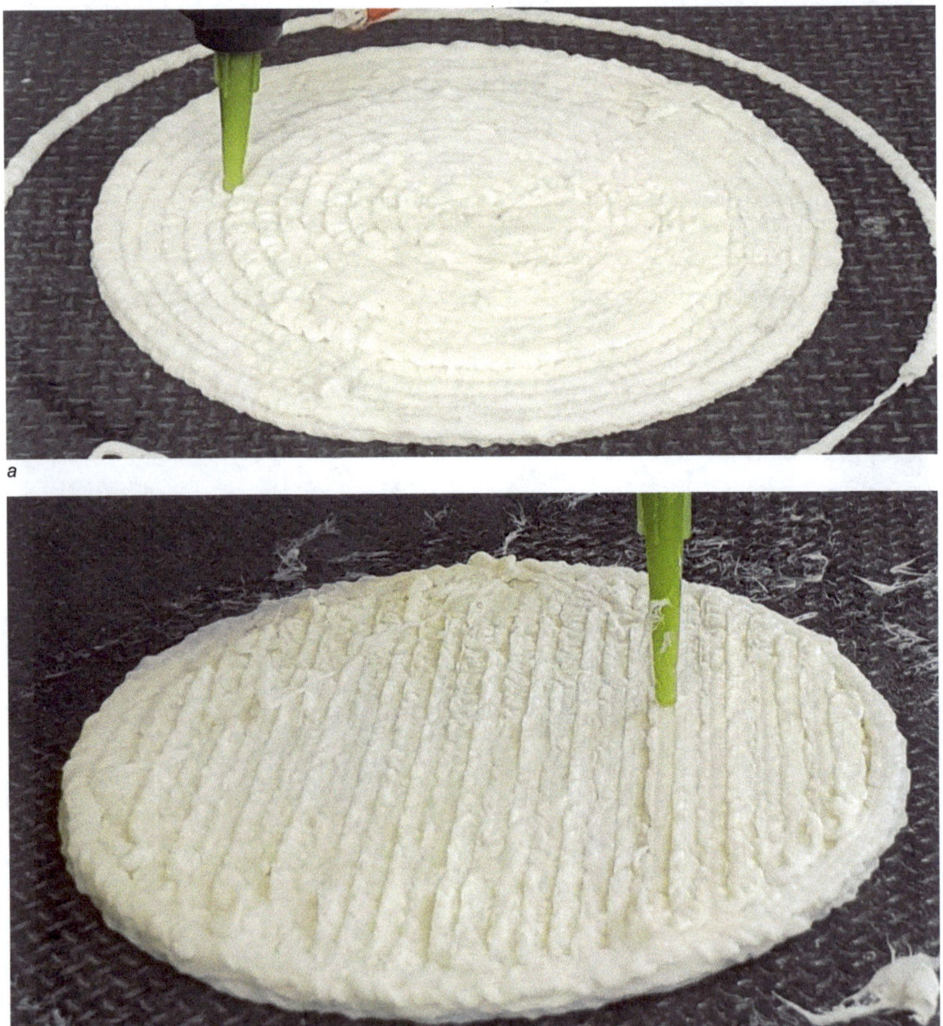

a

b

Figure 3.39 The filling effect on the aesthetics of the part: circular(a) and rectangular pattern (b).

Fig. 3.40 demonstrates the concentric pattern of the infill. This type usually has inside gaps. The amount of these gaps depends on the complexity of printed geometry.

Figure 3.40 Concentric infill for rectangular base geometry (a) and for non-uniform geometry (b).

Fig. 3.41 shows the effect of the infill on the shrinkage process; cross layering is more effective in resisting dehydration during drying. Also, with other types of fills, the amount of material extruded in the same layer is greater than with other types. This gives more confinement (more crossing nodes & bonding & friction action) which helps resist shrinkage and gives greater density for the specimen. The ultimate strength, crack propagation and failure mechanism are also affected by the geometry configuration, particularly the type of filler.

a
b

Figure 3.41 Shrinkage results of specimen influenced by infill type: top view (a) and side view (b).

3.3.6 Air Pressure

It's usually kept constant during printing to control quality, or it can be changed according to the balance of the printing process parameters.

3.3.7 G-code Optimization

Additional settings are made within the PruseSlicer software/expert mode before creating the G-code. It's necessary to adapt the process parameters with other settings to obtain a nice surface finish, to correct the single wall width of the extruded filaments and to obtain an aesthetic overall result. These settings are empirical values; found using the try and error method and set according to the dimensional accuracy measurement and virtual evaluation for the products; as shown in Tab. 2.14.

Table 3.14 Additional settings executed for G-code that are different from the default valves before slicing.

Setting identifier	PruseSlicer Guide	Action	Explanation
Extra perimeters if needed	Print Settings → Layers and perimeters→ Quality	on	Provided extra perimeter we needed to avoid gab; reached 70%.
Avoid crossing perimeter	Print Settings → Layers and perimeters→ Quality	on	Quality on This may affect the reduction of stringing. To avoid this, the maximum detour length should be limited. The cutting time would also be slower.
Avoid crossing perimeters - maximum detour length Print Settings	Print Settings → Layers and perimeters→ Quality	150mm	This value depends on personal choice.
External perimeters first	Print Settings → Layers and perimeters→ Quality	on	This setting would help to get a better dimensional accuracy.
Detect thin walls	Print Settings → Layers and perimeters→ Quality	off	Avoid over volumetric spots
Fill gaps	Print Settings → Layers and perimeters→ Advanced	on	Cover the issue that states in Fig.3.40
Z offset	Printer Settings → General→ Size and coordinates	-0.5	To subtract this value from the z-coordinates in order to compensate for the bad Z end stop position of the nozzle on an uneven print bed.
Infill/perimeters overlap	Print Settings → Advance → overlap	50%	Overlapping of the filling on the parameter

3.3.8 Post Processing

Post-processing is necessary to clean the surfaces immediately after printing by removing the extra filaments. These filaments are generated during the movement of the nozzle to the next position and do not belong to the main structure of the print; as shown in Fig. 3.42. A brush is used with light movements to achieve cleaning.

Figure 3.42 3D printed specimen without post processing- side view.

3.4 Curing Process

Observation of the curing process has shown that slow curing gives a better aesthetic appearance and controls shrinkage problems. Nylon plastic film/transparent is used to maintain the humidity of the specimens as shown in Fig. 3.43. The curing time with nylon film is three days for small objects and five to seven days for massive objects. The samples are then left at room temperature until drying is complete. The drying time depends on the printed volume, its mass, and the exposed surfaces for ventilation.

If it takes more than a day to print, more moisture must be retained to compensate for the loss during the drying process. So wet paper towels are placed over and around the samples, together with nylon plastic sheeting. This strategy is used until the next day when printing starts, as shown in Fig. 3.48 (b). The problem that arises when wet paper towels are not used is that the dimensions shrink due to evaporation and drying. As a result, the new print on top of the old print would not match, see Fig. 3.43(c).

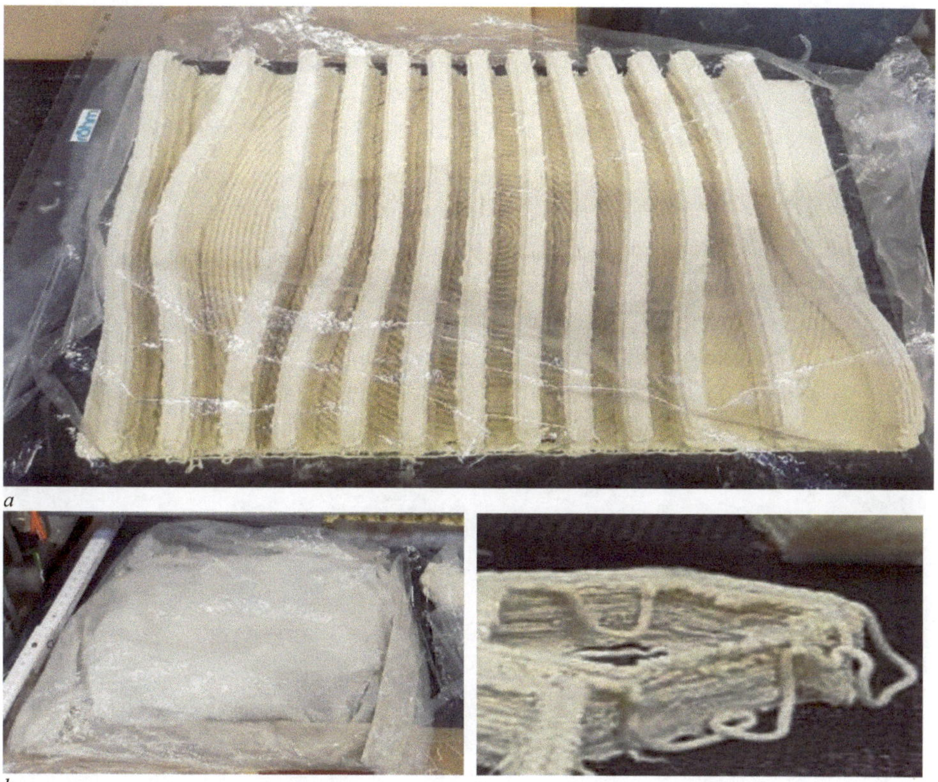

Figure 3.43 Curing process for the specimens(a), when printing exceeded one day (b), and the occurring issues when no curing was used (c).

3.4.1 Shrinkage Outcomes During Drying

Shrinkage is the contraction from the printed dimensions to the final dimensions, due to:

- The inhomogeneous crystallization that occurs in the material components of the mixture, with each component behaving differently and inducing limited shrinkage.
- The cross-linking between the components causes a moderated shrinkage.
- Phase transmission from liquid to solid state during the drying process. This means water evaporation causes a reduction in volume, leading to a noticeable shrinkage. Consequently, controlling the mix formulation by minimizing the water/solid content has a direct influence on this factor. A mix with limited water content will therefore have better shrinkage results.

The shrinkage factor is one of the determinants of print production quality, which is related to dimensional accuracy and aesthetic appearance. It's therefore a limiting factor in the design and optimization of the mixture formulation. It's obvious to notice the un-uniformity of the filament diameter after drying due to shrinkage, which is dependent on the mixture formulation and the ratio of the water content compared to the solid content. Reducing the water content as much as possible and manipulating the other components of the mixture to compensate for the water reduction is the goal to control the shrinkage without affecting the other properties of the paste: such as the extrudability.

Shrinkage Measurements

Shrinkage is measured using a microscope Keyence model VHX-600 Digital, as pictured in Fig. 3.44. More details are shown in Appendix B. It has a magnification of 20x-200x, and the accuracy of the 3D topography mapping is in the microns range. It is therefore used to measure the dimensions before and after drying.

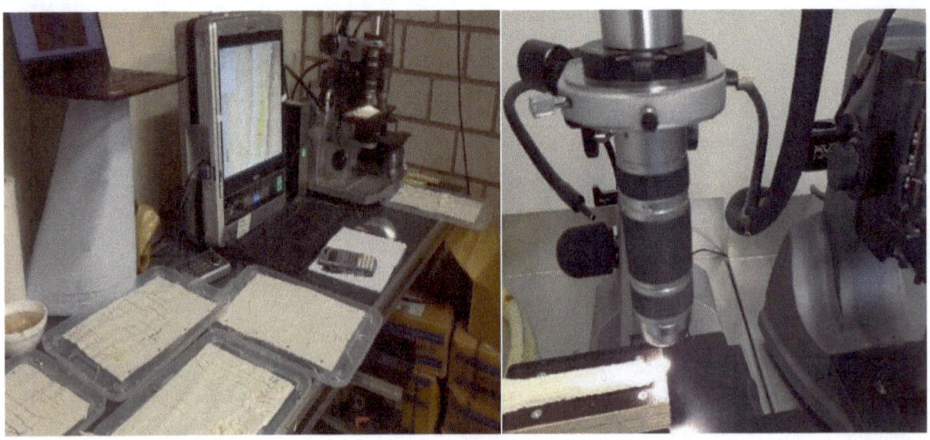

a *b*

Figure 3.44 Keyence Model VHX-600 Digital Microscope(A), Its Lens (20x-200x) (B).

Shrinkage was measured by the longitudinal and radial deformation of the filaments after drying. The radial deformation is assessed by determining the diameter deformation for the filament of each mixture. The result of the deformation is calculated as the percentage between the wet filament immediately after the printing and the dried filament one week after the printing. Each value of the longitudinal and radial deformation in Tab. 3.15, and figures is an average of at least twelve measurements distributed on three marked points inside the filament for each six identical filaments, as shown in Fig. 3.45. Longitudinal deformation is an average calculation for at least six identical printed filaments of 150 mm length; compared with wet filament immediately after printing and with the same filament after one week of drying.

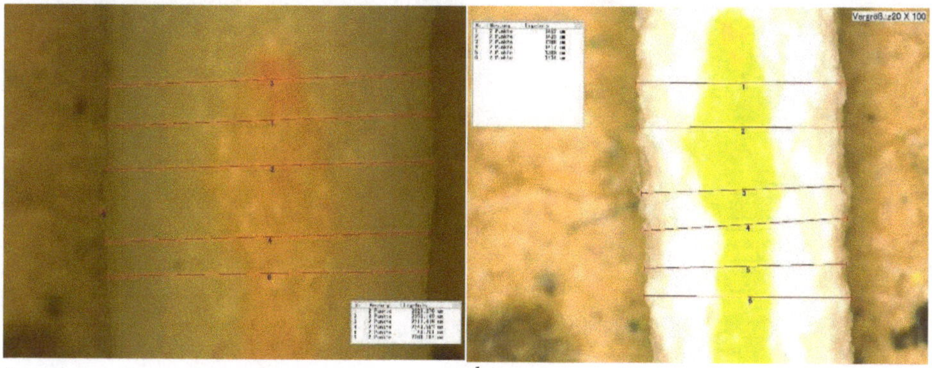

a *b*

Figure 3.45 A spot specified for the radial deformation measurement of a wet filament (a), one-week dried filament of the same spot (b).

3.4.1.1 Effect of Cellulose Type

The effect of using different types of cellulose proves that it has an effective impact on controlling the shrinkage outcomes, due to cellulose properties such as i) the fiber length and width, ii) the percentage of curl and fibrillation within the single cellulose fiber, and iii) the water retention capacity of the cellulose with the other physical properties for each type of cellulose. These factors increase the interlocking within the final skeleton of the mix formulation and help to increase the confinement of the deformation after drying. Fig. 3.46 indicated irregularity of the printed filaments using paste 1. According to Fig. 3.2 and Tab.3.9, the properties of long fiber-cellulose provide a higher potential than short fiber cellulose or powder-cellulose to bound the shrinkage. Comparing paste 2 with paste 1; the shrinkage is reduced by 28.9%. The filament of paste 11 has a lower shrinkage percentage by 60.4% and by 36% in comparison with paste 10 and paste 12; respectively, as shown in Tab. 3.15, Fig. 3.47.

In the experimental investigations, more than one type of short cellulose was used. Each type has different properties and specifications depending on: i) the manufacturing company, ii) its original area, iii) the type of planet, the age of plant when the cellulose is pulped. All these aspects affect the measurement of the properties in Fig. 3.2 and Tab.3.9. In conclusion, Cellulose/ type 3 reduces shrinkage outcomes than the type 2, in terms of using a short fiber. The explanation is due to the physical properties of the cellulose, such as fibers length and width, fibrillation, and curly characteristics, all of which help to increase the interlocking in the network, thus; increasing the confining stresses reduce the shrinkage percentage.

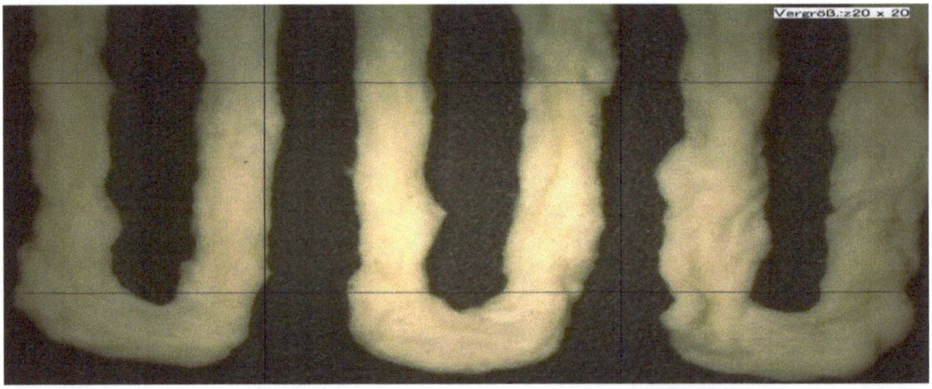

a

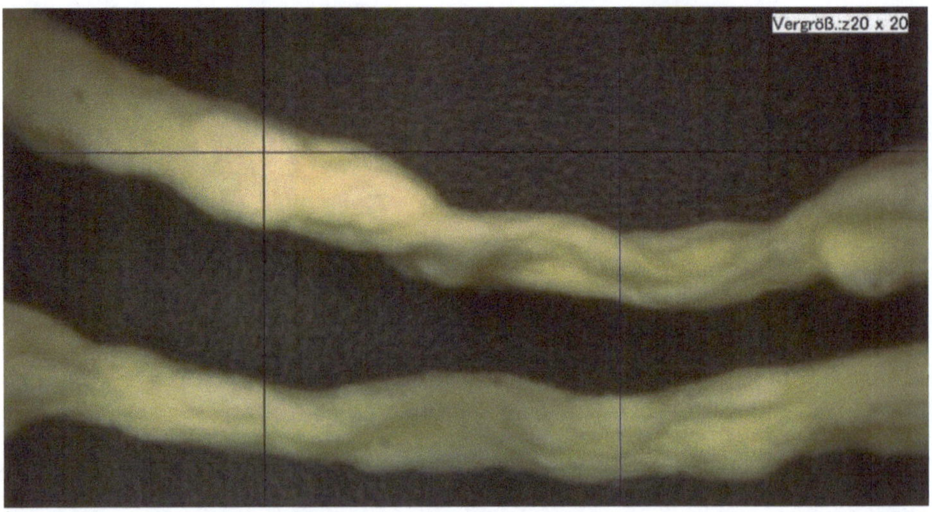

b

Figure 3.46 The uneven line width of the filaments after drying (a), and shrinkage outcomes for the filaments of Paste 1, printed with a 1.6 mm nozzle diameter (b).

Table 3.15 Shrinkages measurements after drying for filaments extruded from the pastes.

Paste From Tab.3.9 , 3.10 and 3.11	Diagonal Strain* of the Diameter %	Longitudinal strain* of the filament %
1	38.52 ± 2.3	2.7 ± 1.5
2	27.37 ± 0.7	1.1 ± 0.4
3	40.2 ± 1.7	3.1 ± 1.2
4	36.4 ± 2.4	2.6 ± 1.3
5	39.12 ± 2.5	2.9 ± 1.5
8	35.7 ± 2.6	2.4 ± 1.8
10	55.6 ± 0.3	2.1 ± 1.5
11	22 ± 2.2	1.2 ± 0.7
12	34.4 ± 1.6	2 ± 1.3
13	32.7 ± 2.1	2 ± 1.7
14	20.8 ± 8.7	1.7 ± 1.8
15	21.6 ± 1	1.4 ± 2.2
16**	27.3 ± 7.3	1.7 ± 1.3
17**	33.4 ± 9.2	2.1 ± 4.2
18**	36 ± 8.5	2.3 ± 3.6
20	26.7 ± 2	1.6 ± 1.5
23	25.2 ± 3.2	1.5 ± 2.2
25	31.2 ± 1.7	1.9 ± 1.3
26	19.1 ± 7.5	1.1 ± 0.8
27	32.3 ± 1.2	2 ± 1.1
28	10.2 ± 1.1	0.9 ± 0.5
29	14.7 ± 1.3	1.1 ± 0.4
30	17.4 ± 0.7	1 ± 0.3
30 No curing process	26.3 ± 2.7	3.5 ± 1.6
31	15.7 ± 0.9	1.1 ± 0.5
32	38.6 ± 1.2	2.8 ± 0.9
33	30.4 ± 0.2	1.9 ± 0.9
37	26.4 ± 0.9	1.2 ± 0.2
38	46.2 ± 1.7	4.4 ± 2.7
40	39.5 ± 2.2	3.9 ± 2.1

* The room temperature is climatized to be 20C°±2, and humidity is to be 60%, which is the same room of the viscosity testing, all the tested filaments are cured as in paragraph "3.4 Curing process" except filament 29- No curing process.
**Due to the rotting, the paste of the mix was inhomogeneous, and had no uniform distribution of the mixture components. It was necessary to remix, but it was used and tested without remixing, that was obvious in the print under microscope, where the cellulose fibers were lumpy and separated from the rest of the mix, that effect on the regularity form of the filament diameter as shown in Fig. 3.47.

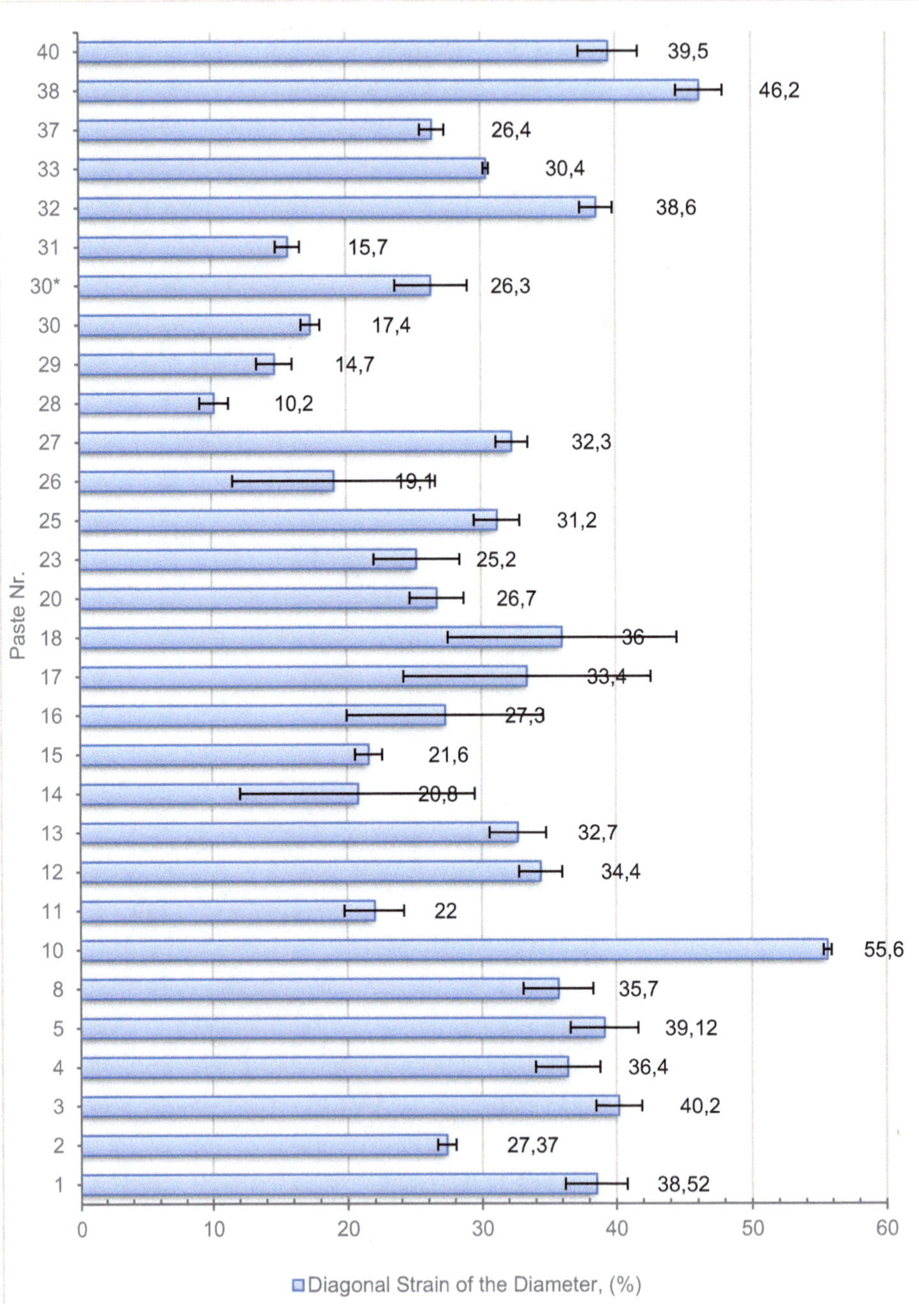

Figure 3.47 Shrinkages measurements after drying for filaments extruded from the pastes.

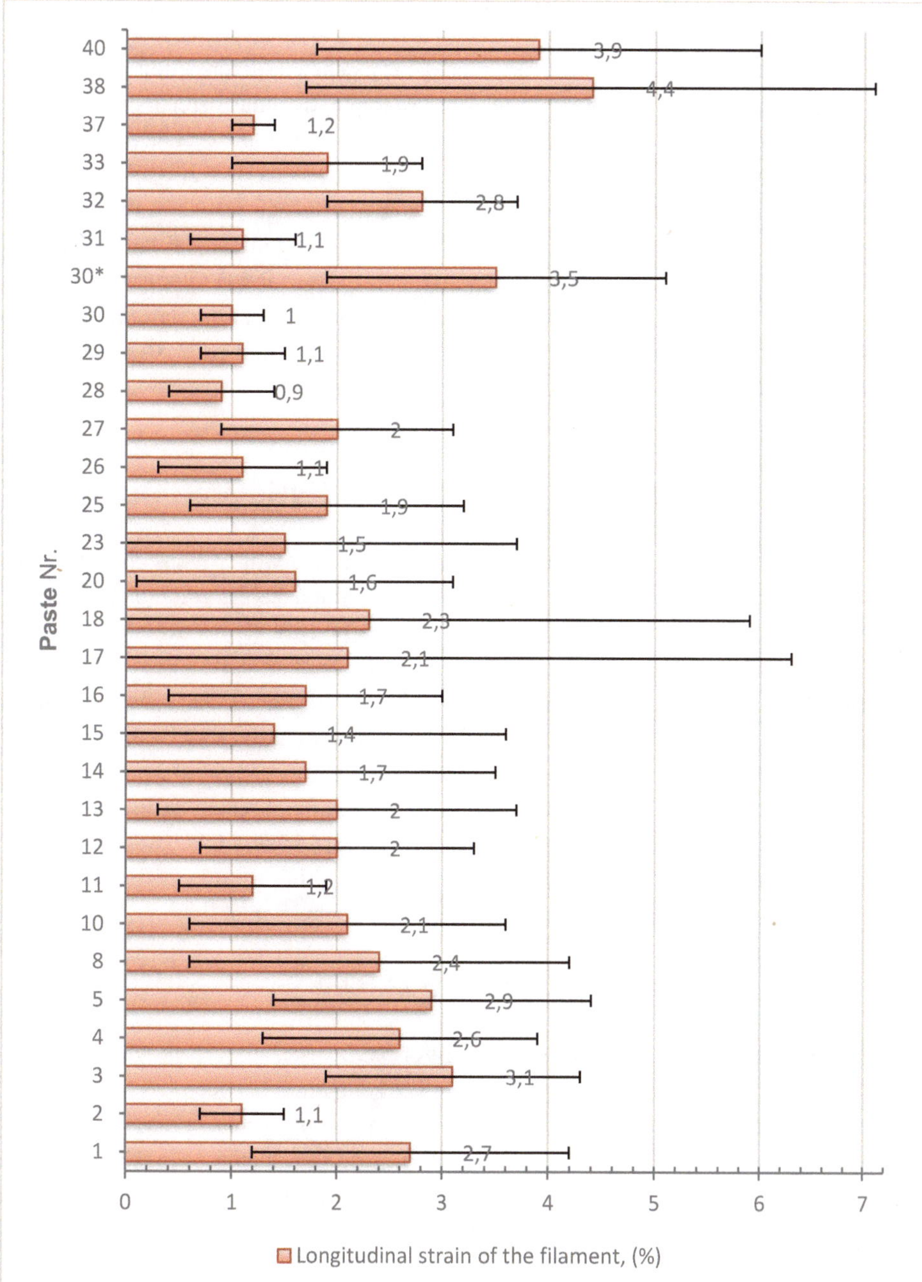

Figure 3.47 Continued.

3.4.1.2 Effect Using Lecithin & Filler

Adding filler and lecithin to mixture increases the resistance to deformation after drying, see the results of pastes 3 to 8 in Tab. 3.15. Manipulating the percentages of these components is important to clarify which components have a major effect on the shrinkage results. The increase lecithin dose in the mix hardly reduces shrinkage, while increasing the filler percentage has an effective influence on the shrinkage results. Tab. 3.3 shows the CMC reaction in the water. This would indicate that the addition of filler and lecithin together would improve the shape regularity and reduce deformation after drying, as they help to stabilize CMC. On average there is an 11.2% reduction in the deformation after drying when comparing to the filaments results for the paste 3 and paste 5. Fig. 3.47 also shows how the use of the filler and lecithin in optimum proportions helps to limit irregular deformation after drying.

3.4.1.3 Water Reduction & Ingredient Ratio Manipulation

Reducing water leads to a reduction in free water content, which helps to reduce the amount of water that evaporates after drying. This would reduce the volume change after drying. However, reducing water influences the extrudability, which is explained in detail in paragraph 3.2.4 Rheology behavior; where it was founded that some types of filler, such as starch, don't react with water, giving the possibility to reduce the water content. In addition, it might be possible to reduce the water content as much as possible by increasing other components that improve the extrudability. This action improves the limitation of deformation after drying. This is demonstrated by the shrinkage results of filament (paste 8) comparing to filaments (paste 1 to paste 7) in Tab. 3.15, where this result is not only related to the reduction of the water content, but also to the use of a different type of cellulose and to modification of the percentages of the components.

3.4.1.4 Effect of Using Rotting Paste

Tab. 3.15: filament series of (paste 11: freshly printed, paste 16: 6 days waiting before printing, paste 17: 6 days in between, and paste 18: 1 month in between) are the mixtures that indicate rotting of the material leading to an increase in shrinkage percentages. The diagonal deformation of filaments from pastes 16, 17, and 18 is increased by 60.6%, 96.5% and 111.8%; respectively compared to filament from paste 11. Fig. 3.48 shows the isolation of the cellulose from the other components due to decay, in the part where the cellulose is concentrated and thicker than the other parts. The other part without cellulose suffers from a greater shrinkage, which proves the scattering and the high standard deviation of the diameter irregularity.

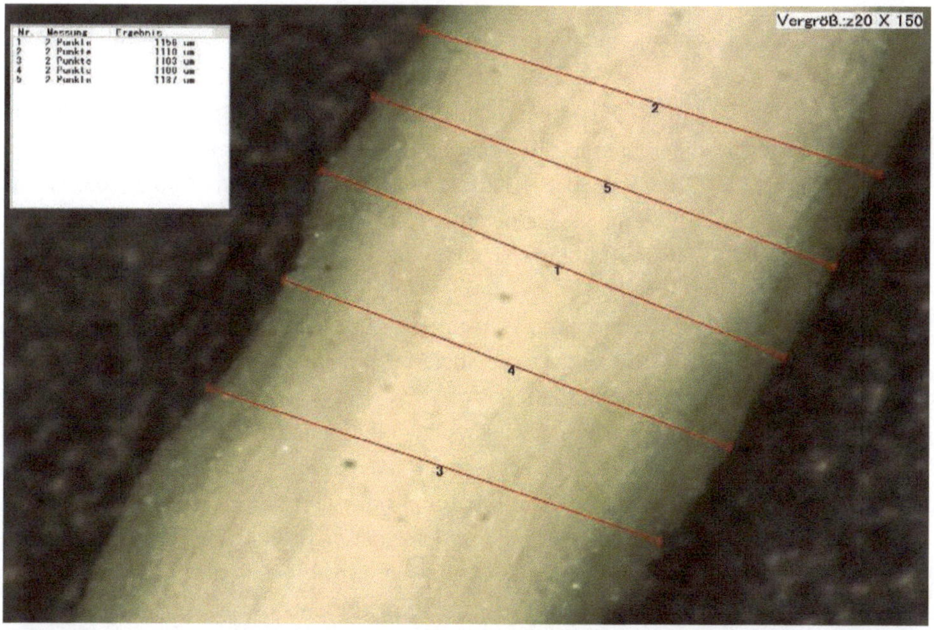

a

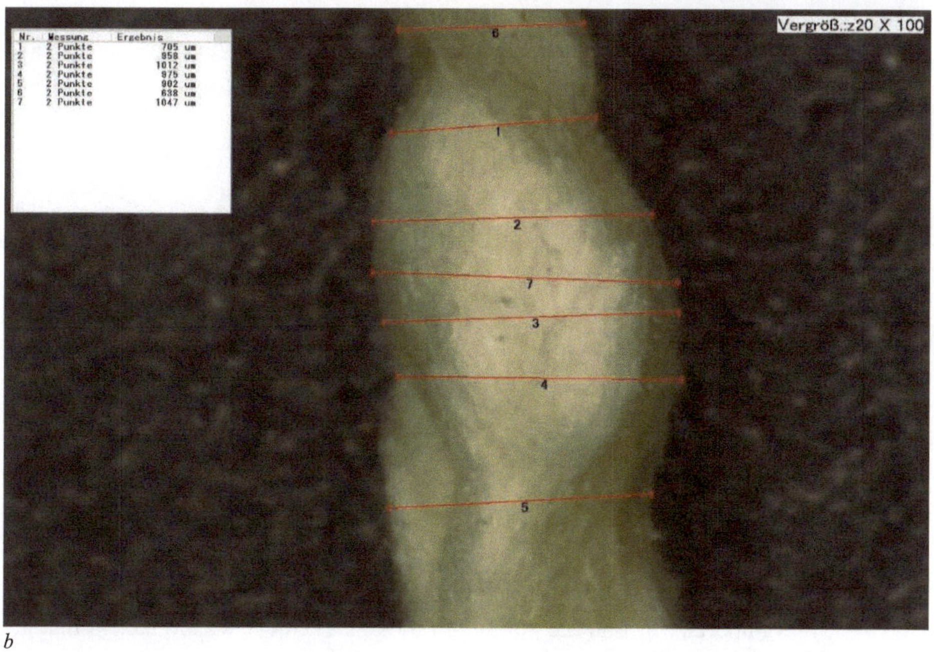

b

Figure 3.48 Regularity of diameter shape after drying (paste 29) (a) and irregularity of diameter shape of rotted paste 17.

3.4.1.5 Effect of the Filler Type

From Tab. 3.15 the results of the filament series (paste 13, 12 and 14) and the series (paste 25, 26 and paste 27) show that the filler type has an effective influence on the shrinkage results. Three types of filler were used. Starch is used as undissolved in water because the water temperature doesn't reach the temperature reactivity of starch for gelatinization. The second filler is chalk. The third filler is Mgso4, which crystallizes with water.
It's noticeable that the filler Mgso4 of (paste 14) reduces the shrinkage of the diagonal deformation by 2% compared to paste 11 (same mixture with starch). On the other hand, Mgso4 affects the recyclability of the paper material, which is a disadvantage for this type of filler since it is a chemical material.
While paste 12 (starch) has a higher shrinkage than the same paste but with chalk, this is due to the free water which does not react with the starch, unlike chalk which causes water to form a colloidal concentration. Therefore, the shrinkage result is 5.2% lower (paste 13 compared to paste 12). Similarly, the diagonal deformation results of paste 27 (with chalk) is smaller by 3.5% than the same paste but with starch (paste 25).
However, the use of starch in the mixture also indirectly reduces shrinkage when the free water is replaced by solid material (less water content), thus stiffening the paste, and reducing shrinkage. What's more, starch is an organic substance and does not prevent the paper material from being recyclable, which is what gives it the privilege of being used.
Fig. 3.49 shows the chemical effect of using Mgso4 on the finished material, where Mgso4 causes the surface roughness compared to the smoothness of the surface using the starch as a filler.

3.4.1.6 Effect of Different Filler Ratios

The potential to reduce the water content in the mixture through starch, as explained above, is the aim for the next step to investigate the correct percentages for the filler (starch) with the other components. Regardless of the effect on extrudability, increasing the filler percentage used gives a significant reduction in the shrinkage results. The results of Tab. 3.15 for Paste 28 (15.6% starch) are 42% and 68.4% lower than those for Paste 29 (9.7% starch) and Paste 27 (3% starch) respectively. This difference is illustrated in Fig. 3.50 where the effect of the filler content on the regularity of the diagonal deformations after drying is very clear. The explanation is that the filler, due to its fineness, fills the areas in a distributed way and stabilizes the CMC, which would limit the shrinkage results. However, it is important to find the optimum ratios that achieve all the factors that are essential for 3D printing in this study, i.e. (extrudability, buildability, shrinkage results and geometric integration & dimensional accuracy after drying).

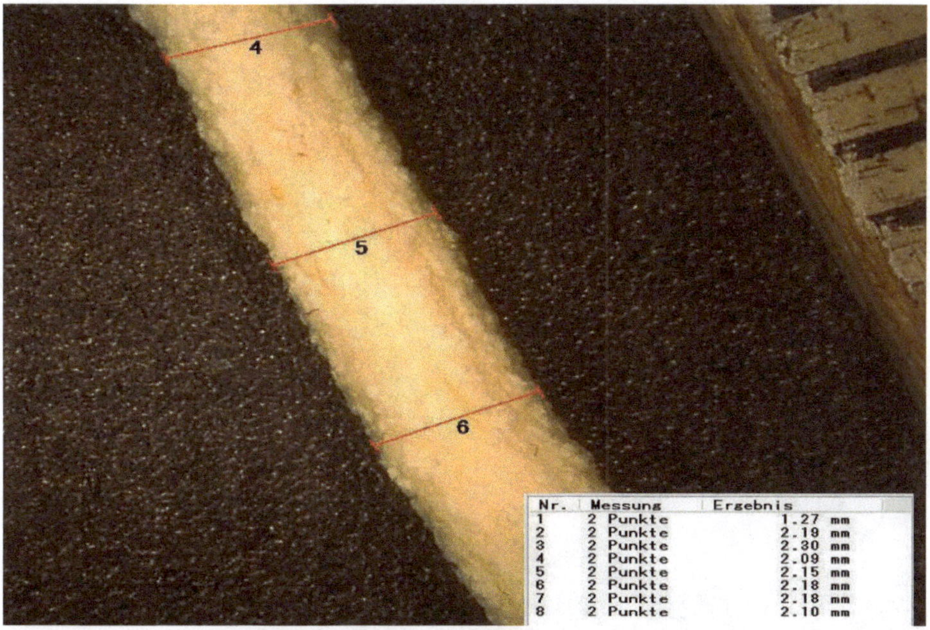

a

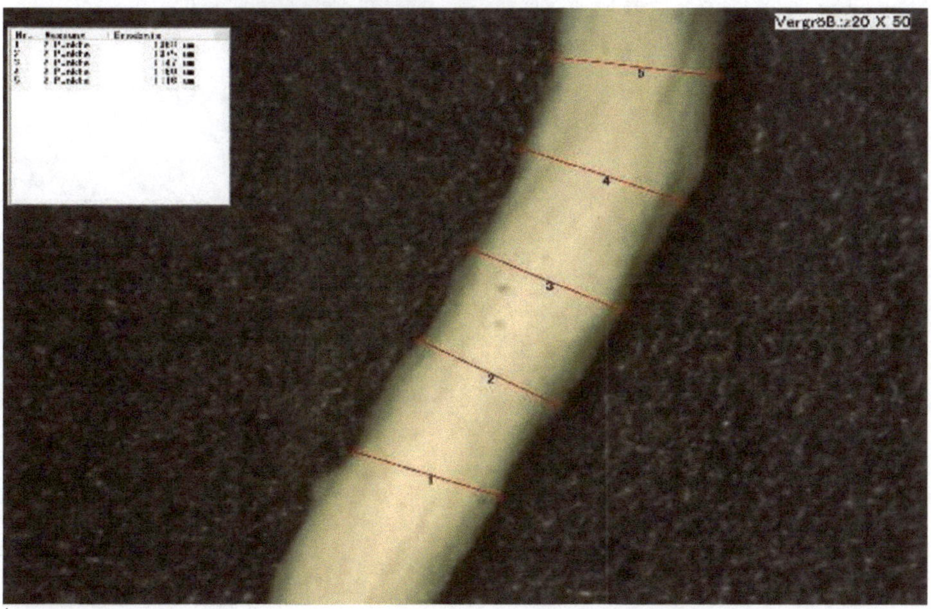

b

Figure 3.49 Surface roughness due to using mgso4 (a), surface smoothness of using starch (b).

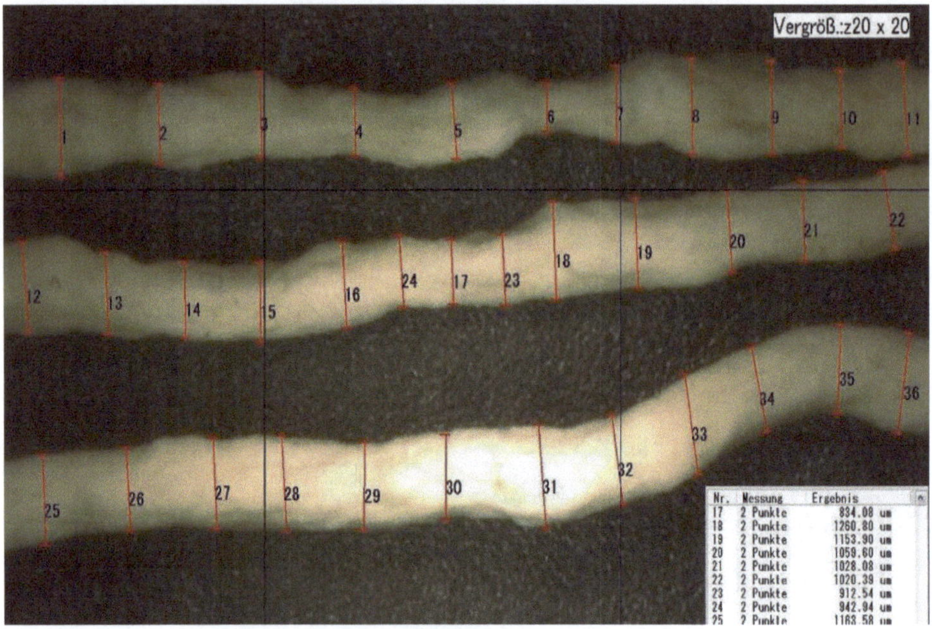

Figure 3.50 Shrinkage outcomes for specimens with different filler percentage (top: paste27, center one: paste29, bottom: paste 28).

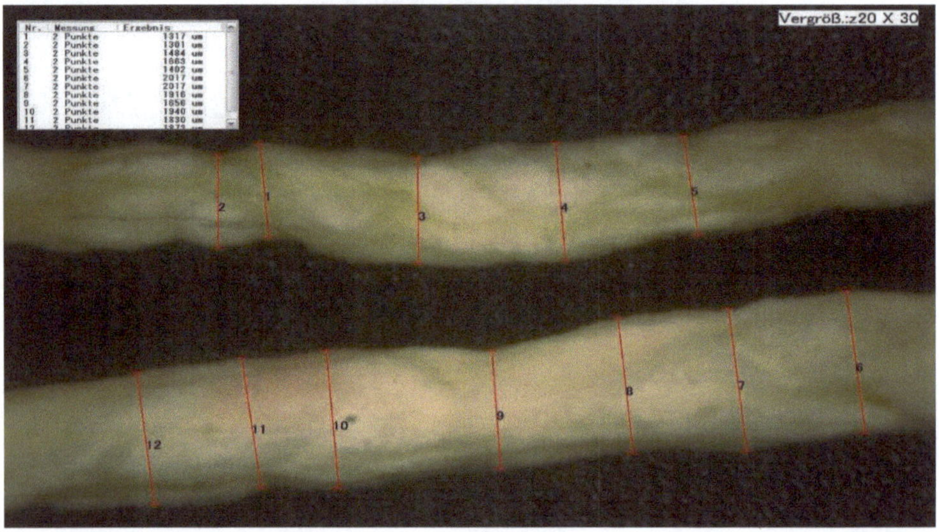

Figure 3.51 Shrinkage outcomes for (top: un-curried specimen, bottom: curing applied according to paragraph 3.4 Curring process).

Finally, after further optimization and reformulation of the proportions of the components, the optimal proportions were obtained that fit the extrudability and limit the shrinkage as much as possible for nozzles 1.6 mm, 1.2 mm, 0.6 mm, indicated by paste 30, paste 31 and paste 32, respectively.

3.4.1.7 Effect of the Curing Process

The shrinkage (diagonal deformation in Tab. 3.15) is reduced by 53.9% when using the curing process, as in section 3.4, we compare the filaments of paste 29 and the identical sample but without the curing process, as can be seen in Fig. 3.51. Fig. 3.52 shows the monitoring of the environment (temperature and relative humidity) from the day of printing to the day of testing for the samples without curing.

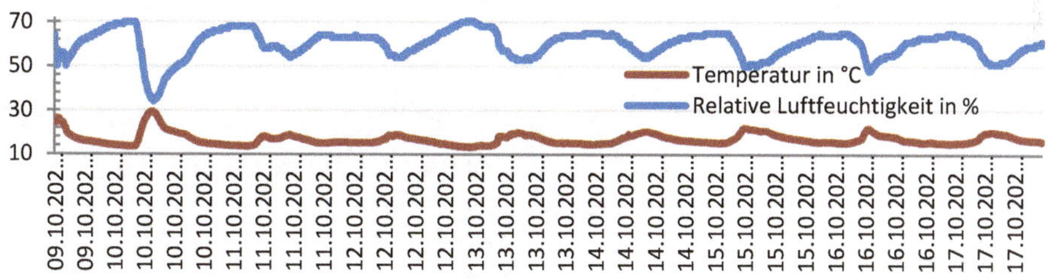

Figure 3.52 Environmental monitoring (temperature and relative humidity) from the day of printing to the day of testing.

3.4.1.8 Effect Using the Wood Fibers

The shrinkage results of pastes 20 and 23 (both with the same wood fibers content of 3%) (Tab. 3.15) show that the deformation after drying is better controlled by using more and shorter wood fibers.

Wood fibers are known to reduce shrinkage. The addition of 1.6% wood fibers to Paste 33 (Paste 37) results in a 15% reduction in diagonal deformation when comparing Paste 33 with Paste 37. Similarly, paste 40 (addition of +0.96 wood fibers to paste 38) reduced the shrinkage by 17%, comparing paste 40 with paste 38.

3.4.1.8 Summery

By reducing the water content of the mix and compensating for this reduction by manipulating the proportion of ingredients, shrinkage can be controlled up to 17.4% (Paste 30), which is the best result for a 1.6mm nozzle.

3D printed hollow triangular pyramid with two lines of thickness and 1.6mm nozzle diameter.

4 Printing Quality Assessment

Until now, 3D printing has no globally recognized standards/specifications or norms for a comprehensive list of performance requirements and test methods for a printing mixture [Ming Xia et. Al. 2019]. Also, until now, there are no guidelines for calculation identification/quantification of print quality that are consistent measurement methods. However, from industry overlook and scientific experimental overview, the term print quality is usually referred to uniform lines with smooth outlines and have acceptable aesthetics with geomatical integration and considerable accuracy. [Julian Schirmer et.al 2018].

In this chapter, the quality of 3D printing performance and production process has been introduced by evaluating the methodology and parameters used in the production process. The quality targeted in this chapter would consider several aspects that are effective, starting from the materials used to the results for the printed geometry. The challenges and solutions are also reviewed to come up with better strategies and effective optimizations to achieve structural integrity with possible accurate dimensions.

4.1 Cellulose Solution

Each type of cellulose has a different structure that creates its characteristics, not only the plant family changes the cellulose properties, but also the location of the cellulose, the age of the plant, seasonal variations with the surrounding available environment such as the abundance of water availability in the soil, and the manufacturing process between mechanical processing and refining of cellulose fibers [Allen, S.T. et. al, 2022]. These factors have a major impact on fiber properties such as strength, morphology, rheology, water retention capacity, etc. The degree of refining can be measured by Schopper Riegler (DIN EN ISO 5267-1) and Fiber Morphology (ISO 16065-2). For these reasons, cellulose products, even for the same company production, have significantly different properties for each production set, this has been noticed from experimental work after experiments using the same type of cellulose from the same company but more than the production set. The results were also different, especially within the shrinkage outcomes.

As a conclusion from the experimental work, the cellulose type that gives better shrinkage results is the cellulose from Mercer's pulp mills/ long fiber, NBSK_ECF, Pine STENDAL/ Germany. However, the short cellulose fiber (Aspen PHK-DMI/Canada) is the better choice from the extrudability ease aspect that fits fine die diameters such as 1.6mmn and 1.2mm, 0.6mm and the second better results in buildability and shrinkage outcomes.

D. Abdullah Agha, *3D Paper Printing for the Built Environment*, Mechanik, Werkstoffe und Konstruktion im Bauwesen 73,
https://doi.org/10.1007/978-3-658-45076-2_4

4.2 Optimal Mixtures

The paste formulation is designed to achieve these main aspects:

- Extrudability/printability: this is the ability to extrude the material homogeneously, smoothly, easily, and fluently without blocking. The paste should be as soft as possible.
- Buildability and stability: ensure the ability to be built in a vertical direction aspect and the bottom layers could hold the load of the top layer without being compressed or changing shape. It should also be stable enough to hold without collapsing until drying is complete. The paste should be as solid as possible.
- Adhesion property: which provides adhesion between the components of the mix and provides a suitable bond between the printed layers. This is achieved by having an appropriate CMC content ratio. This property was tested by printing and visual evaluation. If the CMC content is too low, the top layer will not adhere and will slip.
- Shrinkage results in deformation after drying; it means that the components of the mixture combine with the water. In other words, the free water in the mixture and the water content are limited as much as possible.

The mixture consists of four components (cellulose, carboxymethylcellulose CMC, lecithin, and filler) with water. Each component of the mix contributes to a specific purpose to achieve a compatible paste with all the above considerations.

Cellulose is the main structure of the mix, acting as a skeleton for the paste. CMC forms a gel with hot water for adhesion and as a thickener/rheology. Fillers are used to stabilize the CMC and increase the number of interparticle contacts, which are effective in improving the buildability factor. Lecithin is a fatty substance used to improve flowability and to help the paste deform during extrusion through the die [Abdullah Agha and Knaack, 2022].

Although the cellulose from Mercer's pulp mills/ long fiber, NBSK_ECF, Pine STENDAL/ Germany gives the best results in terms of shrinkage and deformation after drying, it causes blocking due to the use of fine nozzles diameters such as 1.6 mm and 1.2 mm, 0.6 mm. When using this type of cellulose, it is necessary to use pastes with an appropriate content of lecithin and fillers, which would help to soften and ensure the homogeneity of the mixture to avoid blocking, as shown in Fig. 4.1.

For 0.6 mm nozzle diameter, it is better to use powdered cellulose to avoid blocking.

Based on these aspects, three pastes will be investigated and developed for 3D printing, as shown in Tab. 4.1.

Figure 4.1 Blocking caused using long fibrous cellulose with a nozzle diameter of 1.6 mm.

Table 4.1 Mixture formulation suitable for nozzle diameters of 1.6 mm, 1.2 mm and 0.6 mm.

Nozzle used mm	**Cellulose** % by wt.	**Water** % by wt.	**CMC*** % by wt.	**Lecithin** % by wt.	**Filler** % by wt.
1.6	15, Aspen	69.2	7.5	2.8	5.6, Starch
1.2	15.5, Aspen	71.5	5.8	2.9	4.3, Starch
0.6	13.4, Aspen	78	3.6	2.6	2.3, chalk

4.3 Printability & Geometry Calibration

The balance of process parameters can be the main key to 3D printing performance and efficiency indicators. The printing parameters are basically selected depending on the used material and machine efficiency. These are optimized according to printability and extrusion quality, and by determining the accuracy of the dimensions for 3D printed geometries.

To consider the coalescence of filaments and layers, printing parameters such as speed and layer height are set to constant values as mentioned in section (3.3 Process Parameters & Geometry Configuration). The layer height was set to (0.7 and 1) mm for the nozzle diameter of 1.2mm and 1.6mm respectively. The process parameters: all speeds= 18mm/s (circumferential speed, infill speed, first layer speed, gap infill speed and max print speed), nozzle diameter= 1.6mm, layer height= 1mm, air pressure= 0.2Mpa, 100% linear fill density, extrusion width= 300%, extrusion multiplier= are experimented as shown in Fig. 4.2 which illustrates the distributed regions for 3D printing with nozzle diameter equal to 1.6mm. The accuracy of the printed geometry is determined by:

$$\text{Divergence in the accuracy} = \frac{D_{\text{printed}} - D_{\text{Designed}}}{D_{\text{Designed}}} \text{x}100\% \qquad (4.1)$$

where D_{Printed}= measured printed dimension; and D_{designed}= designed dimension.

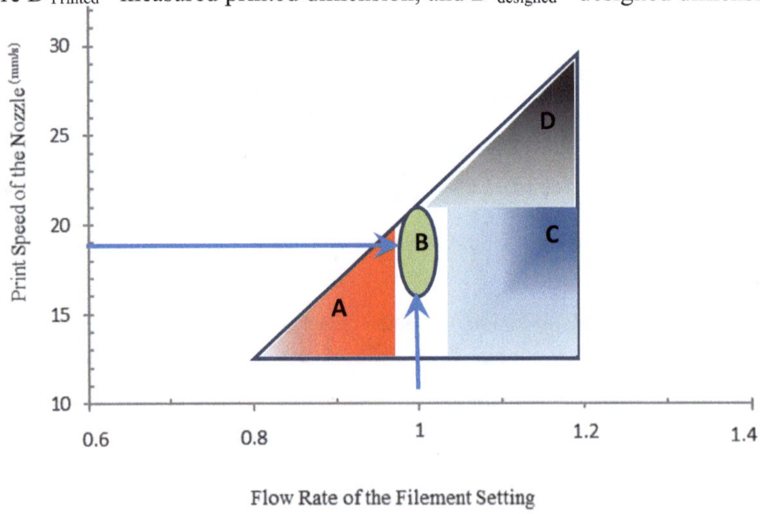

Figure 4.2 The process parameters & printing ability [Abdullah Agha and Knaack, 2022].

Fig. 4.2 and Tab. 4.2 display the regions that give the printed geometry with its forming manufacturing and the distribution of the possible process parameters with 3D paper printing, as: [Abdullah Agha and Knaack, 2022].

Region A: the balance between speed and flow is not quite right, with too much material being extruded at low speed. The result is a bulky volume with a deviation in accuracy (eq. 4.1) of at least 3% when compared with the samples from region B, as shown in Fig. 4.3 and Fig. 4.4.

Region B: a typical balance between speed and flow, where the material extrusion has a nice finish surface quality and correct single wall widths around 100% accuracy in dimensions (eq. 4.1). Fig. 4.5 shows the perfect surface quality and sharper corners, geometric accuracy and especially the aesthetic appearance of the 3D printed geometry.

Region C: the speed is typically suitable, but the flow gives the defect of non-straight "zigzag" lines for the extruded filaments, see Fig. 4.6.

Region D: the speed is barely adequate; it is extremely fast than the average, which does not give enough time for the extrusion to be able to stick and have enough adhesion to the previous layer, so that the extruded filament is pulled with the nozzle movement as mentioned in paragraph "3.3.1 Nozzle movement speed".

Table 4.2 Mixture Formulation that fits the nozzles diameters like 1.6mmn, 1.2 mm, and 0.6 mm.

Limitations	**Nozzle diameter**		
	1.2 mm	1.6 mm	0.6 mm
Region A			
The flow rate and the speed (mm/s) start from: respectively.	0.9 and 7	0.7 and 7	1.2 and 7
and their effect progressively decrease until: respectively	0.95 and 12	0.75 and 12	1.3 and 12
Region B			
The flow rate and the speed (mm/s) are closed to respectively	1 and 18	0.85 and 18	1.4 and 18
Region C			
The flow rate and the speed (mm/s) start from: respectively.	1.05 and 7	0.9 and 7	1.6 and 7
and their effect increase to reach the high effect in respectively	1.1 and 15	0.95 and 15	1.7 and 15
Region D			
The flow rate and the speed (mm/s) start from: respectively.	1.05 and 16	0.9 and 16	1.6 and 16
and their effect ascending reach the peak effect in respectively	1.1 and 26	0.95 and 26	1.7 and 26

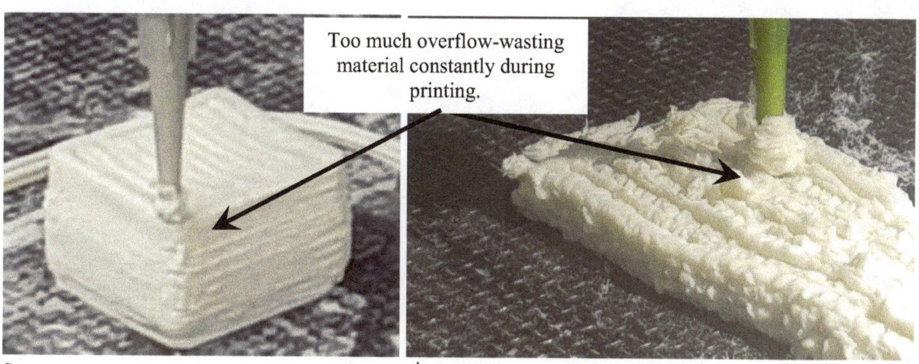

Figure 4.3 Overflow defect (a), and an excessive extrusion causing bulky volume (b).

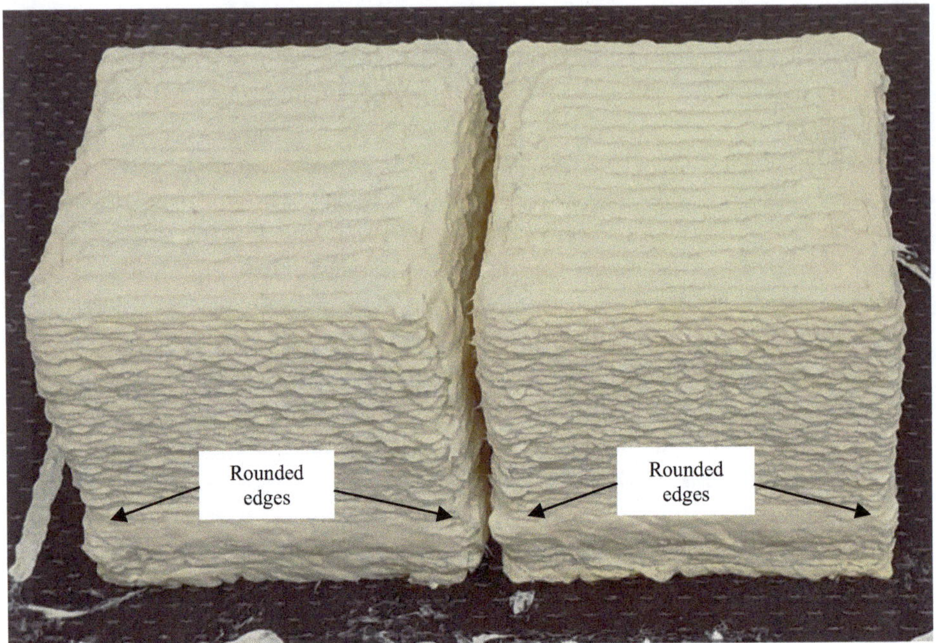

Figure 4.4 Unbalanced printing parameters causing overflow and rounded corner.

Figure 4.5 3D printing with balanced printing parameters.

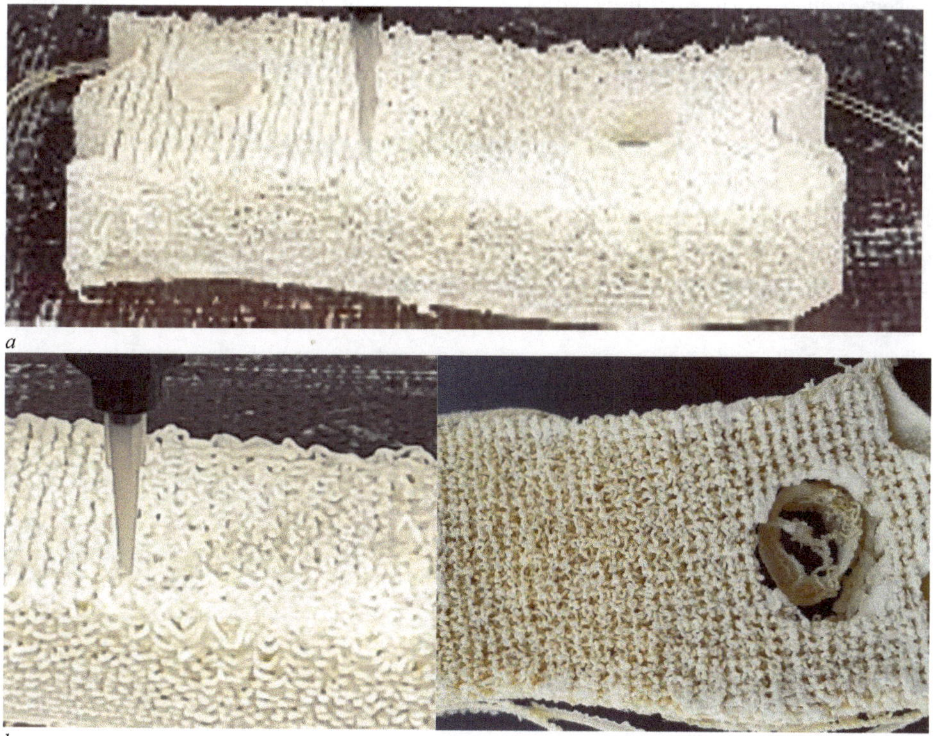

a *b* *c*

Figure 4.6 Zig-Zag Defect: (a)the complete brick during printing(a); closer view(b), and after drying(c).

4.4 Optimized Printing Parameters

Every mixture has a certain consistency and viscosity. More precisely, its consistency and viscosity determine its printing parameters. Therefore, balanced process parameters have been studied and identified for each mixture to improve the printing quality with better aesthetic results, controlled deformation and optimal accuracy. The choice of printing parameters is directly linked to the time required to complete the manufacturing process, as well as to the mechanical properties. 3D printing usually has the production speed in preferences. For example, choosing double size layer height will be approximately double faster to finish printing, but that will also affect the coalescence and wide spreading of pores, and that have direct impact on the adhesion between successive layers due to decreasing the contact area; subsequently the density and stiffness decrease in the direction of the layer height. Conversely, to have a delicate, detailed object, choosing a lower layer thickness improves the resolution of the object and gives smoother surfaces.

Based on the experimental tests with the LUTUM -VormVrij® 3D clay printer version 2.1, the printing parameters for all pastes are listed in Tab. 4.3. It should also be noted that the internal surfaces in contact with the paste should be moistened before printing to avoid changing the water content of the mixture.

Tab. 4.4 shows the evaluation for the pastes based on the rheological results from Tab. 2.13. The ease of mixing index assesses the effort required to mix the paste to achieve homogeneity. Extrudability is a free-flowing indicator of how smoothly the full extrusion quantity will flow out of the die during printing; without blocking the material in the die. Buildability is a measure of the degree of efficiency in building two lines of wall thickness to a height of 100mm without buckling/side-sway or instability, where each step describes 20mm. Stability and shape retention assesses the extent to which the paste can maintain its shape from the load of the overhead layers of at least 15 layers until drying, without compression [Ali Kazemian et al. 2019].

Table 4.3 Balanced printing parameters for each mixture [Abdullah Agha and Knaack, 2022].

Mixture From Tab. 3.9	Used nozzle mm	Printing speed mm/s	Flow rate	Layer height mm	Air pressure MPa	Path width, mm
Paste 1	1.2	18	1	0.75	0.2	1.5
Paste 2	1.2	18	1.2	0.75	0.6	1.5
Paste 10	1.6	18	0.85	1	0.3	2
Paste 11	1.6	20	1	1	0.35	2
Paste 12	1.6	20	0.80	1	0.32	2
Paste 13	1.6	18	0.85	1	0.3	2
Paste 14	1.6	20	1	1	0.4	2
Paste 15	1.6	20	1	1	0.4	2
Paste 25	1.6	20	1	1	0.8	2
Paste 26	1.6	20	1	1	0.8	2
Paste 27	1.6	18	1.2	1	0.8	2
Paste 30	1.6	18	1	1	0.8	2
Paste 31	1.2	18	1.2	1	0.8	2
Paste 32	0.6	15	1.4	0.25	0.8	0.5
Paste 33	0.6	15	1.2	0.25	0.7	0.5
Paste 38	0.6	15	1.5	0.25	0.7	0.5

Table 4.4 Assessing and evaluating pastes based on experimental work by 1.6 mm nozzle.

Mix Tab. 3.9, 3.10 and 3.11	Consistency & Mixability Index*	Extrudability index	Efficiency buildability	Stability** & shape retention
Paste 1	■■■■■	■■■□□	■■□□□	■□□□□
Paste 2	■■■■□	■□□□□	■■■□□	■■□□□
Paste 6	□□□□□	■□□□□	■■■□□	■■□□□
Paste 7	■■□□□	■■■□□	■■■□□	■■□□□
Paste 10	■■■□□	■■■■■	■□□□□	■□□□□
Paste 11	■□□□□	■□□□□	■■■■□	■■■■□
Paste 12	■■□□□	■■■■□	■■■■□	■■■□□
Paste 13	■■□□□	■■■■□	■■■□□	■■■□□
Paste 14	■■□□□	■■■■■	■■■□□	■■■□□
Paste 16	-	■■■□□	■□□□□	■□□□□
Paste 21	■■□□□	■■■■□	■■■■□	■■■■□
Paste 24	■■□□□	■■■■□	■■■■□	■■■■□

* Each filled block expresses the degree of mixing depending on time scale.
** Each step describes 20mm printed stable.

Table 4.4 Continued.

Mixture from Tab. 3.9, 3.10 and 3.11	Consistency Mixing ease Index	Extrudability index*	Efficiency buildability	Stability & shape retention
Paste 25	■■□□□	■■■■□	■■■□□	■■■■□
Paste 26	■■□□□	■■■■□	■■■□□	■■■■□
Paste 27	■■□□□	■■■■□	■■■□□	■■■■□
Paste 28	□□□□□	■■□□□	■■■■■	■■■■■
Paste 29	■□□□□	■■■□□	■■■■■	■■■■■
Paste 30	■□□□□	■■■□□	■■■■■	■■■■■
Paste 31	■□□□□	■■■■■	■■■■□	■■■■■
Paste 32	■■■■■	■■■■■	■■□□□	■□□□□
Paste 33	■■■□□	■■■■■	■■□□□	■■□□□
Paste 37	■■□□□	■■■■□	■■■□□	■■■□□
Paste 38	■■■■■	■■■■■	■□□□□	□□□□□
Paste 40	■■■■■	■■■■■	■■□□□	■□□□□

4.5 Shape-Retention & Stability Ability

In general, dimensional stability (also known as shape retention and green strength) is the ability of a material to retain its original shape after loading or to maintain its original appearance during use. Typically, this property is measured for fibers and fabrics, but it is now also being assessed for 3D printed materials as an evaluation. Where shape retention directly influences the size, shape, surface quality and efficiency of extrusion-based layered forming. [Song, J. et.al, 2019] [ISO 20864:2004] [Ali Kazemian et.al. 2019]. To date, there are no applicable guidelines or proposed methodologies for the measurement of shape retention for 3D printing, or a well-defined approach to acceptance criteria from a norm or standard. Therefore, researchers assess the fresh/wet properties of the printing mixture based on rheological properties with other observations that define the meaning of the property [Ming Xia et. al. 2019].

The shape retention for 3D printing is evaluated by i) rheology properties through the measured shear yield stress and viscosity, ii) by the paste's ability to withstand a load of its own weight to retain its shape until drying and iii) maintain the accuracy compared to the designed geometry without collapse, buckling, sideways bending, or compression.

Figures 4.7 and 4.8 show a specimen that couldn't hold its shape and collapsed during the drying process. This is one of the weaknesses of wet material (green strength). The material of this sample collapses within the time required for drying and couldn't hold its own wet weight. The details of the production are using Paste 8 and the printing parameters are using 100% filling of concentric type, all speeds= 18mm/s (perimeter speed, infill speed, first layer speed, gap infill speed and max printing speed), nozzle diameter=1.6mm, layer height=1mm, extrusion multiplier=0.8, air pressure=0.2Mpa, extrusion width= 250%, the printing time is about 17 hours. This sample is built in four parts over three days as three parts. The explanation is that the mix formulation is not hard enough to build at such a height and maintain buildability and dimensional stability. The specimen was a bite unstable and having buckling and side sway issues. After a height of 100mm the specimen was wobbly during layering and tended to have a side sway, so it needed support during printing. The bottom layers were swollen and had a different consistency to the top layers because they had to be kept wet for four days, making them more susceptible to compression. The curing process is carried out as described in section "3.4 Curing process". The sample is isolated with nylons and wet towels for each printed part due to the time breaks between prints. The actual printing time per day is approximately 6 hours.

Fig. 4.9 shows the monitoring of the environment (temperature and humidity) from the start of printing until the sample is completely dry. The Details of designing and functionality are shown in Appendix D: Design Details.

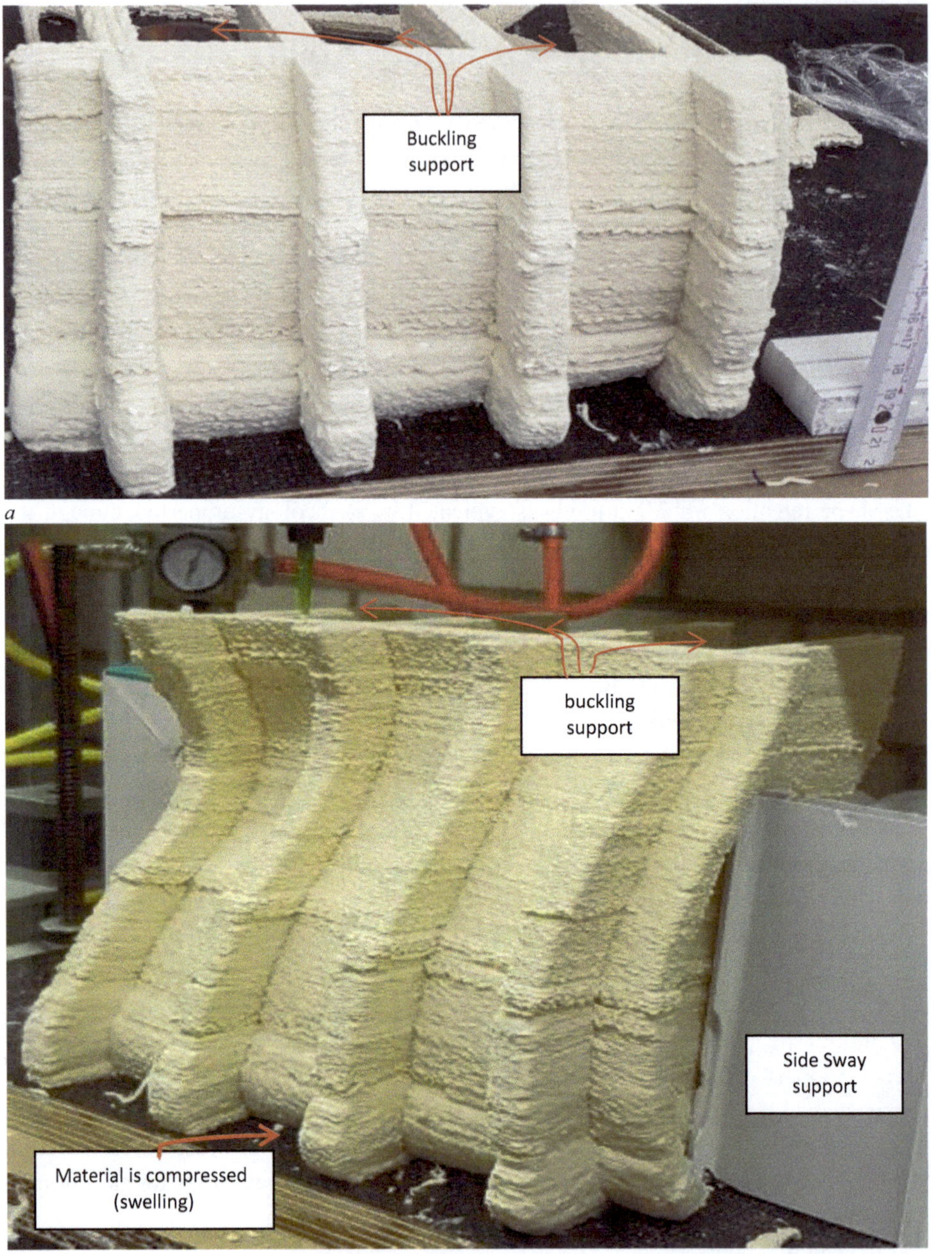

a *b*

Figure 4.7 Elevated geometry during printing at 160 mm height (a), and 220 mm height with side-sway support (b).

4.5 Shape-Retention & Stability Ability

a

b

Figure 4.8 Specimen collapse during drying process (a) and after drying complete (b).

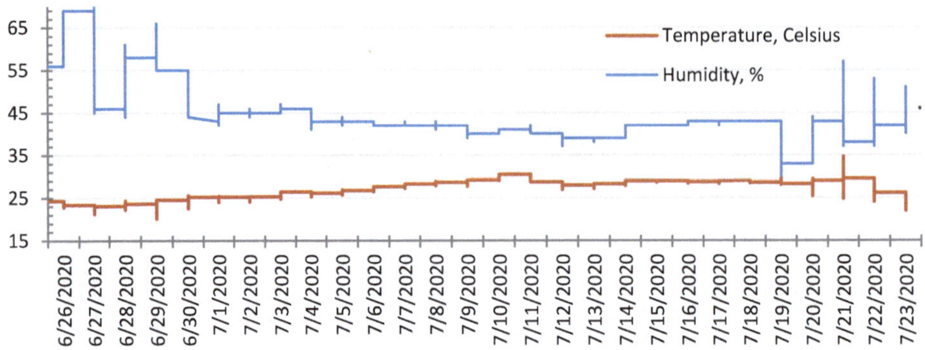

Figure 4.9 Monitoring the surrounding environment (temperature and relative humidity) from the day of printing until the sample is completely dry.

The ability of the specimen to retain its shape and stability throughout the printing process depends on the type and stiffness of the paste used, which is essentially dependent on the mix formulation; if the printing parameters are set and used effectively, the specimen should be able to retain its shape and stability. Figure 4.10(a) shows a model during the printing process which is still stable up to a height of 80 mm. The mixture used is Paste 12 from Tab. 3.9. The printing parameters of the samples are as follows: the infill type is 100% rectilinear. The nozzle diameter is 1.6 mm. All speeds= 18mm/s (perimeter speed, infill speed, first layer speed, gap infill speed and max print speed), layer height= 1mm, extrusion width= 200%, the air pressure= 0.8 MPa and the extrusion multiplier= 1. The time required to complete printing was 13 hours, with the entire sample printed on the same day. The samples were cured under the same conditions as mentioned in paragraph "3.4 Curing process". However, the specimen had to be supported after the height of 80 mm. Due to the height and the type of mixture of materials used, the specimen had a side sway and movement during compression. Despite the use of support for the specimen during printing, the specimen didn't collapse during drying.
Fig. 4.11 shows another case study that were successfully printed without using any support due to the selection of integral suitable paste that have the optimized mix formulation to be a self-supporting material. The manufacturing details are by using paste 30 from Tab.3.9 and the printing parameters are using 100% filling of straight-line type, all speeds= 18mm/s (perimeter speed, infill speed, first layer speed, gap infill speed and max print speed), nozzle diameter=1.6mm, layer height=1mm, extrusion multiplier=0.8, air pressure=0.2Mpa, extrusion width= 250%. The samples were cured under the same conditions as mentioned in Section 3.4 Curing Process.

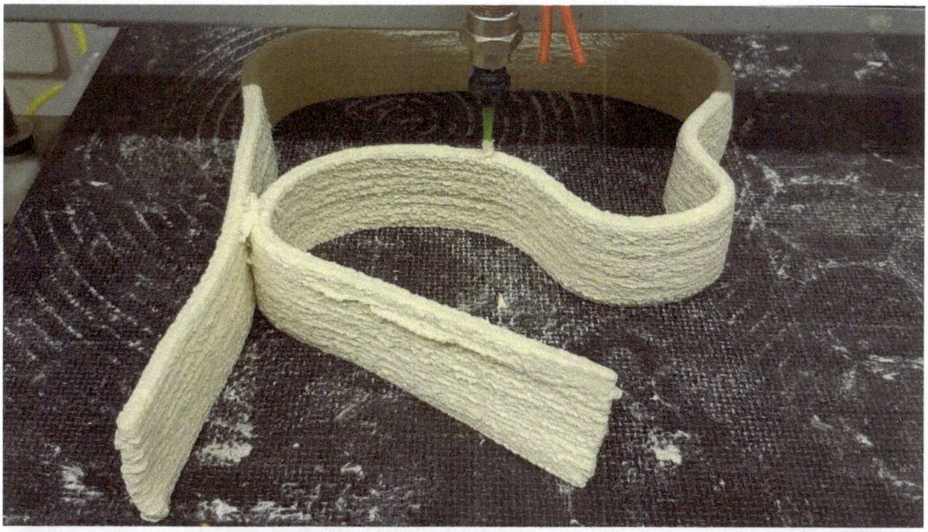

a

Figure 4.10 Good buildability of the specimen while printing (a) and after printing is complete (b).

a

b

Figure 4.11 Good buildability of the specimen while printing (a) and after printing is complete (b).

4.6 Mixture-Validity Time Before Printing

The paste is made of organic materials. The paste decays over time. Experimental observations have shown that the paste is no longer a good material for printing after approximately one week. The shelf life depends on the storage location and the formulation of the mixture.

In general, CMC is a direct spoilage material, so it's better to keep it in the fridge. Usually, it's better if the paste is to be freshly mixed and printed immediately. This was done for this study to avoid the growth of putrefactive bacteria in the mix, which would affect the strength of the material.

Figure 4.12 shows the shape of the rotten paste using powdered cellulose and Fig. 4.13 shows the difference between using rotten paste and fresh paste, paste 16 from Tab. 3.9 is used here to print the rotted paste sample, it is noticeable to see the difference between the two samples in terms of the final color and the deformation and shrinkage results. The deformation results of the rotted paste show uneven behavior and have relatively high shrinkage percentages. The Details of designing and functionality are shown in Appendix G.

a *b*

Figure 4.12 Rotted paste (a) and shrinkage outcomes printing when with rotted paste (b).

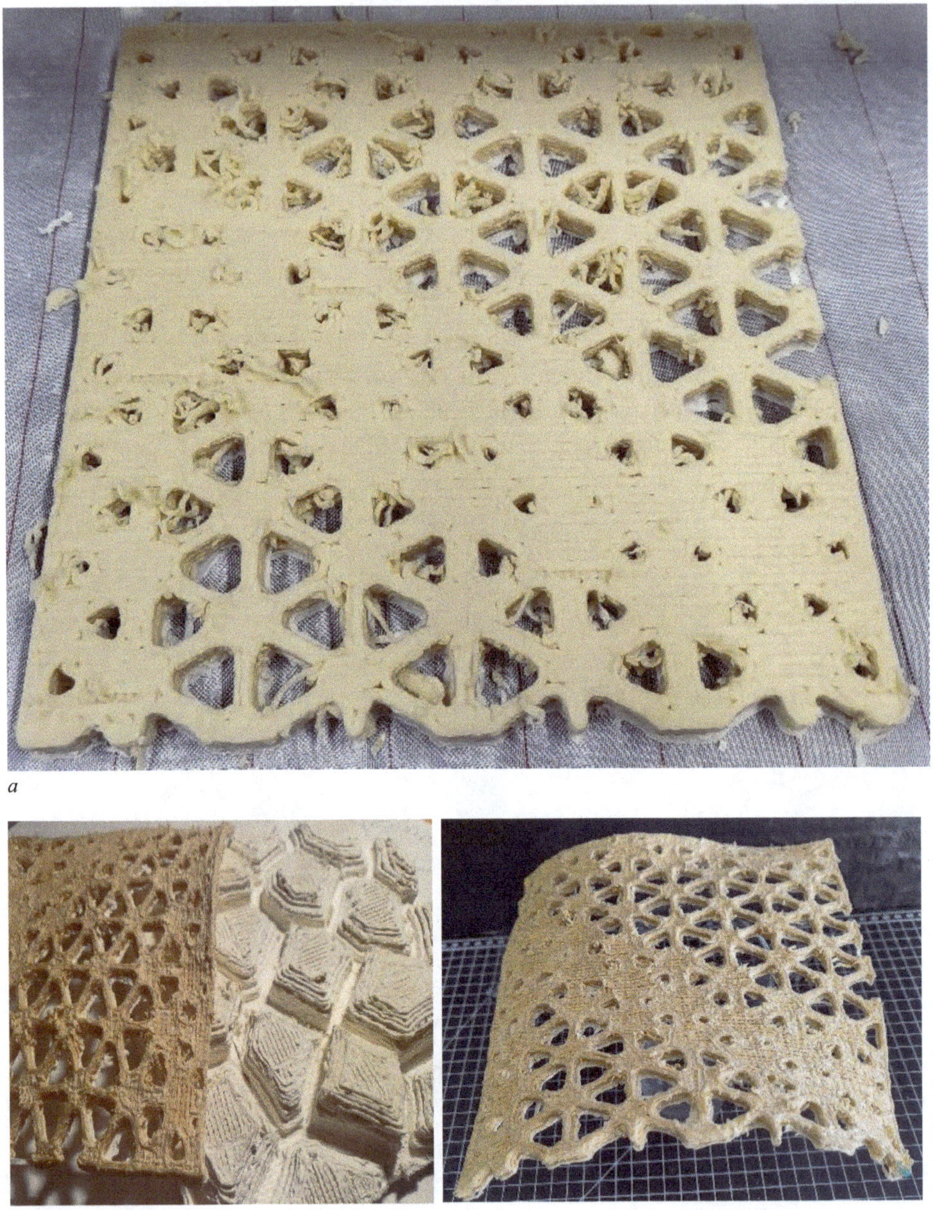

a

b c

Figure 4.13 Printing with rotted paste (a) after complete printing in comparison with another model printed with the same fresh paste (b); and the uneven shrinkage of the model after drying.

4.7 Recyclability

Recyclability is an important tool for designing products with the environment in mind. Thus, it is one of the environmental evaluation tools that integrates both important aspects such as the cost of reuse and waste management [Zhichao Wang and Fang Liu, 2021]. The recyclability assessment is adopted in this study by placing a sample with the same dimensions for all pastes in water and calculating the time needed to complete hydrolysis, as shown in Fig. 4.14 and Tab. 4.4. The recyclability of a mixture depends on the components of the mixture, whether they are organic or chemical. If the mixture had a chemical component, such as a paste with the filler (magnesium sulphate), the mixture wouldn't be recyclable. While the filler component (starch or chalk) delays the time needed for complete hydrolysis.

Table 4.4 Behavior assessments for each mixture using a 1.6 mm nozzle diameter.

Mixture From Tab. 3.9, 3.10 and 3.11	Recyclability	Disintegration/ Hydrolysis Time hr
Paste 1	✓	≥ 40
Paste 2	✓	≥ 40
Paste 3	✓	≥ 40
Paste 4	✓	≥ 40
Paste 5	✓	≥ 40
Paste 6	✓	≥ 24
Paste 7	✓	≥ 24
Paste 10	✓	≥ 40
Paste 11	✓	≥ 40
Paste 12	✓	≥ 40
Paste 13	✓	≥ 40
Paste 14	×	-
Paste 15	✓	≥ 40
Paste 16	✓	≥ 40
Paste 19	✓	≥ 40
Paste 20	✓	≥ 40
Paste 21	✓	≥ 40
Paste 24	✓	≥ 40
Paste 25	✓	≥ 40
Paste 26	×	-
Paste 27	✓	≥ 40
Paste 28	✓	≥ 45
Paste 29	✓	≥ 45
Paste 30	✓	≥ 45
Paste 31	✓	≥ 45
Paste 32	✓	≥ 45
Paste 33	×	-
Paste 37	×	-
Paste 38	✓	≥ 40
Paste 40	✓	≥ 40

b

b *c*

Figure 4.14 Recyclability measurements for all pastes (a), incipient hydrolysis for one sample (b) and complete hydrolysis for one sample (c).

4.8 Overhang Slope

The slant is one of the limitations that constrains the complexity of the design with free creativity when it depends on the ability of the machine that implements the architect's design. However, this is the reality of 3D printing when it is done unsupported.

In additive manufacturing, the general concept of unsupported printing is to consider geometry designed with a maximum inclination of 45-degree. The explanation is that the 45-degree tilt provides support for each layer, with about 50% of the contact lying down around the layer below. This provides enough support to build on and therefore the print is good. This is also true for overhangs that have a slant of more than 45 degrees, but the challenge here is when the slant is less than 45-degrees, where support is needed.

Fig. 4.15 shows several specimens in which pressure overhang slops were investigated. The slopes are 42-, 45-, 47- and 50 degrees as shown from left to right. The overhang slopes equal to and/or greater than 45-degrees are found to be printed normally, while the samples with overhang slopes greater than 45-degrees are not printed well.

Fig. 4.15(b) shows that the material bends after drying because it is not sufficiently supported by the material. It's also worth noting that the 45-degrees specimen printed perfectly, and the material was able to support its own weight with the slant during drying. The details of the production of these samples in Fig. 4.15 are made using Paste 30.

The nozzle used is 1.6 mm and the printing parameters are set to 100% filling of the concentric type, all speeds= 18mm/s (perimeter speed, infill speed, first layer speed, gap infill speed and max print speed), layer height= 1mm, extrusion multiplier= 1, air pressure= 0.8Mpa, extrusion width= 200%. These samples are dried in three days. The specimen in (Introductory Figure for Chapter Four) is an example of a 45-degree overhang; it's a hollow pyramid geometry with two lines of wall thickness, but it's also confined by the three corners. The challenge with this specimen is that not only the overhang slop, but also the effective buildability and stability, the type of mix formulation is the solution here to control these issues.

a

b

Figure 4.15 Overhang Slopes from 3D paper printing (A) and for specimens after drying.

The production details are using Paste 30. The nozzle used is 1.6 mm and the printing parameters are set to 100% filling of linear type, all speeds= 18mm/s (perimeter speed, infill speed, first layer speed, gap infill speed and max print speed), layer height=1mm, extrusion multiplier=1, air pressure=0.8Mpa, extrusion width= 200%.

However, there are other tricks that are followed to avoid overhangs slops, like example using 3D plastic model as support structure. These could also be used for 3D paper printing with consideration that material needed a significant duration for wet material to dry from the water content; like:

- Changing the orientation in the other direction could avoid the overhanging slope. It should consider choosing the better orientation that reduces the number of overhanging slopes as much as possible.
- Splitting the model into parts and within the design of the part, an interlocking function could be made in the geometry to fix the parts together.
- Modify the G-code using the 'Shell' slicing settings in the PrusaSlicer software to print the inner layers before the outer. In addition, the infill/perimeter overlap could be increased to (75 - 100) %. This would improve the surface quality of the overhanging layers and could reduce stringing of the filament. This would help the overhanging paths to adhere to the inner paths. It is achieved: Print Settings → Infill → Advance → Infill before perimeter.
- It would also be possible to provide an external support structure. In this case, however, the G-code would have to be specially programmed to ensure that the print path was precisely matched to the dimensions and location of the external support structure. Figure 4.16 shows 3D printed plastic used as a base support for 3D printed paper.

Figure 4.16 3D paper printing using an external support structure.

- The wet material does not help either. Therefore, a slower print speed will help to increase adhesion and improve stability during printing.
- Adjusting the extrusion width would be an effective factor to increase the contact area between the previous and subsequent layers. This means that it increases the overlap area of this 50% lying contact. However, this would only work up to a certain limit.
- It's also possible to print the backing if the printer is capable of double extrusion, but this is not the case in this study.

4.9 Removal from the Print-Bed

This defect occurs when the base area of the model is too large and after drying, as shown in Fig. 4.17. To maintain the quality and dimensional accuracy of the model, a solution should be adopted. Here we're using a screen-printing plate/plywood with a peel ply fabric. This means that printing is carried out on a rough surface to ensure that the first layer adheres to the print bed. The advantage of using a peel ply fabric is that it is easy to remove the model from the print bed and the fabric can then be easily removed from the model.

Figure 4.17 Defect of the specimen base due to sticking to the printing-bed.

4.10 Surfaces Topography

The surface topography is specified by the profile shape and the surface roughness, which is evaluated by the waviness and the asperity or finish, which refers to the tip radius or sharpness [Qianfan Xin, 2013] and [Waghmare, Ajay & Sahoo, Prasanta., 2017].

In additive manufacturing, the surface topography defined by the extrusion layer is not evaluated by the asperity or harshness of the surface material, but rather by the local deviation of a surface from its original plane, which is characterized by the statistical parameters such as the variance of the original dimensions, the slope, and the waviness and curvature. The surface topography, measured by calculating the peak heights and the range of curvature, is extended and evaluated with a 3D surface profiling system at different locations (side surface, top end surface and upright) to assess the effect of process parameters (build orientation, layer height and flow rate) on the geometric properties [Eustaquio García Plaza et. al, 2019].

In this study, the focus is generally on showing the aesthetics of the finish for the pastes using different factors that affect the surface topography, without measuring the parameters of the surface topography. For 3D paper printing, it's obvious that the surfaces should be layered, and this depends on the mix formulation and the type of cellulose used, as shown in Fig. 4.18. Here, the long fibers with overflow rate significantly influence the surface topography, as shown in Fig. 4.18 (a). The manufacturing details are by using paste 2. The nozzle used is 1.6 mm, the printing parameters are set to 100% filling of linear type, all speeds= 18mm/s (perimeter speed, infill speed, first layer speed, gap infill speed and max printing speed), layer height= 1mm, extrusion multiplier= 1.2, air pressure= 0.8Mpa, extrusion width= 200%. Meanwhile, the manufacturing details of the sample in Fig. 4.18 (b) are made using paste 1. The nozzle used is 1.6 mm, the printing parameters are set to 100% filling of straight-line type, all speeds= 18mm/s (perimeter speed, infill speed, first layer speed, gap infill speed and max printing speed), layer height=1mm, extrusion multiplier=1, air pressure=0.2Mpa, extrusion width= 200%.

The infill type determines the surface finish in some geometry. It's considered as one of the profile shape parameters which clearly influences the surface topography, as shown in Fig. 4.19.

a

b

Figure 4.18 The surface topography for the same geometry printed with the same nozzle but different printing parameters using long fiber cellulose (a), short fiber cellulose (b).

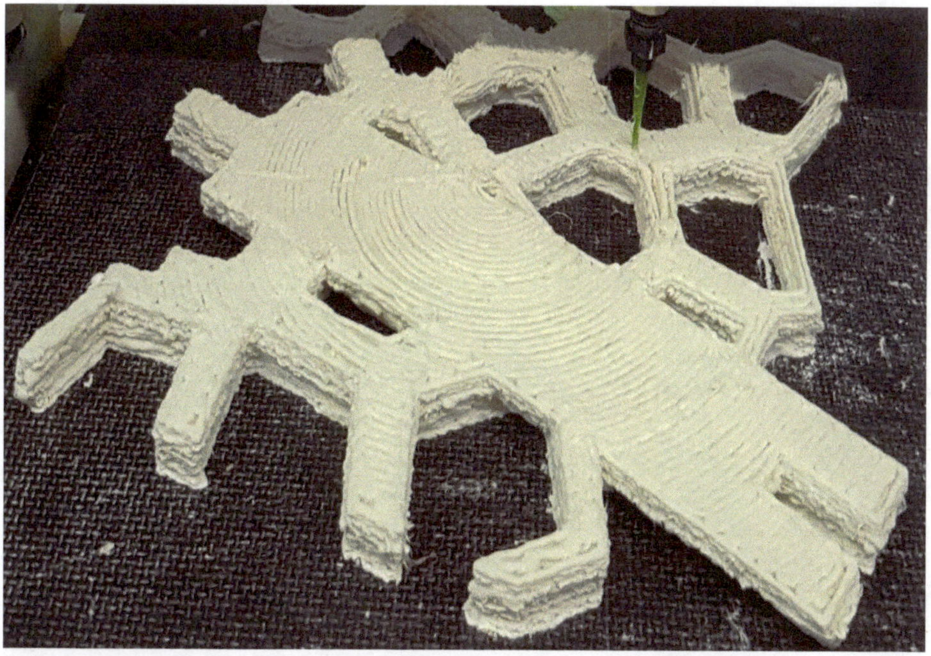

a

b

Figure 4.19 Specimens with concentric infill type, which influences the results of the surface topography: one center point (a) and multiple center points (b).

4.11 Dimensional Accuracy

According to the terminology of ISO/ASTM 52900, accuracy is the closeness of agreement between an individual result and an assumed reference value. It refers to the identical agreement between the measurements in the CAD design and the model once it's printed, expressed by tolerance. It's a key consideration for production quality and accurate geometric integration. In 3D paper printing in general, the main factors influencing dimensional accuracy and tolerances are the manufacturing process and the manufacturing materials.

4.11.1 Production Process Effect

The balance of printing parameters for the required resolution and curing conditions. The dimensions of the sample are checked and optimized several times by manipulating the printing parameters until the balance printing parameters are found, then the dimensional tolerances are specified as the more reasonable accuracy. More details about the balance printing parameters are given in section "4.7 Printability & Geometry Calibration".
The printing parameters affect the size of the model after printing, especially the flow rate, the width path, the air pressure, and the type of filling, etc. Fig. 4.20 shows this effect on the dimensional accuracy when printing a designed cube with dimensions of (30 x 30 x 30) mm. The production settings are by using Paste 1, the nozzle used is 1.6 mm and the printing parameters are selected as follows: 100% filling of linear type, all speeds= 18mm/s (perimeter speed, infill speed, first layer speed, gap infill speed and max printing speed), layer height= 1mm, air pressure= 0.2 MPa, extrusion width= 200%. The different setting between the samples in Fig. 4.24 is the extrusion multiplier, which is 0.75 for sample (a), 0.8 for sample (b) and 0.85 for sample (c).

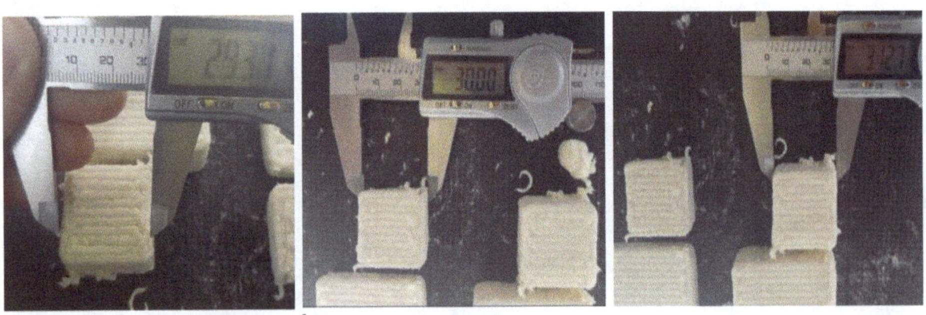

a *b* *c*

Figure 4.20 Printing parameters affect dimensional accuracy when results are less than design (a), matched (b), and printed dimensions greater than design (c).

Coalescence affects the dimensional accuracy of the sample after drying. A complete coalescence is formed by balanced printing parameters and if it is not perfectly adjusted, the filaments would lay separately, causing a weak bond between the filaments, correspondingly each printed line would shrink individually, as shown in Fig. 4.21. Where the shrinkage resulted in this sample being waisted in the middle due to the partial coalescence of the filaments and the lack of material to fill the space of the width. Fig. 4.21 shows a straight infill with a lack of fusion causing the shrinkage behavior shown.

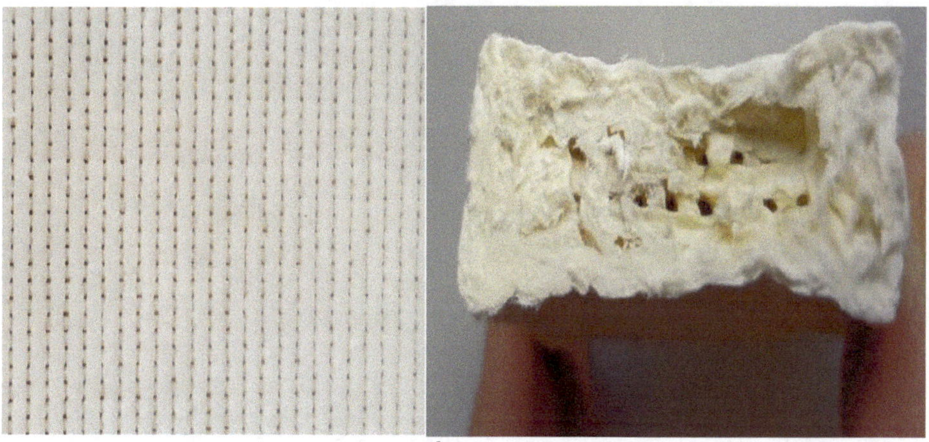

a *b*

Figure 4.21 Figure 4.22 The effect of coalescence on dimensional accuracy depending on the infill type (a) and causing a waisted specimen (b).

4.11.2 Material Behavior Effect (Feedstock)

Material behavior after drying due to shrinkage depends on the composition of the material, i.e., the mixture formulation of a 3D printed paper part. Figure 4.22 shows an example of the different results after drying of two different formulations. This would directly affect the deformation and shrinkage after drying, along with the environmental conditions provided for the sample. The mechanical properties of this material have an effective impact, such as the strength of the material could identify the capability to hold the shape retention then it would avoid the material being compressed under the material's own weight load of successive layers. Mechanical properties also affect the tendency of a print to deform, and to warp. Here the parameters used to quantify the dimensional accuracy of 3D paper printed geometries are the optimization of the mixture Further details in this regard are provided in Tab. 3.15, which shows the quantitative values of deformation after drying for the filament diameter. The minimum value from this table gives the most acceptable result for dimensional tolerance than the other mixtures.

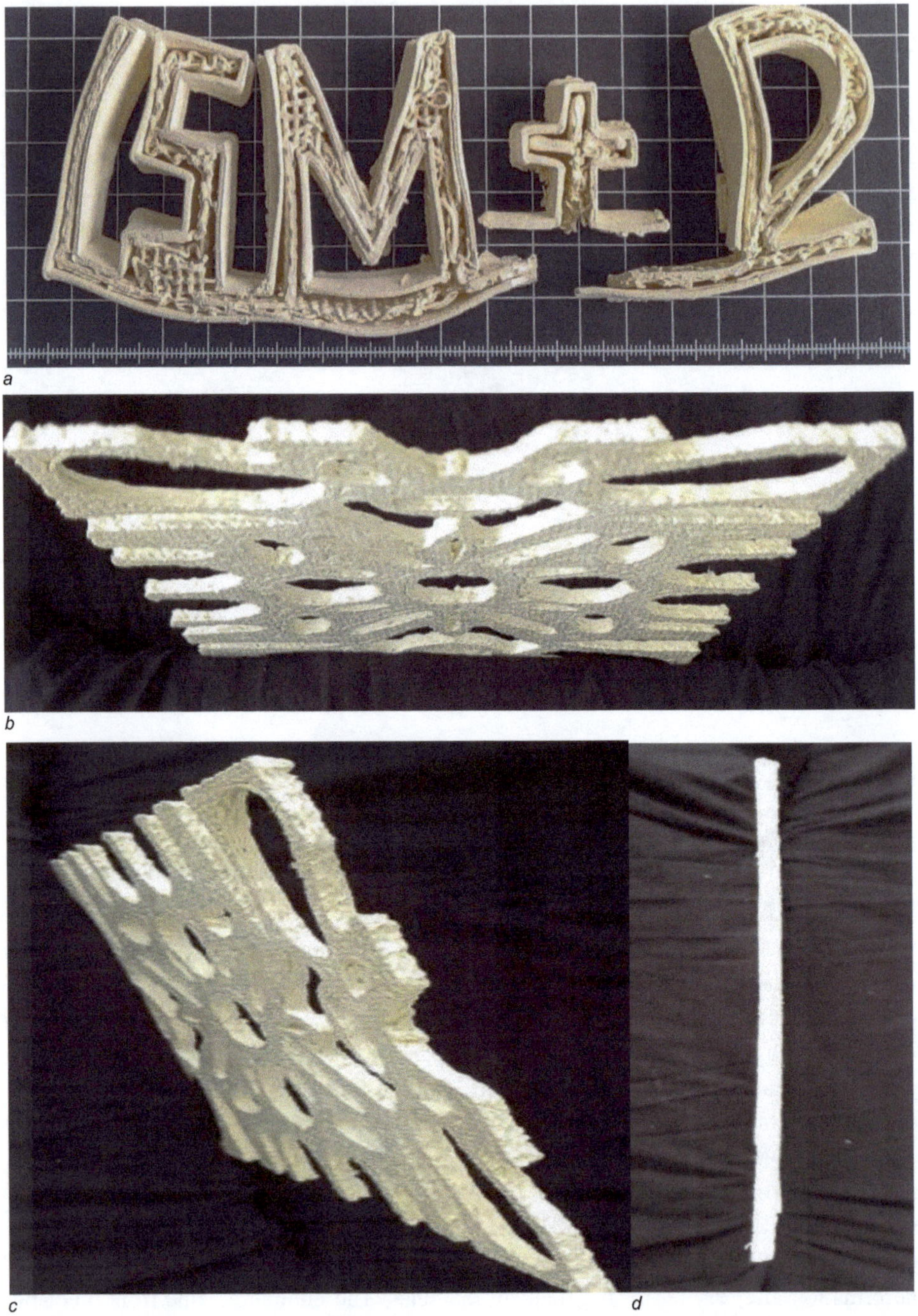

Figure 4.22 Powdered cellulose sample with bad shrinkage behavior (a), other sample with eliminated shrinkage top view (b), side view of the sample (c) and sectional view of the sample (d).

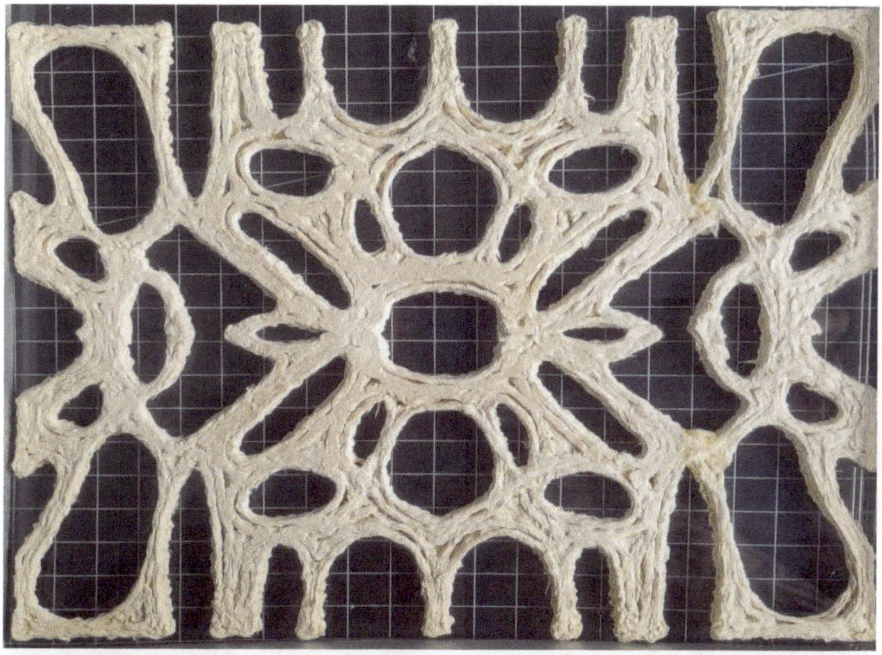

a

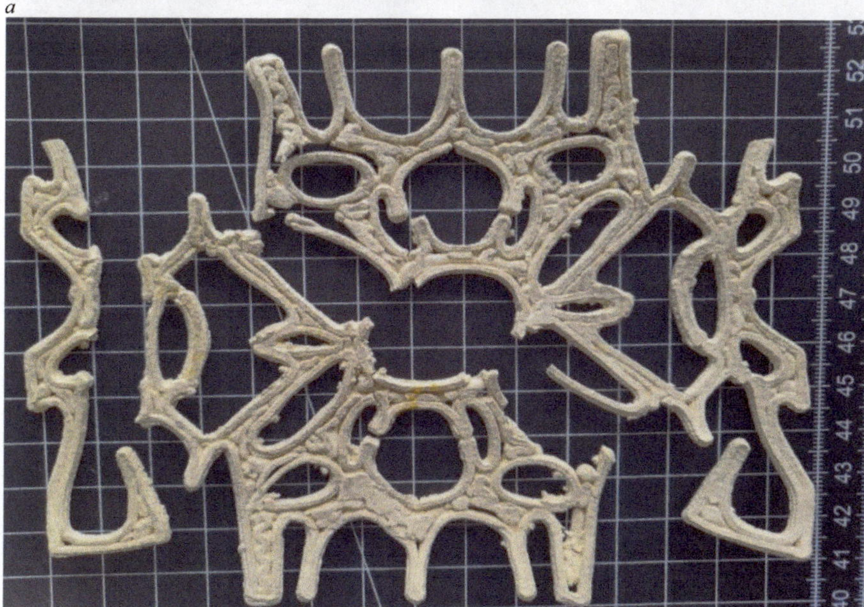

b

Figure 4.23 Model from Figure 4.23b (a) compared with the same printed geometry but with powdered cellulose (b).

The type of cellulose used can affect more than just the shrinkage ratio, as it affects the dimensional accuracy, production quality and warping behavior of the printed sample, as shown in Fig. 4.23. The type of pulp used in the sample in Fig. 4.22 (a) is powdered pulp. The warping behavior is highly deformed and distorted, with the range of values of the curvature reaching about 68.16 degrees compared to the original flat surface before drying. The drying shrinkage and warping for the specimens in Figures 4.22 (b) and (c) have been eliminated by using fibrous cellulose and optimized paste. Fig. 4.24 shows some samples printed to observe the shrinkage behavior of different types of cellulose.

Figure 4.24 Samples of different types of cellulose.

4.11.3 Shrinkage Measurement for 3D Volumes

One of the biggest influences on the quality of 3D paper printing is shrinkage, which depends on the formulation, the proportion of the mix design and the selectivity of the materials. Shrinkage measurement for solids is different from that for filaments. For filaments, shrinkage was evaluated by measuring the diagonal strain of the filament diameter and the longitudinal strain of a given filament length, as shown in Tab. 3.15. Shrinkage for a 3D volume was measured by calculating the change in the three dimensions - length, width, and height - for the standard shape or the change in volume for an irregular shape. The shrinkage for a 3D volume is affected not only by the material behavior (feedstock) but also by the manufacturing printing parameters, the geometry configuration & shape, the drying conditions & drying rate, the model volume and the exposed area of the surfaces for ventilation or confinement such as corners or intersections. In order to specify the volumetric shrinkage in this study, the dimensional change was measured for a standard 3D shape for a given filling and for several sizes. Shrinkage was measured in all three directions due to the orthotropic behavior of additive manufacturing, which depends on the variety of geometry configuration.

Cubic and cylindrical shapes were chosen to represent the change in volume after drying. When evaluating the volumetric shrinkage for a cubic shape, the diagonal measurements were also included in the average of all dimensional measurements and documented within the standard deviation of the given values. Four size groups were printed for the cubic form, namely (20 x 20 x 20 mm), (30 x 30 x 30 mm), (40 x 40 x 40 mm) and (50 x 50 x 50 mm). The infill patterns tested to investigate the differences between them are rectilinear and concentric. Measurements were taken at 8 points (4 corners and 4 points in the middle of the edges). The average of three identical samples is shown in Tab. 4.5.

Shrinkage for a cylindrical specimen is assessed by taking the average of the measurements for both (diameter & height of the cylinder) at the octagonal points of the circle with equally spaced divisions. The average of four identical specimens are then recorded in Tab. 4.5 and Fig. 4.25. The three size groups printed are (20 x 20 mm), (40 x 40 mm) and (60 x 60 mm). The printing parameters used for the samples in Tab. 4.5 are constant whether they are cubes or cylinders. The fill types were 99% and 100% concentric and rectilinear respectively. The nozzle diameter is 1.6 mm. All speeds= 18mm/s (circumferential speed, infill speed, first layer speed, gap infill speed and max print speed), layer height= 1mm, extrusion width= 200%, the air pressure= 0.8 MPa and the extrusion multiplier= 1. These parameters were chosen to achieve better print quality and full coalescence results. The paste type is 30 from Table 3.9. This paste has the better result in limiting the shrinkage of the filament, as shown in Tab. 3.15. The samples were cured under the same conditions as mentioned in paragraph 3.4 Curing Process. The samples were tested one month after printing.

Each group is printed with all the sizes listed in Tab. 4.5 is printed together to ensure that each group has the same configuration conditions, as shown in Fig. 4.26. The environment was monitored from the day of printing to the day of testing as shown in Fig. 4.27.

Table 4.5 Shrinkage measurements for the cubes and cylinders.

Specimen Size* mm3	Y-axis strain** %	X- or Z- axis strain*** %	Volumatic shrinkage %
Cubic (20x20x20), concentric infill	27.99 ± 3	2.97 ± 3.2	32 ± 6.9
Cubic (30x30x30) concentric infill	24.65 ± 2.4	2.86 ± 3.1	28.9 ± 3.7
Cubic (40x40x40) concentric infill	23.64 ± 2.8	2.71 ± 2.8	27.72 ± 4
Cubic (50x50x50) concentric infill	42.03 ± 2.3	4.74 ± 3.17	47.47 ± 1.3
Cubic (50x50x50), rectilinear infill	37.67 ± 1.2	2.78 ± 1.4	41.1 ± 1.5
Cylinder (20 X 25.5), concentric infill	31.13 ± 2	3.9 ± 0.7	36.48 ± 2.4
Cylinder (30 X 38.2), concentric infill	30.64 ± 2.2	3.39 ± 1.7	35.27 ± 1.8
Cylinder (40 X 51), concentric infill	27.8 ± 2	3.18 ± 0.9	32.3 ± 2.8

* Specimens are from paste 15 from Tab. 3.10, except for the (50x50x50) mm size specimens which are from paste12.
** The height of the sample is in the Y direction, i.e., the direction of the Y axis is perpendicular to the build surface.
*** X or Y direction is the plane of the build surface.

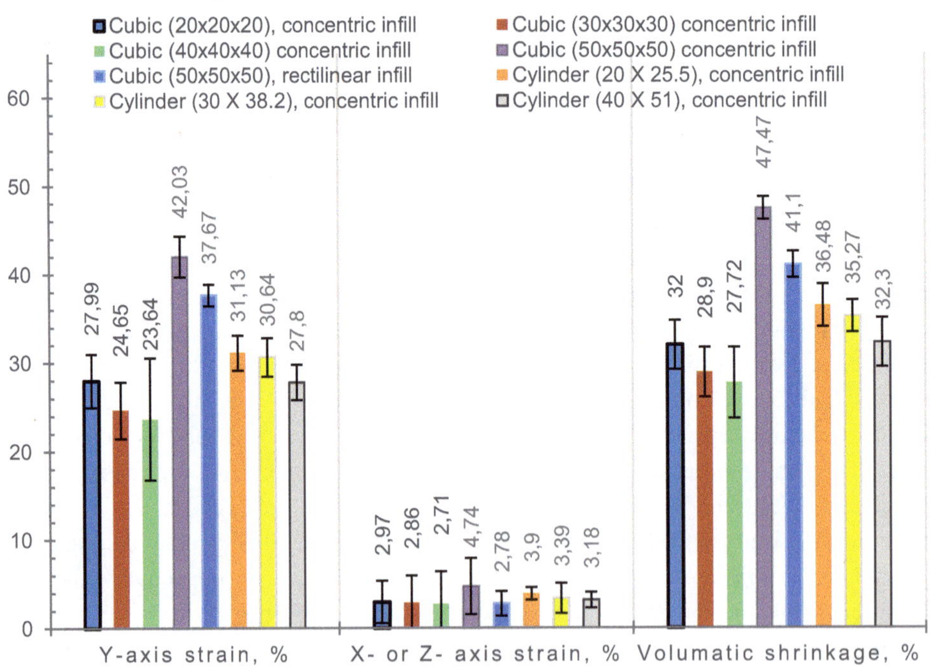

Figure 4.25 Shrinkage measurements for the cubes and cylinders

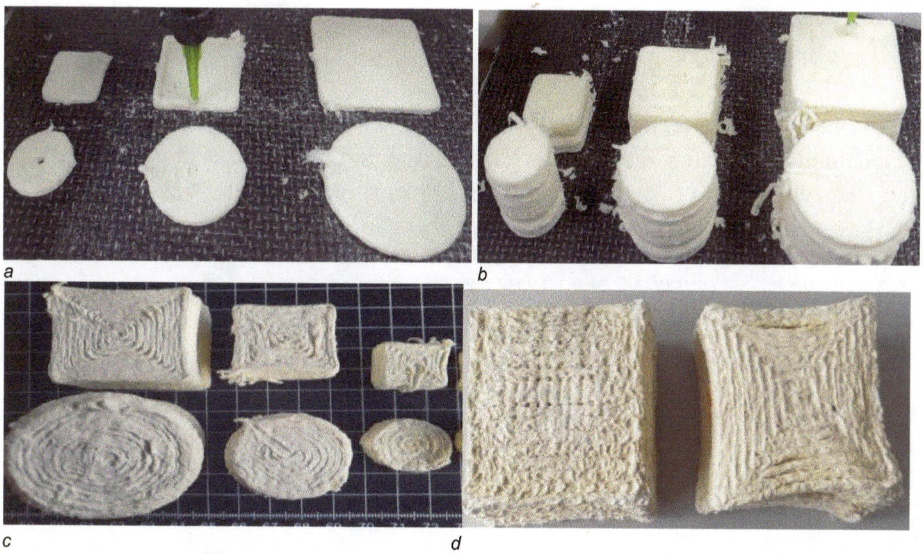

Figure 4.26 Shrinkage measurement specimens during printing (a), after printing (b), after drying (c), and comparing the shrinkage behavior of two types of infills (linear and concentric) (d).

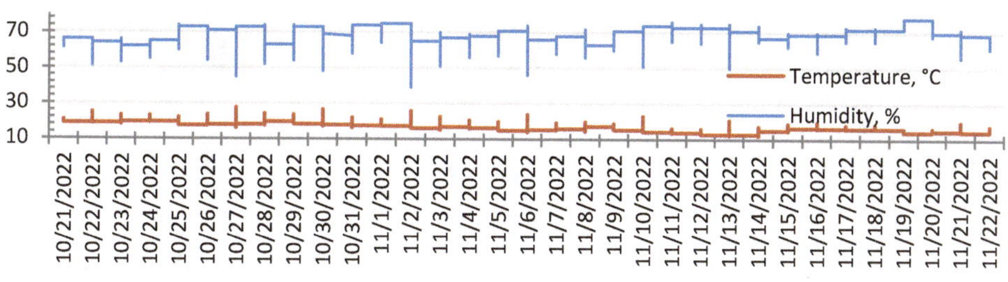

Figure 4.27 Environmental monitoring of samples printed.

4.11.3.1 Printing Direction (Y-axis vs. X- or Z-axis Strain)

There is a significant clear case in 3D paper printing due to the configuration orientation of the additively manufactured layers. This behavior has been explained as the difference in shrinkage results between the matrix direction (Y-axis) and the longitudinal direction (X- or Z-axis). This difference decreases with the use of cellulose fibers and according to the formulation of the mixture in terms of adhesion strength, cellulose used and the ratio of solids to water content. From Tab. 4.5 shows that the shrinkage in (Y-axis) is much greater than in (X- or Z-axis), which is more than 7 times. Therefore, due to the orthotropic behavior of additive manufacturing, it is important to calculate the drying shrinkage in the Y-direction separately from the volumetric shrinkage and drying deformation in the X- and Z-directions.

The resistance to shrinkage in the Y direction is provided by the width of the filaments and the adhesion strength between the layers. This strength is much less than the resistance provided by the length of the filaments in the X or Z direction, as shown in Fig. 4.28 (a) and (b). In another sense, the fibers were relatively forced to lay in a direction parallel to the longitudinal direction of the extruded filament during extrusion, as shown in Fig. 4.28. Here, the length of the fibers constrains the shrinkage more than the width of the fibers; therefore, the Y-axis shrinkage is greater than the X- or Z-axis shrinkage.

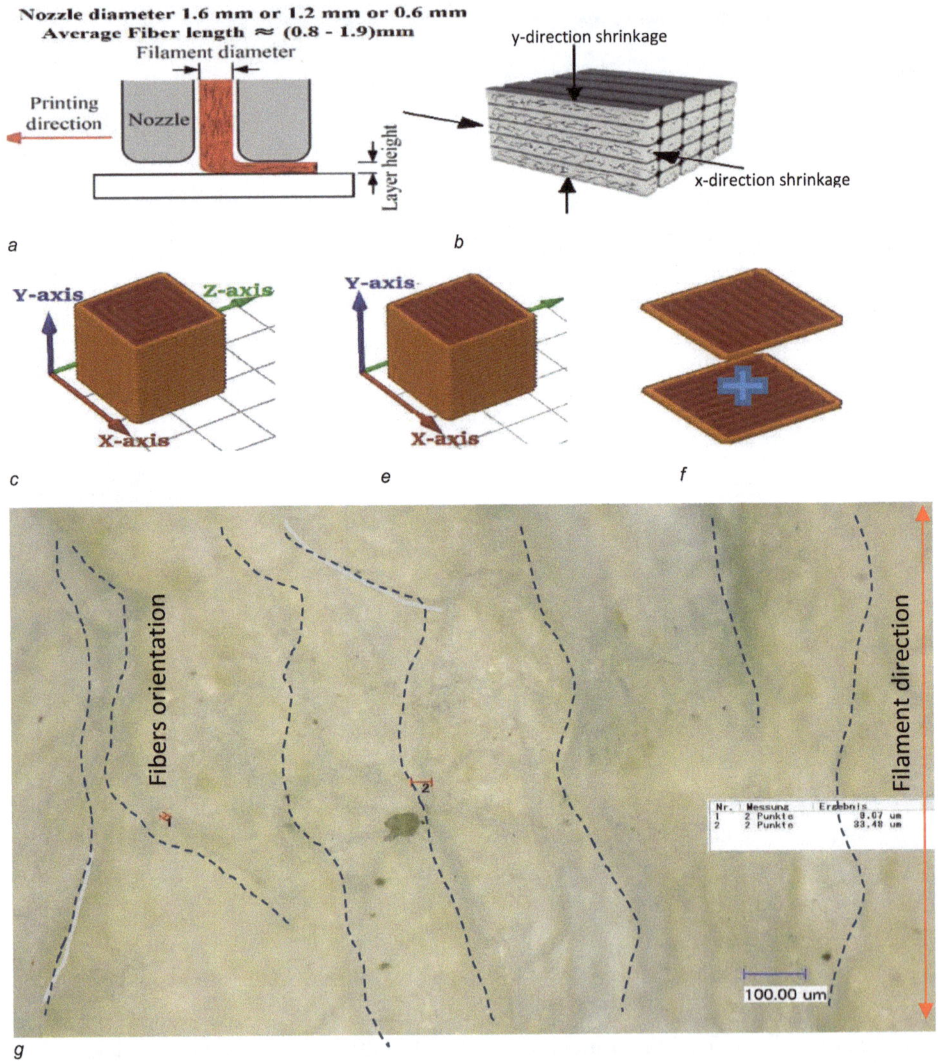

Figure 4.28 Fibers orientation during extrusion (a); layering effect (b); concentric filling of a (20x20x20) mm (c); rectilinear filling (e); two consecutive layers of the rectilinear filling pattern(f), and a 1.6mm filament under the microscope shows fibers orientation inside (g).

4.11.3.2 Specimen Sizing Effect

From Tab. 4.5, it's noticeable that larger specimens have less shrinkage due to their lower ratio of exposed surface area to specimen volume. It preserves a lesser water evaporation rate within the core. These measurements would help to evaluate shrinkage for other specimens. It is evident from Figures 2.29, 2.30 and 2.31 that the increasing rate of shrinkage for cylindrical specimens is higher than for cubic specimens and the shrinkage measurement has been plotted as a fitted curve as follows:
For cubic specimens,

$$S_y = 88{,}333x^2 - 10{,}75x + 23{,}267 \tag{4.1}$$

$$S_{XZ} = 3{,}48\, x^{0.1287} \tag{4.2}$$

$$S_V = 47{,}098\, x2 + 7{,}3477x + 25{,}561 \tag{4.3}$$

For cylindrical specimens

$$S_y = 21{,}232x + 25{,}586 \tag{4.4}$$

$$S_{XZ} = 12{,}353x^2 + 0{,}529x + 2{,}8703 \tag{4.5}$$

$$S_V = 27{,}591x + 29{,}135 \tag{4.6}$$

Where S_y is the y-axis shrinkage, S_{XZ} is the x- or z-axis shrinkage, S_V is the volumetric shrinkage, x is the surface area to volume ratio of the specimen.

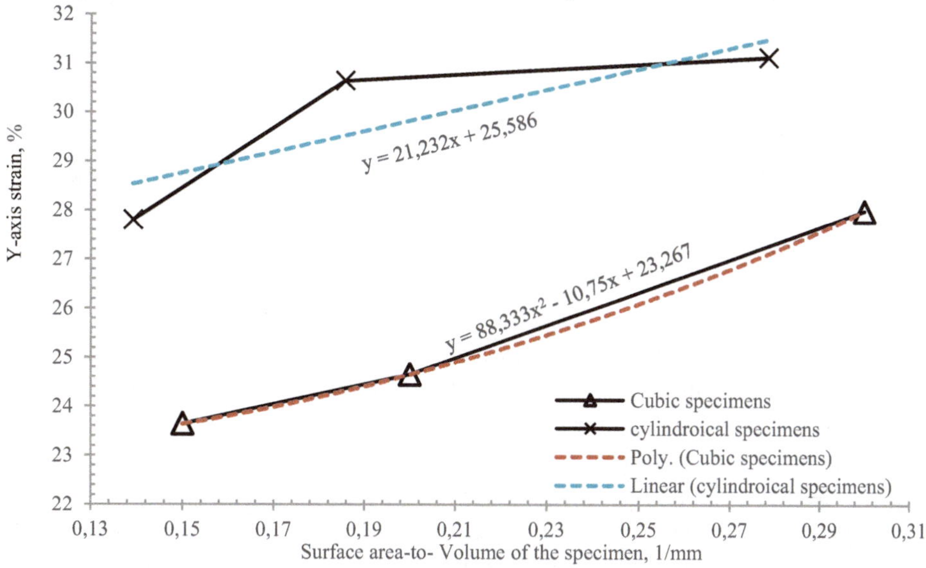

Figure 4.29 Y-axis shrinkage results for cubic and cylindrical samples in terms of ratio (surface area to volume).

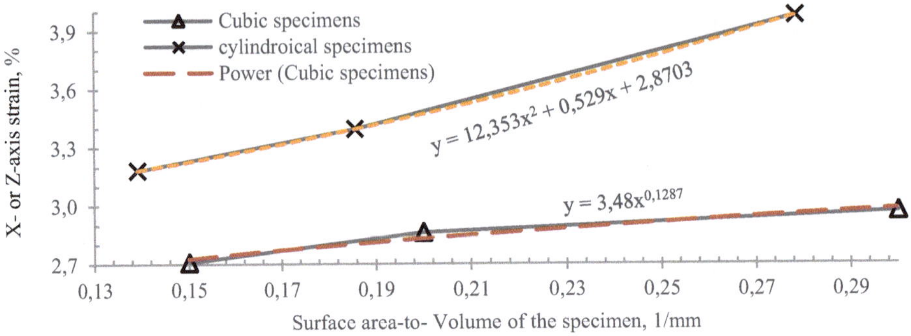

Figure 4.30 X or Z axis shrinkage results for cubic and cylindrical samples in terms of ratio (surface area to volume).

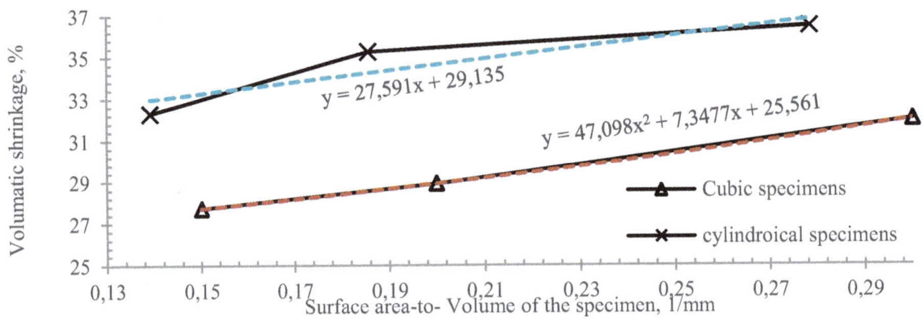

Figure 4.31 Volumetric shrinkage results for cubic and cylindrical samples in terms of ratio (surface area to volume).

4.11.3.3 Shape Changing Effect

From Tab. 4.5, the cylindrical shape has higher shrinkage results by a range (17.9%, 18.7% and 20.5%) than the cubic shape with the same surface to volume ratio as (0.15, 0.2, 0.3)1/mm respectively. The explanation lies in the geometric shape and height of the sample. The cylindrical shape has a greater height than the cubic shape and we know that the (Y axial shrinkage) shrinks more than the width (Z or X axial shrinkage). In addition, the cubic shape has corners which act as a cross-connection, confining the shrinkage and changing the stress distribution and strain behavior compared to the cylindrical shape which has strain distribution in the radial direction and no confinement. These reasons justify the divergence of the results calculated within a higher standard deviation for cubic shape. Therefore, to evaluate the shrinkage for other specimens, it is useful to use the shape transformation factor to transform the cubic shape into an equivalent cylindrical shape. This equivalent shape could have a different volume, which will be calculated accordingly:

$$Y_{cl} = f \cdot \frac{vol_{cb}}{vol._{cl}} \cdot Y_{cb} \tag{4.8}$$

Where, Y_{cl} is the shrinkage for cylindrical shape, Y_{cb} is the shrinkage for cubic shape, $f = \frac{y_{cl}}{y_{cb}}$, which is the transform factor according to shrinkage direction within range (1.11-1.24) for the Y-axis shrinkage, (1.18- 1.34) for the x-axis shrinkage, and (1.14-1.22) for the volumetric shrinkage.
For cubic shape:

$$y_{cb\,1} = 58{,}431x^{-0{,}083} \tag{4.9}$$

$$y_{cb\,2} = 3E\text{-}11x^2 - 6E\text{-}06x + 3{,}0166 \tag{4.10}$$

$$y_{cb\,3} = 59{,}753x^{-0{,}07} \tag{4.11}$$

For cylindrical shape:

$$y_{cl\,1} = -9E\text{-}10\,x^2 + 7E\text{-}06\,x + 31{,}132 \tag{4.12}$$

$$y_{cl\,2} = 10{,}444x^{-0{,}108} \tag{4.13}$$

$$y_{cl\,3} = -3E\text{-}10\,x^2 - 5E\text{-}05\,x + 36{,}927 \tag{4.14}$$

Where cl is a symbol for cylinder, cb is a symbol for cube, y1 is the y-axis shrinkage, y2 is the x- or z-axis shrinkage, y3 is the volumetric shrinkage, x is the specimen area to volume ratio, vol. is the volume of the specimen.

4.11.3.4 Infill Effect

The type of infill and the setting of the process printing parameters control the drying process and the shrinkage results. The type of infill has a significant effect on limiting shrinkage results because the amount of material extruded varies depending on the pattern. If more material is extruded in the same layer, the limitation of volumetric shrinkage is increased. This means that more material within the duplicate layers fills the space and compacts the area. The amount of material extruded in the same layer depends on the pattern. This pattern plays a major role in the configuration and mechanical behavior of the sample, particularly the drying shrinkage and drying process, which helps to limit shrinkage when forming knots or cross contacts. It is also mentioned in section 3.3.5 Filling and Figure 3.41. Thus, when comparing the concentric and rectangular fills, the linear fill is preferred in terms of improving shrinkage results by 10.4%, 41.43% and 13.43% in terms of Y-axis shrinkage, X/Z-axis shrinkage and volumetric shrinkage respectively. The resistance of the concentric infill to shrinkage stresses depends on the coalescence and adhesion properties between the printed filaments. The rectilinear infill in particular has intersections within the pattern that reinforce the confinement and constraint. This type of pattern provides more intersecting nodes and junctions that sufficiently constrain the boundaries, as shown in Figures 4.32, 4.33, 4.34.

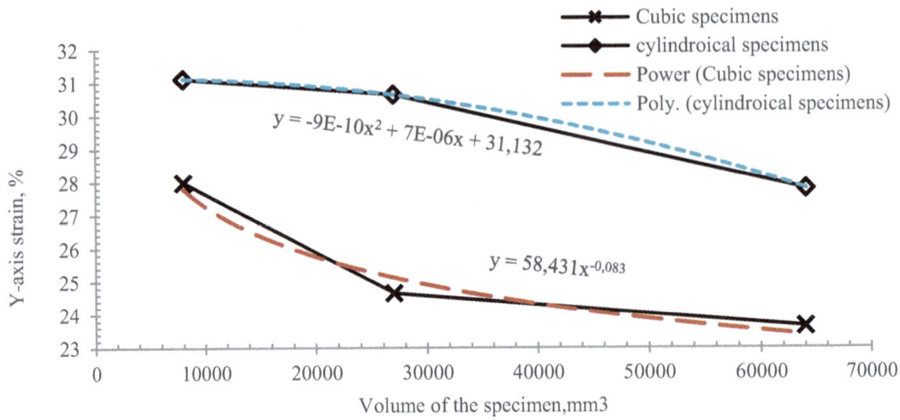

Figure 4.32 Y-axis shrinkage results for cubic and cylindrical samples as a function of volume..

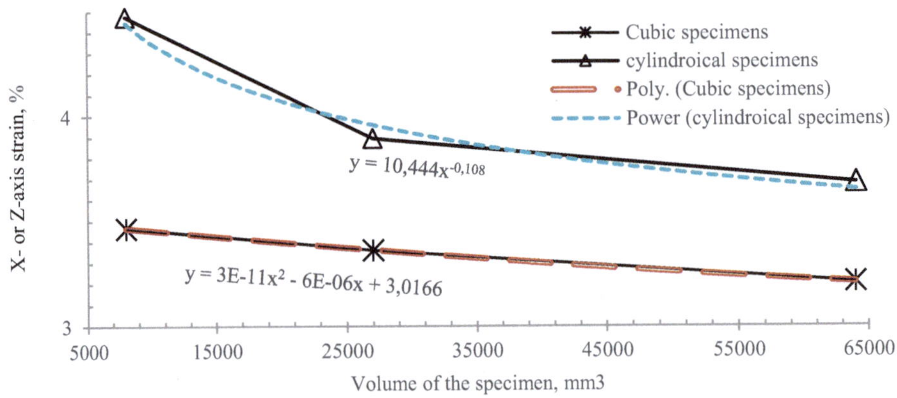

Figure 4.33 X or Z axis shrinkage results for cubic and cylindrical specimens as a function of volume.

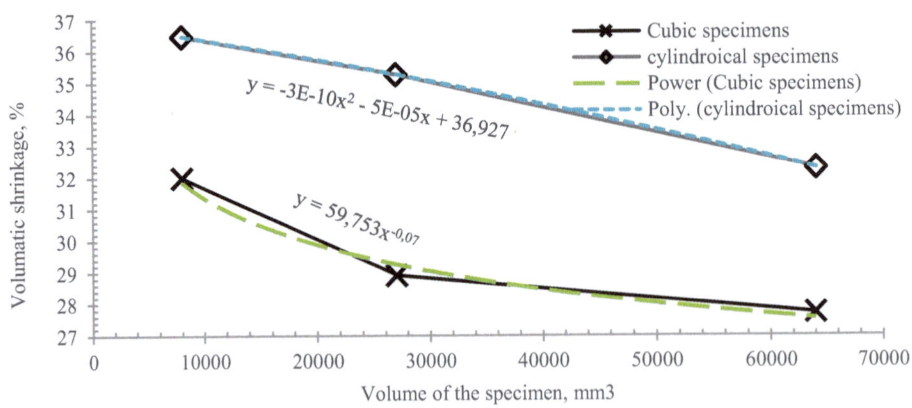

Figure 4.34 Volumetric shrinkage results for cubic and cylindrical samples as a function of volume.

4.11.3.5 Summery for Volumatic Shrinkage

Principally, geometry and surface area to volume ratio have a significant impact on how quickly a material will shrink as it dries.
There is not a directly proportional increase in shrinkage as the volume increases. The principles of moisture diffusion, heat transfer and evaporation kinetics explain these phenomena. In larger volumes, the evaporation process has a longer distance to travel from the interior to the surface, resulting in uneven moisture distribution when comparing two different sample volumes. This results in a slower water evaporation rate and more time, which facilitates cross-linking between components and water when drying is slower. The curing method selected, as described in section "3.4 Curing Method", results in slower drying. This allows sufficient time for moisture to migrate from the interior to the surface. This promotes relaxation and uniform moisture content. This uniformity reduces differential shrinkage. As a result, internal stress development decreases, reducing the probability of shrinkage-related damage.
When the geometry is ununiform cause ununiform shrinkage due the smaller part dries faster causing uneven strain distribution between two parts that causing cracks. This difference creates a non-uniform moisture distribution within the material, causing the material to experience stress gradients. These stress gradients can result in cracking, warping, or distortion of the material's shape, as shown in Fig. 4.35.

4.11.4 Drying Rate & Uniformity Issues

Drying speed, which is controlled by the environment and drying conditions, is a critical factor affecting shrinkage. Drying uniformity is affected by environmental conditions (drying temperature and humidity), the type of curing used and the complexity and bulk of the printed geometry. The thinner part dries faster than the solid part because the ratio of exposed surface area to volume is greater, which means it contains less moisture. So, it dries faster. All of this leads to uneven drying when parts dry faster than other parts, one after the other, causing cracks to form between them. Cracks appear between two parts printed at different times (e.g., when printing is completed the next day due to the end of working hours, which means long breaks), where the old-printed part somehow doesn't stick like the fresh print, even though it is kept moist, as in Fig. 3.36 (a).
Some defects occur during drying such as uneven/irregular, wrinkles, creases, and folds as concave and convex, delamination/debonding within the coalescence between successive layers as shown in Figures 4.36 and 4.37. However, the shrinkage deformation could be controlled by a combination of efficient curing methods and appropriate mix designs [De Schutter G et.al, 2018]. This was also achieved in this research to maintain accuracy and reduce shrinkage as much as possible, as shown in Fig 4.38.

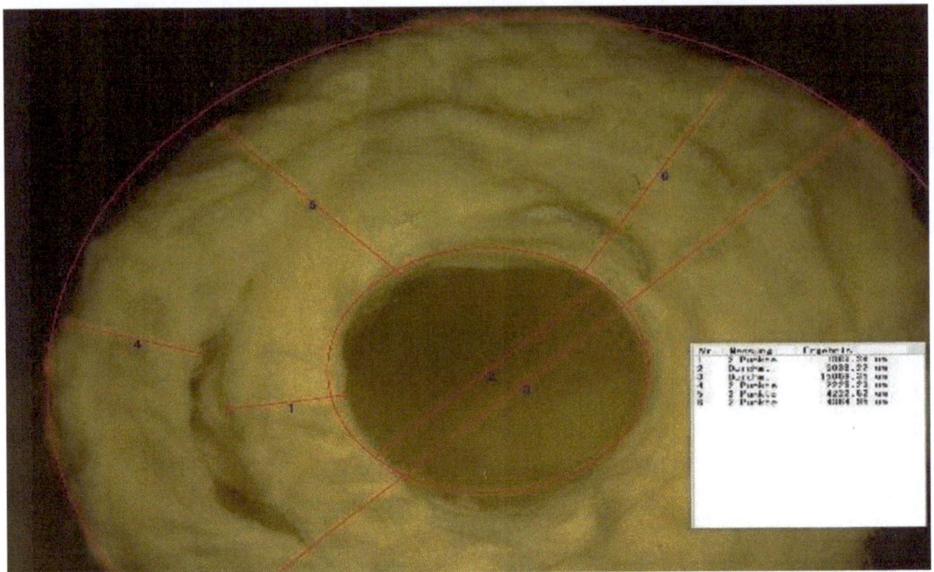

Figure 4.35 Irregularity in drying deformation causes dimensional divergence due to the use of non-optimized paste.

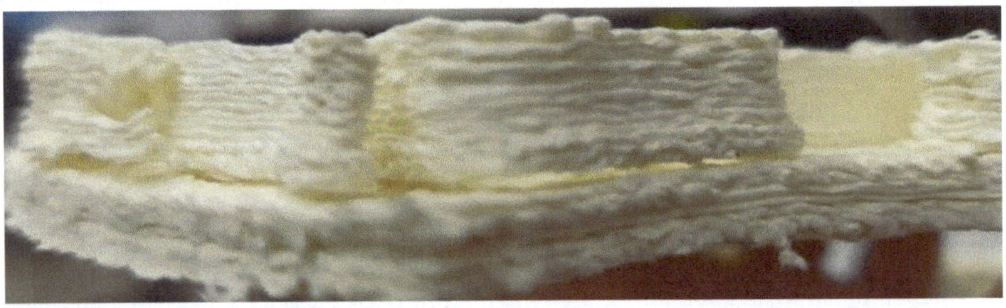

a

a

b

Figure 4.36 Defects caused by drying deformation, splitting between two printed parts in different time zones (a) splitting and warping (b), layer delamination (c).

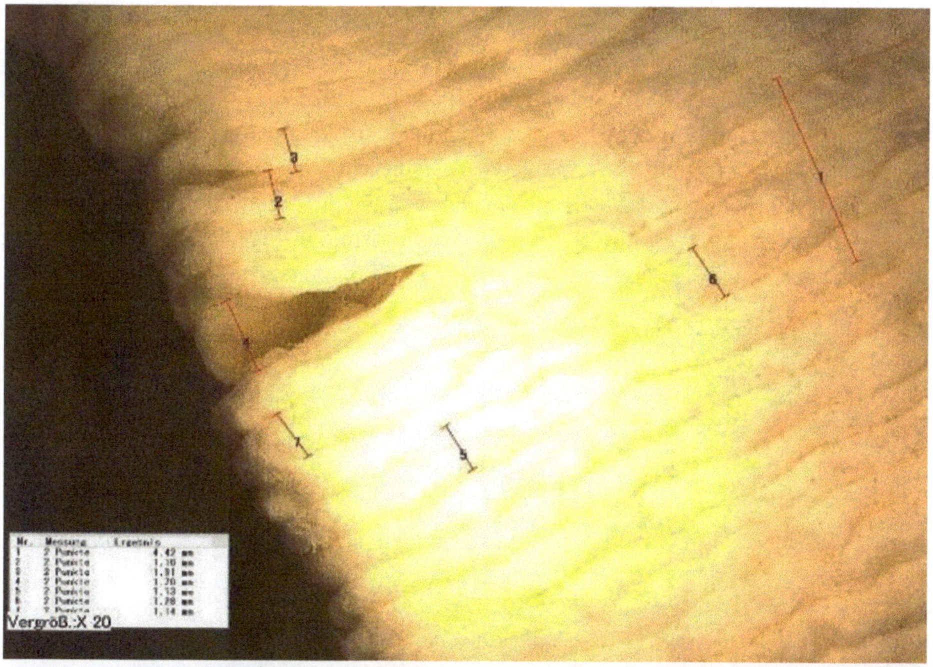

Figure 4.37 Drying deformation causing delamination between the successive layers at the corner.

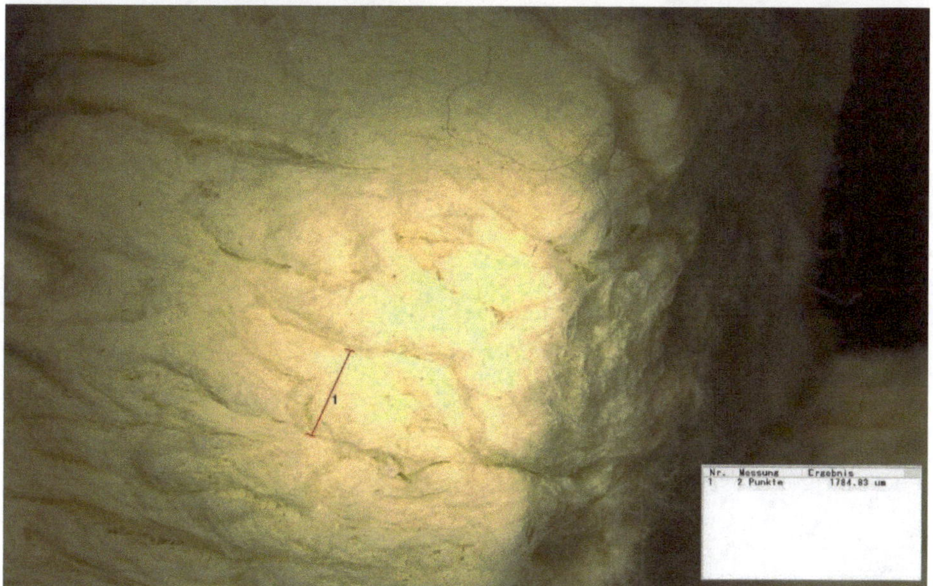

Figure 4.38 The specimen with limited shrinkage, controlled by the material formulation and proper printing parameters.

4.11.5 Zits & Blobs

It's a defect that affects the print quality, especially the dimensional accuracy and the aesthetic appearance. This defect occurs when there are long pauses in printing in hours between the fresh print and the old-printed part. In this case, the old-printed part has shrunk a little, i.e., the dimension has become smaller, and the new printed part has dimensions that no longer fit together. The problem is clearly visible in Fig. 4.39 if the old-printed part hasn't been properly cured at the end of the printing process.

a

b

Figure 4.39 Samples have zits & blobs defect (a) and (b)

4.11.6 Corner Angle

During the evaluation of the 3D paper printed models, it was generally noticed that the corner angles of the filaments are more likely to be routed as sharp, which depends on the printing speed and the extrusion rate. For a speed equal to 18mm/s (perimeter speed, infill speed, first layer speed, gap infill speed and max print speed) and an extrusion multiplier equal to 0.85, the result of corner printing is as shown in Fig. 4.40 and 4.41, where the radius has a range of (2.34-3.96) for right/acute/obtuse angle.

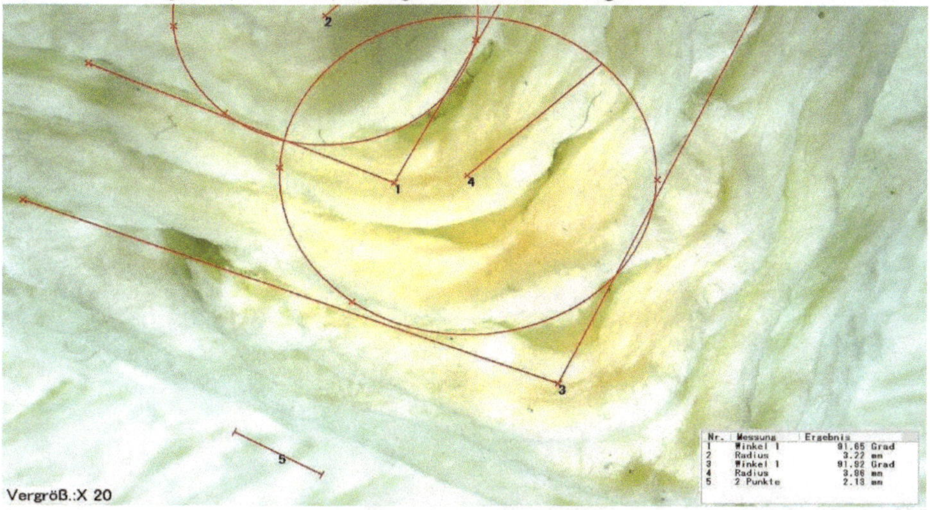

a

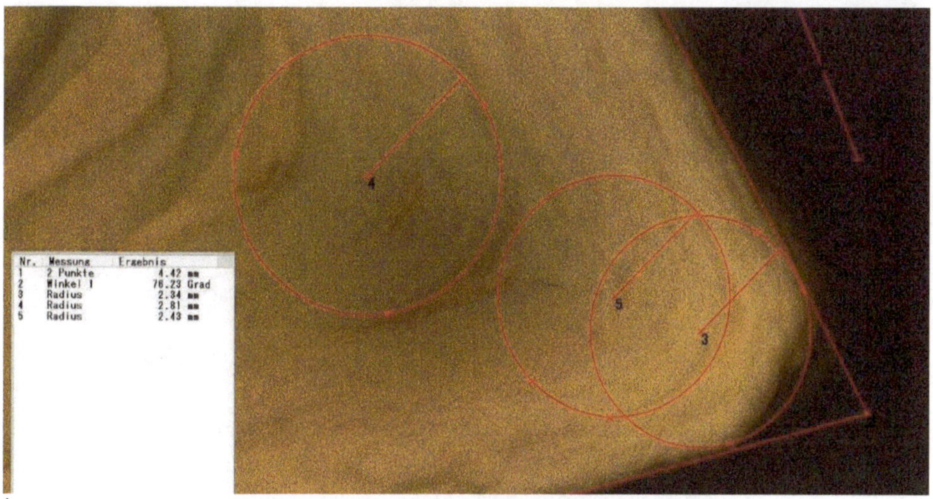

b

Figure 4.40 Outside corner edge for right angle(a), and for acute angle(b).

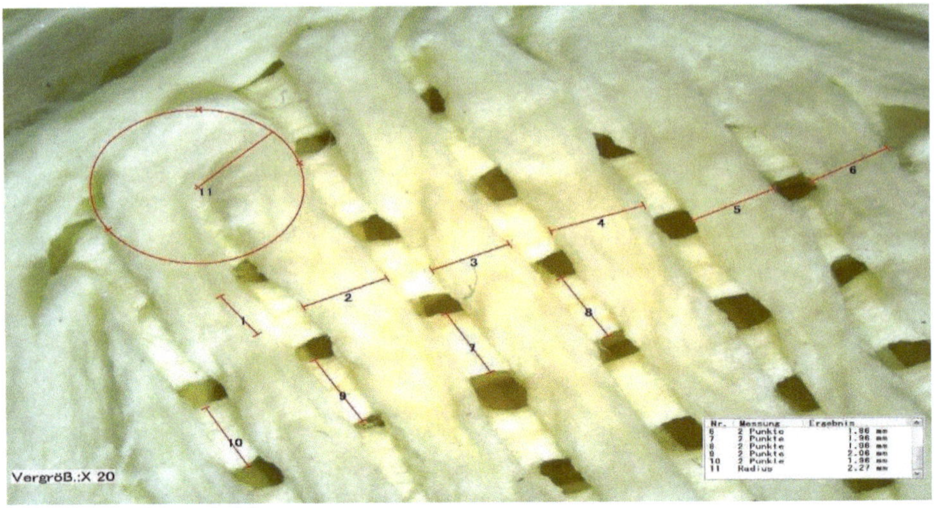

a

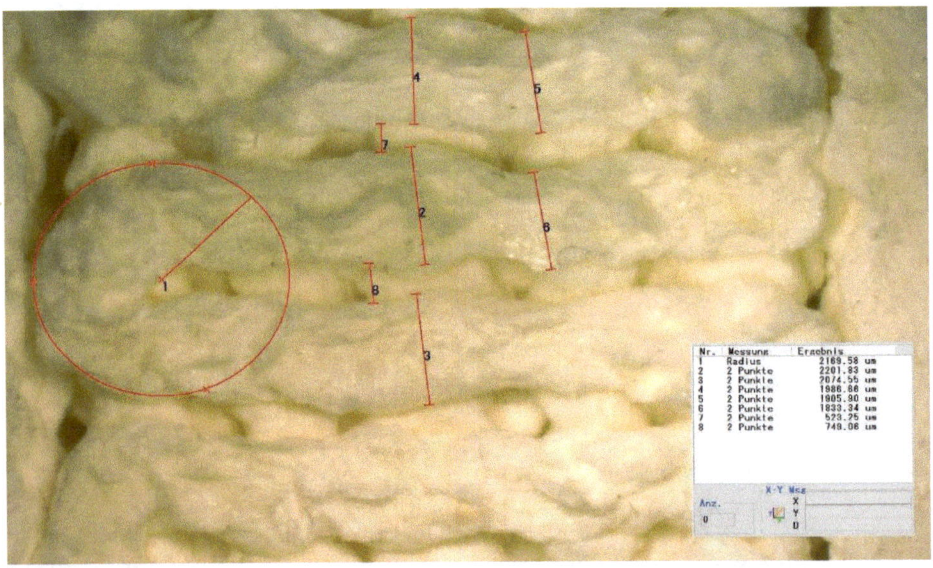

b

Figure 4.41 Inside printed obtuse angle (a), and from filament / perimeter attachment.

The type of cellulose also affects the shape of the pointing end of the corner, where the fibrous cellulose needs to be considered in this issue compared to using powder cellulose, because fibrous cellulose needs more time to complete the extrusion of the whole fiber rather than powder cellulose, as shown in Fig. 4.42. To optimize this issue, either use a smaller nozzle size or reduce the print speed. Fig. 4.43 shows the right angle printed well with a pointed angle when the speed is reduced to 10 mm/s.

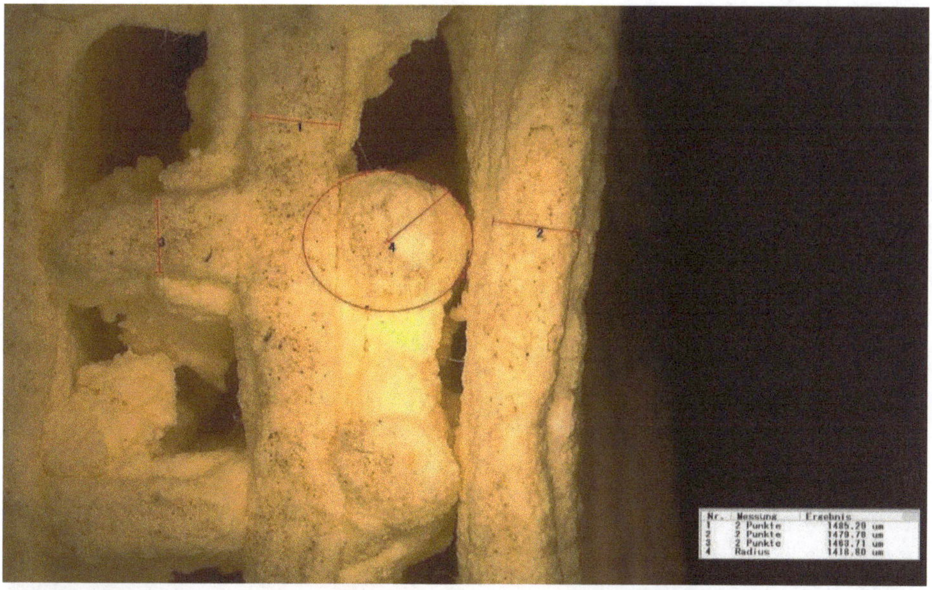

Figure 4.42 Outside corner for powder cellulose using a print speed equal to 18m/s.

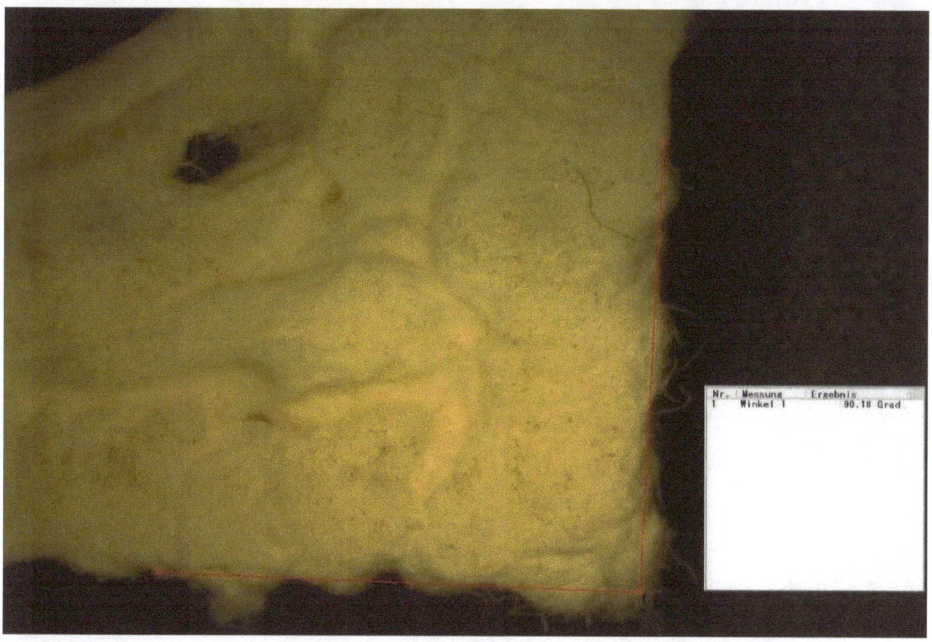

Figure 4.43 Pointed edge of a right angle corner using a print speed of 15mm/s.

4.11.7 Layering Shifting

This could be happened during printing, basically caused extremely from high print speed. Fast speed causes the shifting or creeping of each layer within succussive printed layering. This error happened when print speed exceeded 25 m/s and travel speed exceed 35mm/s due to the speedy movements for the printing which causing vibration in the machine then it loses the accurate coordinates. Fig. 4.44 shows one case for this defect. It is observed that the height of the specimen shifted ununiformly.

a

b

Figure 4.44 Sample with layer shift error in perspective view (a) and front view (b).

Descending or layer creep is measured using a levelling laser and a digital angle finder protractor. In Fig. 4.44, the shifting corner is randomly shifted in the x-y direction. It should be perpendicular.

Fig. 4.45 shows a different type of shift caused by the timing of the two-part print. This causes the entire top part to be altered at a distance from the original coordinates. This error is caused by a mismatch between the start coordinates of the new part and the old part, due to a mismatch in the origin point. This leads to this defect as an individual during the production process. Human error therefore caused this defect.

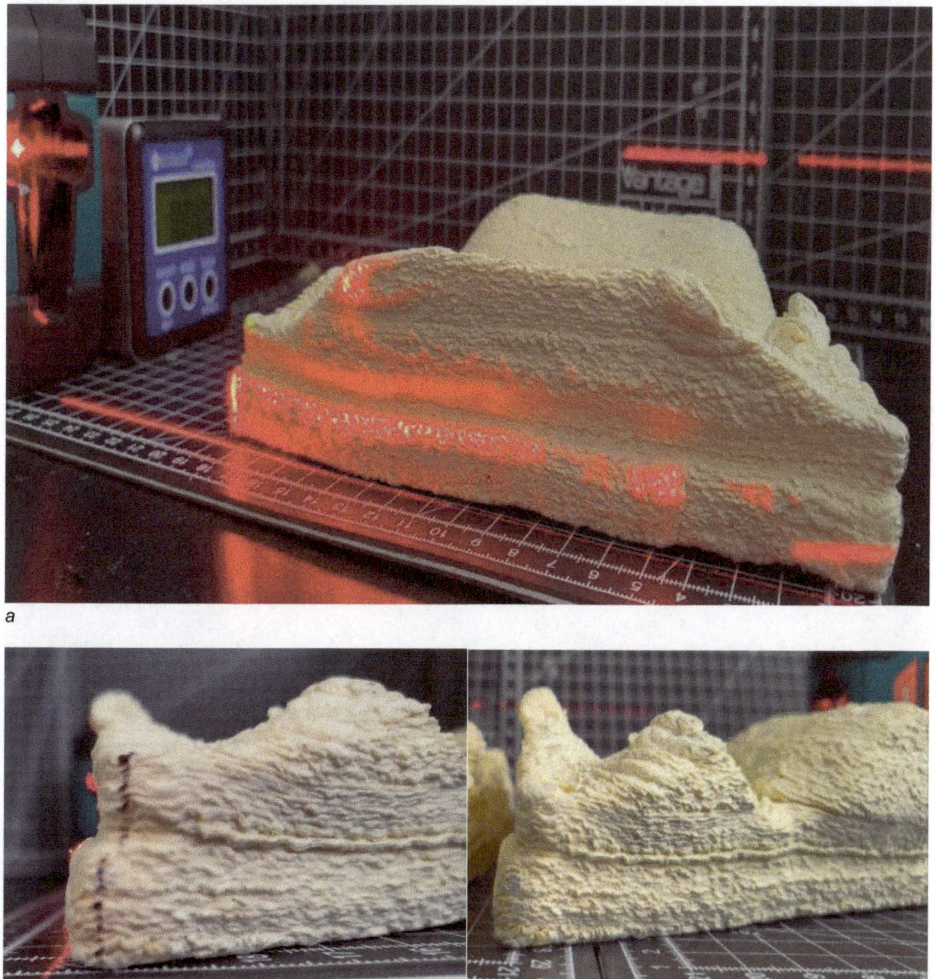

a b c

Figure 4.45 Sample with misaligned origin between new and old part (a), side view of end corner (b), and another side of the same corner.

I-bones specimens' preparation for tensile testing.

5 Material & Mechanical Characteristics

In the development of a new material, it is important to identify the material properties, understand the mechanical response and specify its ability to transmit and resist deformation and forces (internal or external). Mechanical properties are therefore essential for predicting material behavior, identifying weaknesses, and providing solutions. The response and behavior of any structural member under load depend on the type of construction, the nature of its constitutive materials, properties, dimensions, the type of connection between the construction units and the type and magnitude of the applied load [EN 1991-1-1: Eurocode 1(2002)].

The study of the mechanical properties of any developed material such as 3D printed paper paste is important, it is considered part of an integral quality of the product and shows the ability of a material to be molded in a suitable configuration. For example, mechanical properties can be used to estimate the adhesion properties of the paste and to indicate them after drying by testing the strength and stiffness between the layers.

However, standards for testing 3D printed paper samples are not directly available. The following specifications have been adopted for AM in general and for plastic materials, as both are considered as polymeric materials within the standards: ISO/ASTM 52910, 2018(E), ISO/ASTM 52900, 2015(E), and ASTM F2971- 13.

This chapter presents the mechanical properties and material response of 3D paper printed specimens. These comprehensive characterizations are concluded in terms of stress-strain behavior, Young's modulus, fracture strength, toughness strength, and elongation at break.

5.1 Density

The density of paper material is generally low. Engineers are always looking for alternatives to have low weight components that do not have or add extra load to the structure. Paper fulfils these desires in some specific applications where it is one of the lightest materials within the structural components.

The wet density of 3D printed paper paste is measured as a viscous material according to ISO 2811-2:2011. The wet density has a range of (7.265-8.074) kN/m3. The bulk density of 100% infill after drying is measured for a standard shape of geometries by quantifying the weight of the sample divided by the dimensions of the sample after drying, the mean considering the effect of shrinkage. The samples measured are the same as those used to calculate the shrinkage results.

D. Abdullah Agha, *3D Paper Printing for the Built Environment*, Mechanik, Werkstoffe und Konstruktion im Bauwesen 73,
https://doi.org/10.1007/978-3-658-45076-2_5

The density for cylinder is lower than the density calculated for cube, although the printed paste is the same, due to the differences in shrinkage behavior for geometry shape and Y-axis shrinkage as explained in paragraph 4.11.3.3 Shape change effect.
Tab. 5.1 shows the calculated density for a 3D paper printed sample depending on the geometric shape and the type of filler. The paste type from Tab.3.9 has an impressive effect on the density results due to the mix formulation and the reduction in water content, when comparing the density of Paste 1 with Paste 30.

Table 5.1 Density measurements for the cubes and cylinders.

Specimen Size*	Cubic (20x20x20)	Cubic (30x30x30)	Cubic (40x40x40)	Cylinder (20X20)	Cylinder (30X30)	Cylinder (40X40)	Cubic (50x50x50) rectilinear infill	Cubic (50x50x50) concentric infill
	kN/m³	kN/m³	kN/m³	kN/m³	kN/m³	kN/m³	kN/m³	kN/m³
Density	5.658 ± 0.3	5.411 ± 0.2	5.372 ± 0.2	5.062 ± 0.1	4.891 ± 0.1	4.696 ± 0.2	4.618 ± 0.1	4.017 ± 0.1

* The specimens are made of Paste 30 from Tab. 3.9 and the filling is concentric, except for the specimens of size (50x50x50) mm3 which are made of Paste 12.

5.2 Compression Strength

The test was performed in accordance with ISO/ASTM 52910, 2018(E), ISO/ASTM 52900, 2015(E), ASTM F2971- 13 and ASTM Ref: D695 - 15. The tests were performed in the laboratory ISMD Institute /TU Darmstadt using Zwick/Roell Z050 and VIC-3D™ measurement system as shown in Fig. 5.1. The specimens were tested with strain rate of 5 mm/min [Gao et al. 2018]. Each result in Fig.5.2 is average of at least three specimens. The specimens tested were produced using the following printing parameters: 99% and 100% concentric and rectilinear respectively. The nozzle diameter is 1.6mm. All speeds= 18mm/s (perimeter speed, infill speed, first layer speed, gap infill speed and max print speed), layer height= 1mm, extrusion width= 200%, the air pressure= 0.8 MPa and the extrusion multiplier= 1. The paste used is 1 from Table 3.9. The specimens were cured under the same conditions as mentioned in paragraph 3.4 Curing process. The samples were tested one month after printing.
From Fig. 5.2, the material behaves as a ductile material under compression. The failure could be considered as a large strain-shape change, that exceeded a certain acceptable limit, as shown in Fig. 5.3. During the test, the specimen began to bulge outward at the sides and became barrel shaped. As the load increases, the specimen flattens out, providing greater resistance to further shortening. Due to the use of fibrous cellulose, this type of behavior is greatly influenced by the surface roughness and surface area of the fiber.

This resistance is not only due to the strength of the individual fibers, but also to the bonding of the network fibers, which adhere to each other to form a mat interlocking between the surfaces of the adjacent filaments. [Chen T. et al, 2016]. Concentric infill has more stress concentration in the corner. Therefore, the specimen fails as a fracture in the corner at a later stage of loading. This causes this slip in the curvature of the curve. The pattern is shown in Fig. 5.3. The environment was monitored from the day of printing to the day of testing, as shown in Fig. 5.4.

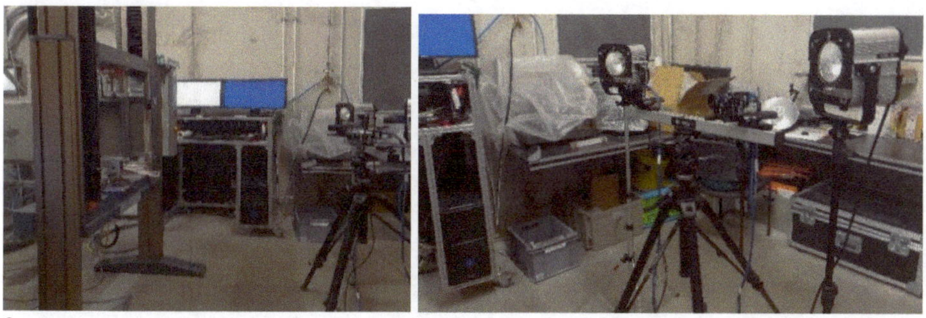

a

Figure 5.1 The testing machine (a) with the DIC system (b) at the ISMD Institute / TU Darmstadt.

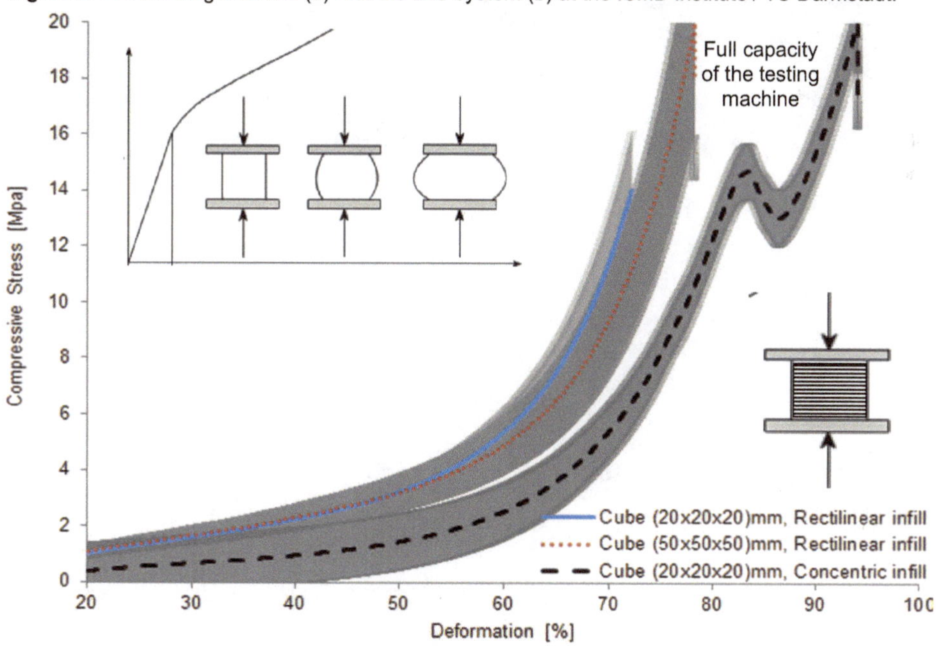

Figure 5.2 The stress-strain results of the compression test for 3D paper-printed specimens and the drawn analogue behavior of steel under compression [NIST NCSTAR1-3D 2005].

a

b

Figure 5.3 Tested specimens after compression failure top view(a), and side view (b).

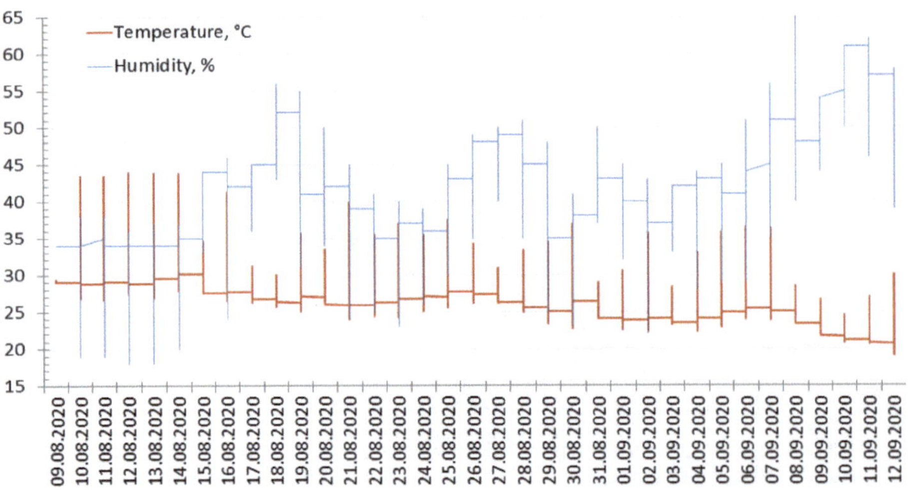

Figure 5.4 Monitoring the specimen environment from printing to the day of testing.

5.3 Tensile Strength

Tensile testing is a method of determining many of the most important mechanical properties of a material. [Jonas Schmidt et al, 2022]. The test was carried out in accordance with "ASTM Designation: D638 - 14 Standard Test Method for Tensile Properties of Plastics". The tests were carried out in the laboratory of the ISMD Institute /TU Darmstadt using the Zwick/Roell Z050 and the VIC-3D™ measuring system, as shown in Fig. 5.5.

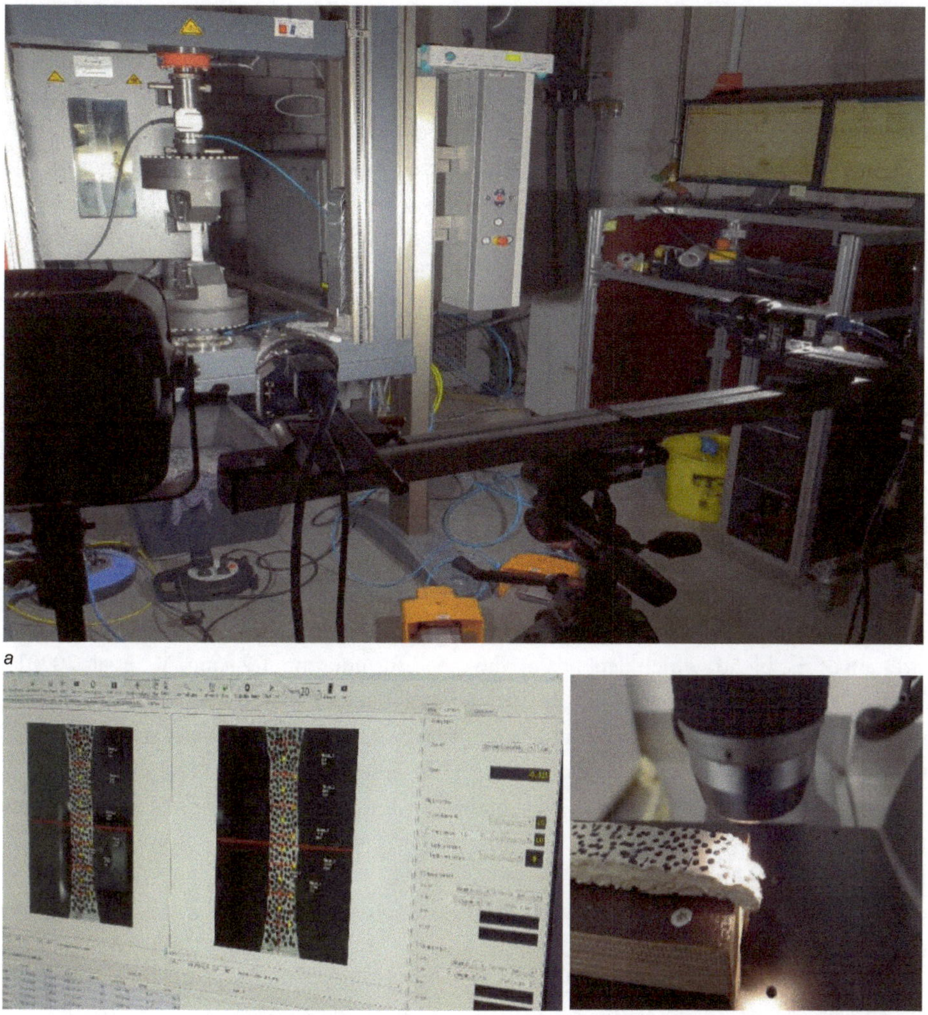

Figure 5.5 The machine preparations for the tensile test (a), the specimen preparations tested (b) and the specimens examined under the microscope (c).

This test was performed to determine the ultimate strength of the material and its deformation. Mechanical properties were also calculated for 3D printed paper samples. Fiber's involvement and post-break bonding behavior were also investigated using the microscope, as shown in Fig. 5.5(c). More details about the microscope are given in section 3.4.1 Shrinkage results during drying.

5.3.1 Filaments Orientation Effect (0- , 90- , 45-) Degree

The infill identifies the strength of the configuration. The orientation of the infill is used as an internal design for the part to optimize strength and involves orienting in a specific direction to maximize the strength of the part. The properties of any 3D printed part or model are identified by effective layer bonding and raw filament performance.

Bonding quality and material adhesion properties determine the mechanical properties and integrity of parts produced using the FDM process. Bonding is the cohesive force between the components of the mixture forming the filaments within a layer and between different layers in 3D printed parts [Brenken, B. et. al, 2018] and [Polyzos, E.et al, 2020]. This bonding process is commonly referred to as [Polyzos, E. and Pyl, L. 2022]:

- Intra-layer bonding: bonding between deposited filaments of the same layer.
- Inter-layer bonding: bonding between successive deposited layers.

It's important to examine the strength and stiffness behavior of the filament orientation of the 3D printed paper part in different directions, such as the stresses in the axial direction of the filaments or perpendicular to see the adhesion properties and coalescence strength between the layers. The axial strength of the filament depends on the strength of the cellulose fibers and the matrix strength, which is the network strength of the interlocking and adhesion between the fibers. While the strength of the filaments bonded together in the direction perpendicular to the direction of the force axis depends on the adhesion strength of the blend, which depends on the blend formulation. [Zhang, Y., et.al, 2020].

Tab. 5.2 shows the ultimate tensile strength and secant modulus of elasticity for the specimens. The specimen (A1.6) is tested by a force direction lies in the same direction of the filament orientation for the specimen with the direction. Specimen (A1.6_45) is tested by applying an axial force, but the orientation of the printed filaments is at 45 degrees from the axial force. This means that the resistance strength comes from the cellulose fibers and the adhesion force between the filaments. Specimens (A1.6+f-90D) and (A1.6_No f_190D) are tested with axial force lies perpendicular to the filament orientation. One is with frame (perimeter) and the other is without. The tensile strength of these specimens is only due to the adhesion force between the filaments.

Fig. 5.6 shows the tensile behavior for the samples with different filament orientations with respect to the axial force. The Young's modulus and tensile strength of specimen A1.6 are 12% and 24.5% higher than those of specimen A1.6_45D.

The printing parameters of the samples are constant. The infill type is 99% rectilinear. The nozzle diameter is 1.6mm. All speeds= 18mm/s (perimeter speed, infill speed, first layer speed, gap infill speed and max print speed), layer height= 1mm, extrusion width= 200%, the air pressure= 0.8 MPa and the extrusion multiplier= 1. The paste type is 30 from Table 3.9. The specimens were cured under the same conditions as mentioned in paragraph 3.4 Curing Process. The results are shown in Tab.5.2 and Fig. 5.7.

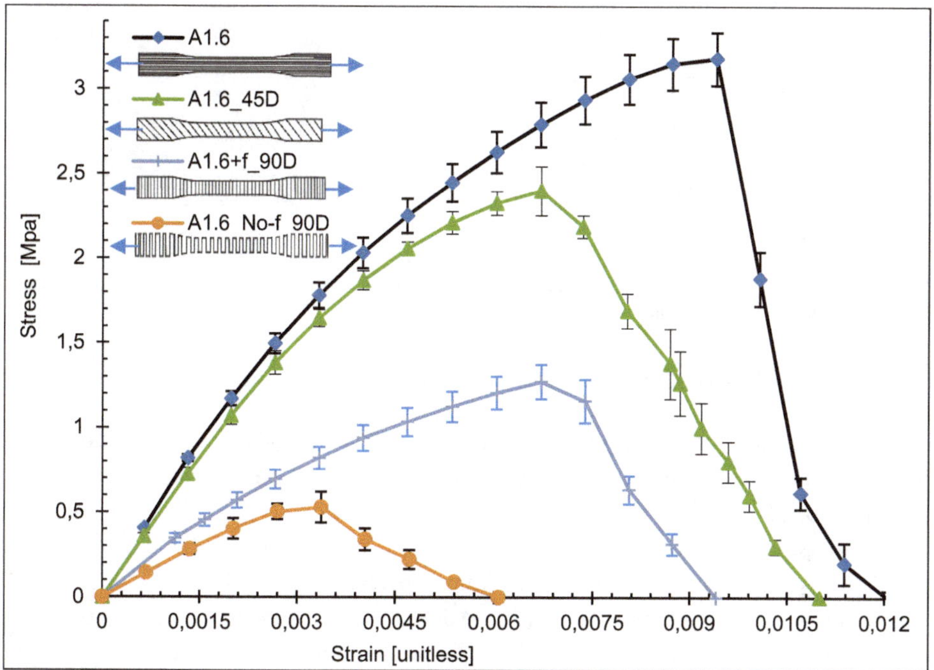

Figure 5.6 Tensile strength as a function of the angle between force direction and filament orientation.

Table 5.2 Secant Young's modulus and ultimate strength for the specimens with different filament orientations.

Specimen	Secant Young Modulus Mpa	Ultimate strength Mpa	Failure Mode
A1.6	589.25 ± 22	3.18 ± 0.2	Fracture failure by cellulose fibers, as shown in Fig. 5.8.
A1.6_45D	518.69 ± 25	2.4 ± 0.2	Fracture failure, the adhesive strength contributes to the resistance as shown in Fig. 5.9.
A 1.6 +f_90 D	246.05 ± 20	1.273 ± 0.2	Fracture failure
A 1.6 No f_90 D	178.53 ± 15	0.53 ± 0.05	Delimitation/ Adhesion failure between the extruded filaments.

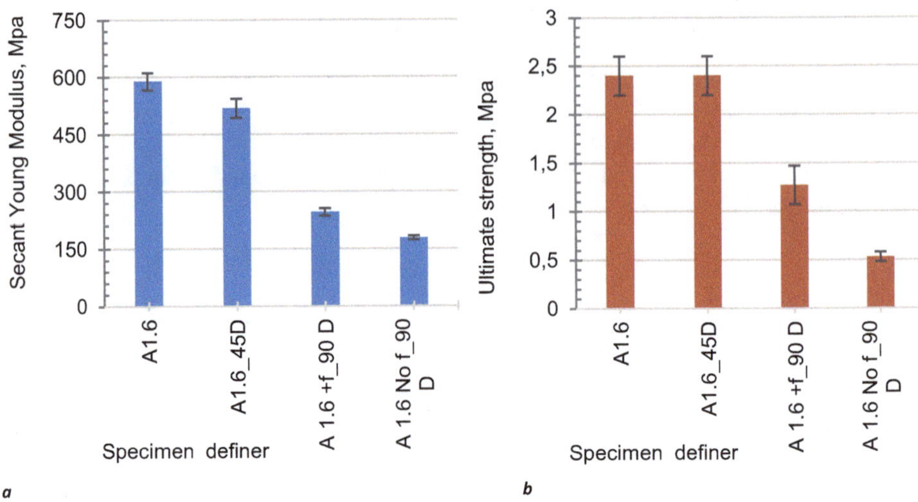

Figure 5.7 The results of tensile tests: secant Young's modulus (a) and ultimate strength (b) for the specimens with different filament orientations.

5.3.2 Adhesion Properties

Adhesive properties of a 3D printing mixture refer to the ability of the material to adhere to itself and to other surfaces. This is an important factor in determining the mechanical properties of 3D printed objects. It affects the strength and durability of the object. Poor adhesion can lead to weak bonds between layers, which can lead to a weak and brittle object that is prone to delamination or cracking, especially between layers. Good adhesion can result in strong and durable objects. Adhesion also influences and improves the dimensional accuracy and surface finish of the printed object. It also plays an important role in limiting shrinkage results. The adhesion strength of 3D printing with paper material is indeed influenced by i) the type and properties of the cellulose fibers used, ii) the formulation of the blend, iii) and the printing configuration, which controls the coalescence between the filaments. The cellulose fibers are the main component of the paper pulp and play a crucial role in determining the properties of the paper objects, including their strength, flexibility, and texture. For example, longer fibers tend to have better adhesion than shorter fibers because they provide more points of interlock between the layers. In addition, the thickness and orientation of the fibers can also affect the bond strength. Thicker fibers may provide more strength but may also make the mixture more difficult to extrude, while fibers oriented in the same direction may provide weaker adhesion than randomly oriented fibers.

- The formulation of the mix can also affect the bond strength of the 3D printed object. The amount and type of binder used, such as CMC or filler, as well as the ratio of cellulose fibers, can all affect adhesion strength.

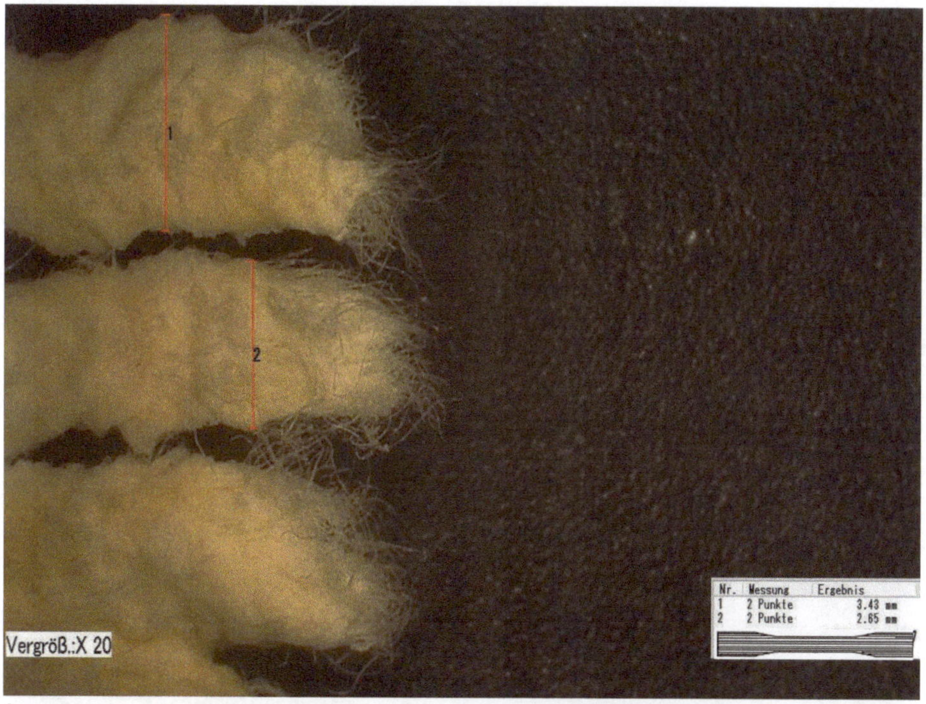

a

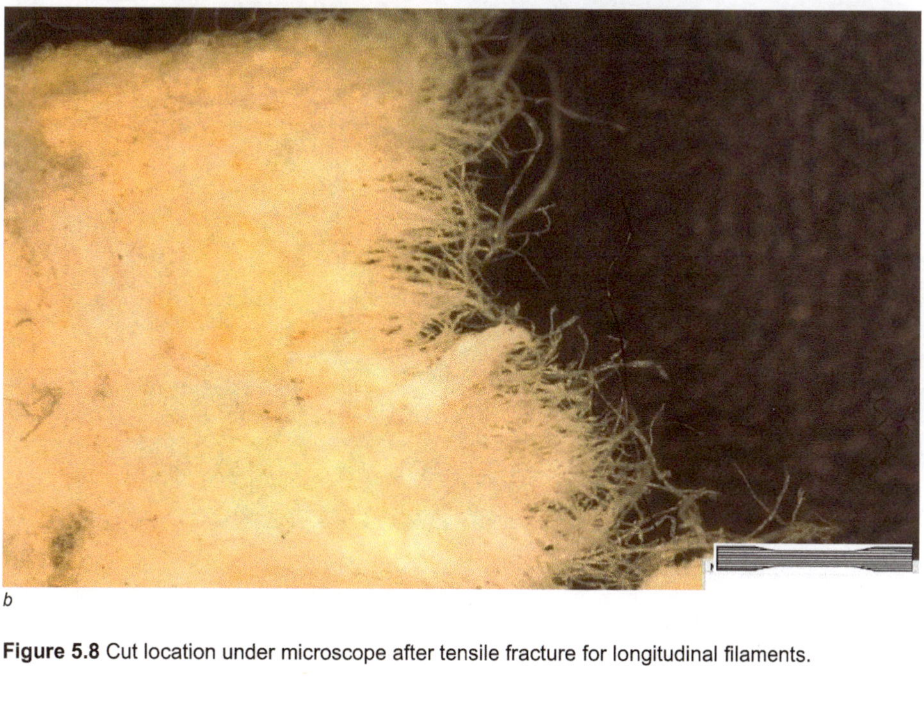

b

Figure 5.8 Cut location under microscope after tensile fracture for longitudinal filaments.

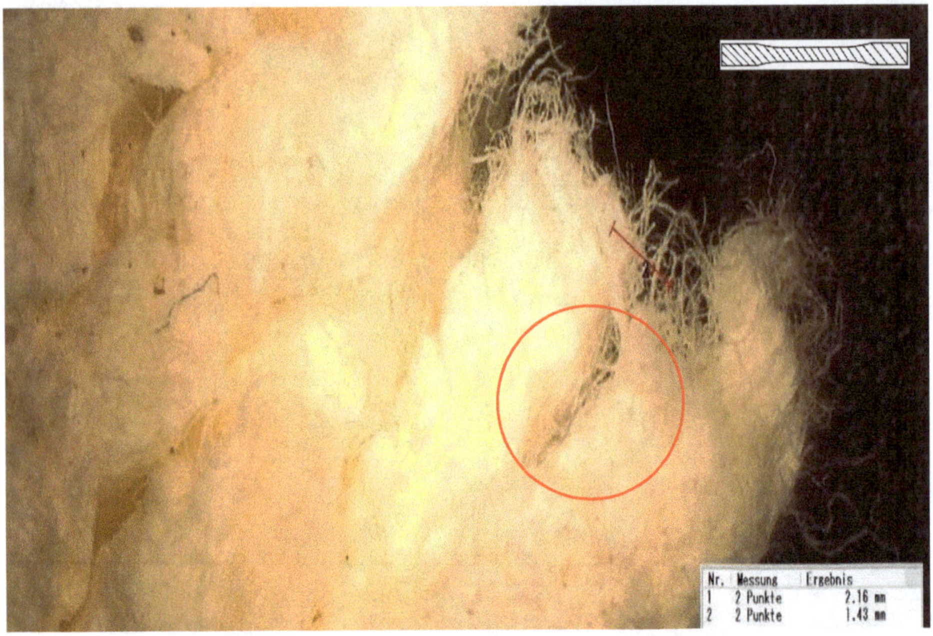

a

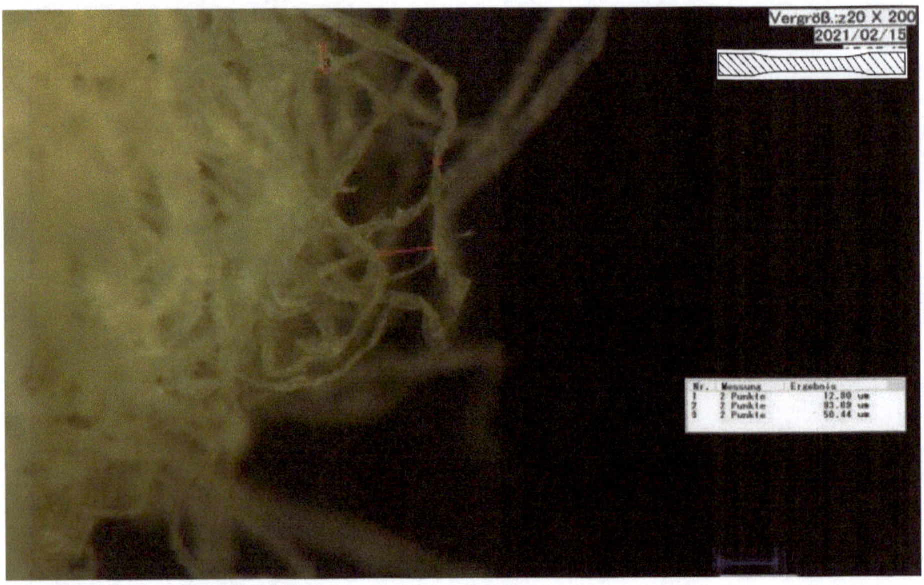

b

Figure 5.9 Location of the cut under the microscope after tensile breakage of the longitudinal filaments (a) and zoomed area located under the red circle (b).

- The formulation of the mix can also affect the bond strength of the 3D printed object. The amount and type of binder used, such as CMC or filler, as well as the ratio of cellulose fibers, can all affect adhesion strength.
- Layer adhesion is a critical aspect of 3D printing quality as it affects the strength and stability of the final print. Interlayer adhesion is influenced by several factors, including [Abolhasani, M. M., Salari, et al, 2019], [Hu, Q et. al, 2017], and [Li, Y., Li, S., & Zhang, Z., 2020].
- Surface preparation: The cleanliness and roughness of the built plate can affect adhesion, and it's important to ensure that it is free of dust and debris and has the appropriate roughness for good adhesion.
- Paste formulation: the proportions of the compounds and their type/properties have varying levels of adhesion, as well as depending on the water/solids content ratio.
- Sample stability and dimensional stability during printing. Adhesive strength helps to increase the sticking and collapsing or buckling problems.
- Drying and shrinkage: Adequate drying of the filament as it is laid down is crucial to ensure good adhesion between layers without uneven shrinkage leading to cracking, as it helps to prevent warping and other problems that can affect the final print quality.
- Bed adhesion is the force that holds the first layer of the print to the build plate and is critical to ensuring good adhesion between layers.

This property is tested by measuring the resistance between the filaments during the tensile test. The results are shown in Tab. 5.2 and Fig. 5.6.

Two modes are used on the samples (A1.6+f-90D and A1.6_No f_90D). The extruded filaments of these specimens are oriented perpendicular to the axis of the force applied. There will be two types of test specimen, one with a frame and one without a frame. The unframed specimen tests the pure effect of adhesion behavior between filaments. On the other hand, the specimen with external frame (parameter) is tested to study the behavior of solid objects with same filling. In both modes it is shown that the adhesion strength can maintain the resistance, as shown in Fig. 5.10.

The Young's modulus and ultimate strength of the specimen with 90-degree filaments without frame (A1.6 No f_90 D) are 30.3% and 16.67% preforming respectively from the specimen with 0-degree filaments (A1.6). These values are relatively adequate. This indicates that this formulated mixture has acceptable adhesion strength and coalescence, as shown in Tab. 5.2.

The Adhesion property for the mixture to other surfaces is investigated in Appendix E.

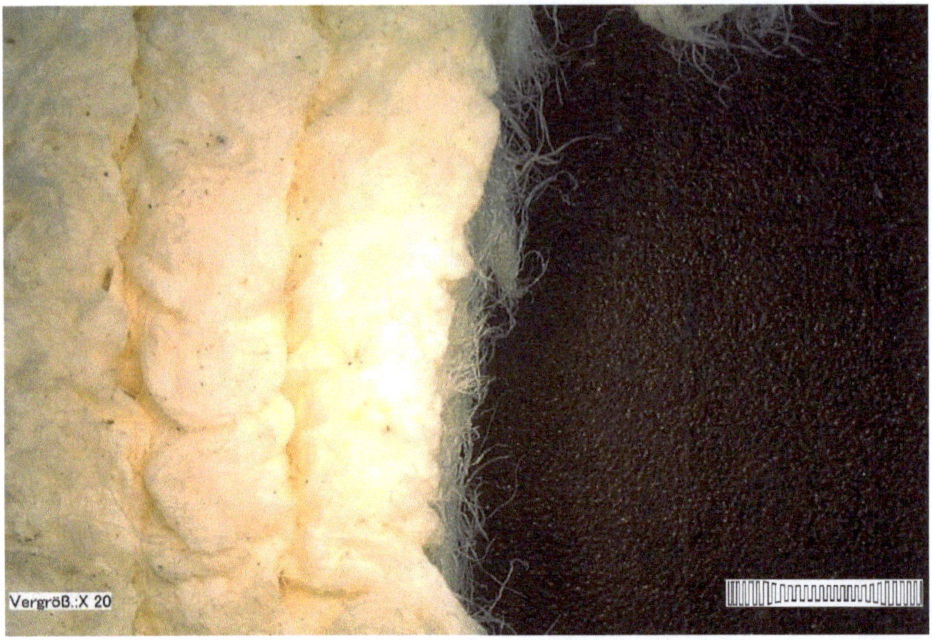

a

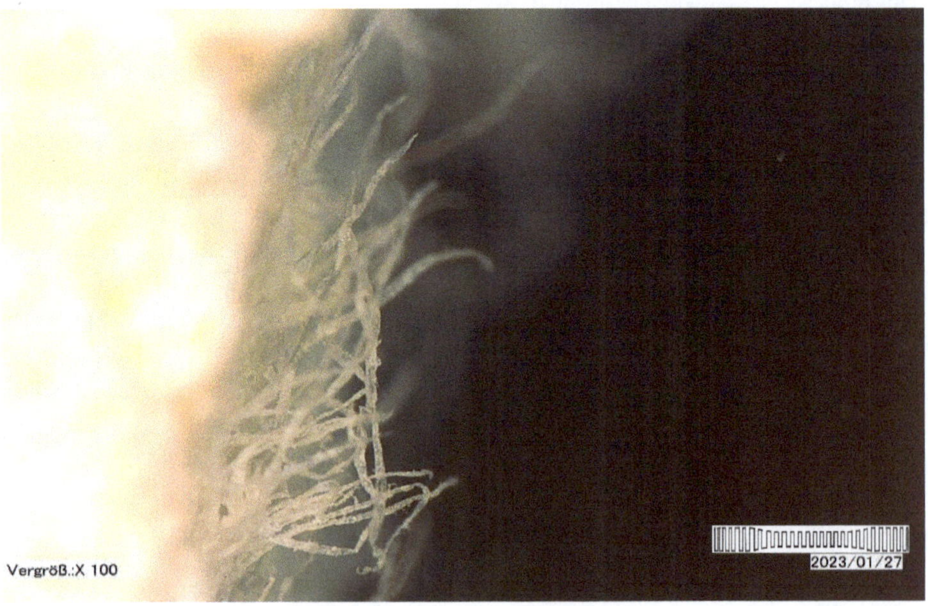

b

Figure 5.10 Adhesive failure of the 90-degree filament orientation (a) and the magnified area of the adhesive failure (b).

5.3.3 Effect of Nozzle Size

Nozzle size is beneficial for objects that need to be close together or require a high level of detail. Nozzle size is often directly related to layer height, as a smaller nozzle size allows for a smaller layer height and more detail in the print. A smaller layer height can result in stronger and more rigid prints because there are more layers, and the individual layers are stronger. It also allows for a more consistent, uniform texture and structure. A smaller nozzle size results in smaller and more closely spaced infill lines, thus increasing the layering effect and contact area more than a larger nozzle, thus increasing the strength and stiffness of the print, and increasing the adhesion strength between the layers and the filaments. However, it is important to note that very fine nozzle sizes have a clogging effect. This can have a negative effect on the mechanical properties of the print if the printing parameters are not set up well. Nozzle size has a significant effect on 3D printing mechanical properties. Based on the desired final properties of the print, it is important to select an appropriate nozzle size. Two nozzle diameters (1.2 mm and 1.6 mm) are used to study this effect. To characterize this effect, filament orientation and adhesion strength are studied. The specimens are defined as in Section 5.3.1 "Effect of filament orientation (00 - 90o-450)". The printing parameters of the samples are the same. The fill type is 99% rectilinear. The nozzle diameter is either 1.6 or 1.2 mm, all speeds= 18 mm/s (perimeter speed, infill speed, first layer speed, gap infill speed and max print speed), layer height= 1 mm, extrusion width= 200%, the air pressure= 0.8 MPa and the extrusion multiplier= 1. The paste type is 30 from Table 3.9. The specimens were cured under the same conditions as those mentioned in the paragraph "3.4 The curing process". The environmental data are given in Fig. 5.11. Tab. 5.2 and Figures 5.12 and 5.13 confirm that the ultimate strength and stiffness of the smaller nozzle (1.2 mm nozzle) are higher than those of the larger nozzle (1.6 mm nozzle). From Fig. 5.12 and Tab. 5.3, the ultimate stress and modulus of elasticity of the 1.2 mm nozzle are 11.9% and 7.3% higher, respectively, than those of the 1.6 mm nozzle when the filaments are oriented in respect of the axial direction of the force. When the filaments are oriented at 45 degrees to the direction of the force, the ultimate stress and modulus of elasticity of the 1.2 mm nozzle are 5% and 6.9% higher, respectively, than those of the 1.6 mm nozzle.

On the other hand, when the filaments are oriented at 90 degrees to the direction of the force, the tensile strength and modulus of elasticity of the 1.2 mm nozzle are 18.1% and 22.67% higher, respectively, than those of the 1.6 mm nozzle (in the case of the providing perimeters) and 24.5% and 56.7% higher, respectively, than those of the 1.6 mm nozzle (without perimeters). It's interesting to note that the ratio difference in the ultimate stress between the specimens (A1.2 No f_90 D and A1.6 No f_90 D) is higher than the ratio difference between the specimens (A1.2 and A1.6). The explanation is that the matrix worked effectively when the adhesion strength plays a role for a smaller nozzle size and the cellulose fiber had no direct effect on the behavior, as shown in Fig. 5.14.

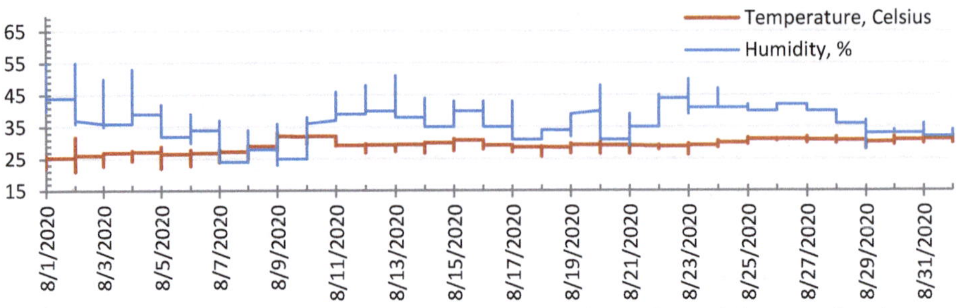

Figure 5.11 Monitoring the surrounding environment for specimens from printing to testing day.

Table 5.3 Secant Young's Modulus & Ultimate Strength for different nozzle size specimens.

Specimen	Secant Young Modulus Mpa	Ultimate strength Mpa	Failure Mode
A1.6	589.25 ± 22	3.18 ± 0.2	Fracture failure, see Fig. 5.7.
A1.2	632.1 ± 20	3.56 ± 0.2	Fracture failure
A1.6_45D	518.7 ± 25	2.4 ± 0.2	Fracture failure
A1.2_45D	554.3 ± 40	2.52 ± 0.18	Fracture failure
A 1.6 +f_90 D	246.05 ± 10	1.27 ± 0.2	Fracture failure
A 1.2 +f_90 D	301.85 ± 35	1.5 ± 0.1	Fracture failure
A 1.6 No f_90 D	178.53 ± 5	0.53 ± 0.05	Delimitation between the filaments.
A 1.2 No f_90 D	279.73 ± 45	0.66 ± 0.17	Delimitation between the filaments.

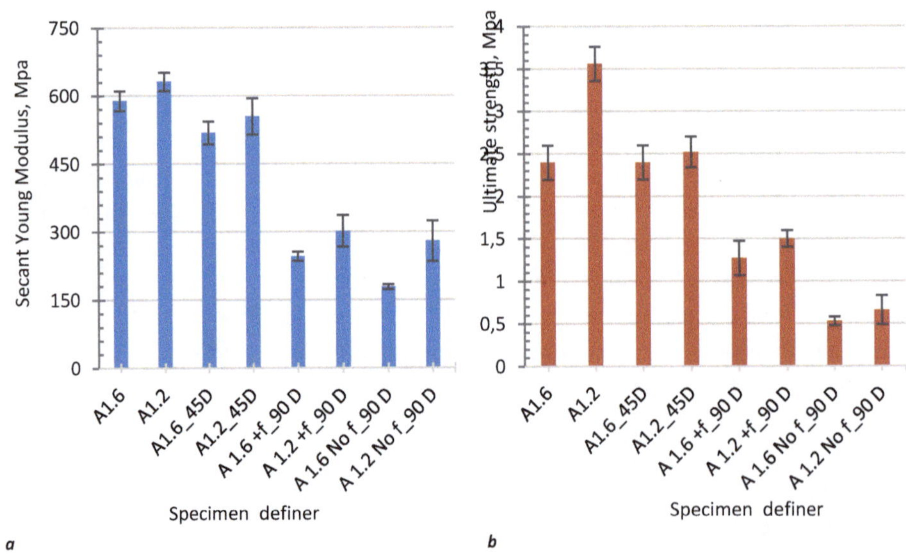

Figure 5.12 The results of tensile tests: secant Young's modulus (a) and ultimate strength (b) for the specimens of different filament diameters of with different filament orientations.

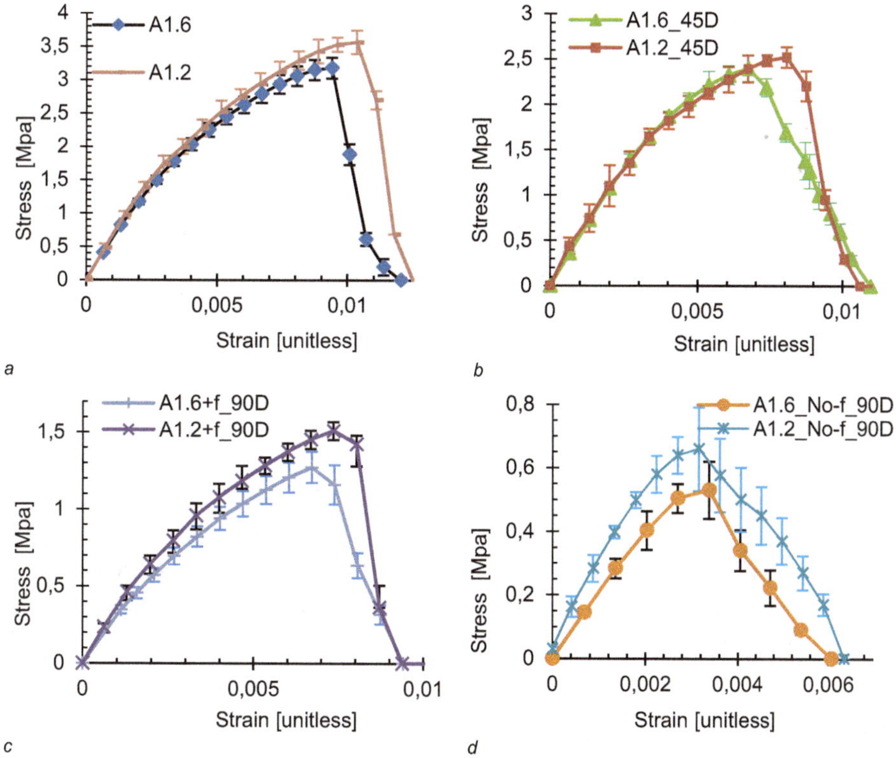

Figure 5.13 Stress-strain behavior of the effect of different nozzle size.

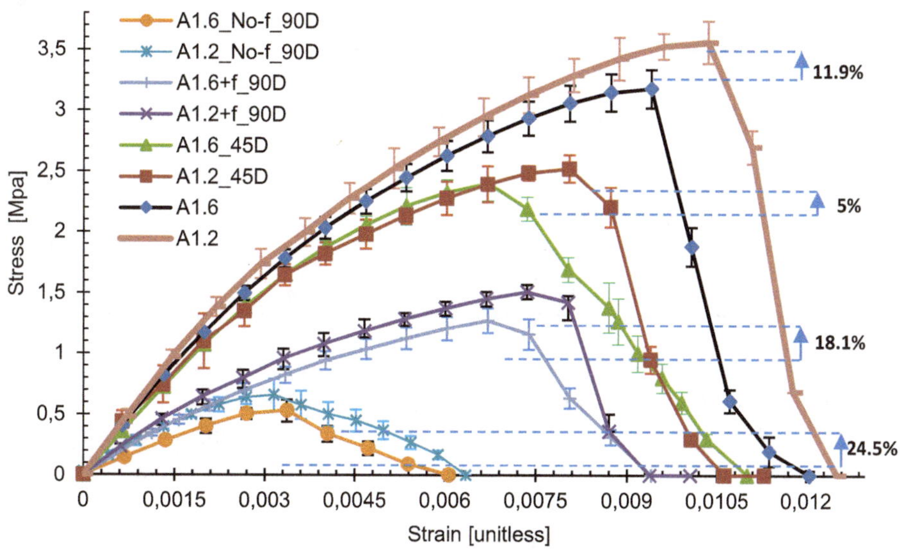

Figure 5.14 Stress-strain behavior of different nozzle size and different orientations.

5.3.4 Effect of Wood Fiber

The addition of wood fibers to the mixture changes the mechanical behavior. This depends on the proportions and types of wood fibers added to the mixture, as listed in Tab. 5.3 and Fig.5.15. Wood fibers are a popular choice for reinforcement in composites because they are cellulosic fibers, naturally abundant and relatively inexpensive. If added to the mix, wood fibers can improve the mechanical properties including tensile strength, flexural strength, and impact resistance of the finished part. The improvement in mechanical properties is attributed to the reinforcement provided by the wood fibers. Fiber size, shape and orientation, matrix type and processing conditions are factors that can affect mechanical performance. The specimens in Tab. 5.4 are named according to the percentages used and the length sizes that were sieved using sieves with dimensional holes that kept the fiber sizes in between. A1.6 +2%wf (0.5-0.4) mm means that the nozzle used is 1.6 mm, the percentage of wood fiber added is 2% by weight of the total weight of the mixture, and (0.5-0.4) refers to the sizes currently used. It was studied to find out the possible percentages that could be added to the mix, which could influence the properties in a positive way. The best improvement in terms of ultimate strength and modulus of elasticity is the specimen (A1.6 +2%wf (1-0.5) mm) that is shown in Fig. 5.16. It noticed that (1-0.5) mm is the better wood fiber size than (0.5-0.4) mm.

Table 5.4 The Secant Young's Modulus & Ultimate Strength for specimens of different fiber proportions and sizes.

Specimen	Secant Young Modulus Mpa	Ultimate strength Mpa	Toughness Mpa	Failure Mode
A1.6	589.25 ± 22	3.18 ± 0.2	0.022073 ± 0.00087	Fracture failure
A1.6 +2%wf (0.5-0.4) mm	525.48 ± 50	3.71 ± 0.1	0.033333 ± 0.00092	Fracture failure
A1.6 +3%wf (0.5-0.4) mm	474.52 ± 148	3.69 ± 0.17	0.036492 ± 0.01463	Fracture failure
A1.6 +5%wf (0.5-0.4) mm	521. 45 ± 40	3.72 ± 0.08	0.046397 ± 0.00356	Fracture failure
A1.6 +2%wf (1-0.5) mm	529.45 ± 77	4.49 ± 0.1	0.046711 ± 0.00974	Fracture failure
A1.6 +3%wf (1-0.5) mm	401.6 ± 45	4.62 ± 0.15	0.132072 ± 0.00429	Fracture failure
A1.6 +5%wf (1-0.5) mm	352.3 ± 30	4.36 ± 0.2	0.148466 ± 0.00681	Fracture failure

Fig. 5.16 shows that when using (0.5-0.4) mm wood fibers, the better percentage is 2% (sample A1.6 +2%wf (0.5-0.4) mm). Concerning the modulus of elasticity, the percentage 2 % is 10.74 % and 0.77 % higher than the percentages 3 % and 5 % respectively. Meanwhile, the values of the ultimate strength of 2%, 3% and 5% remain remarkably close. However, the specimen with 5% wood fibers has higher toughness stress than the specimens with 3% and 2% wood fibers of (0.5-0.4) mm fibers length, which absorbed energy higher by 27.1% and 39.2% respectively.

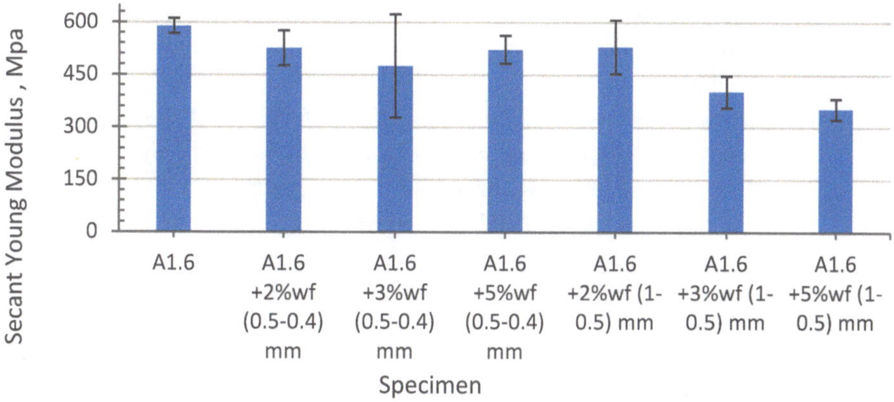

a

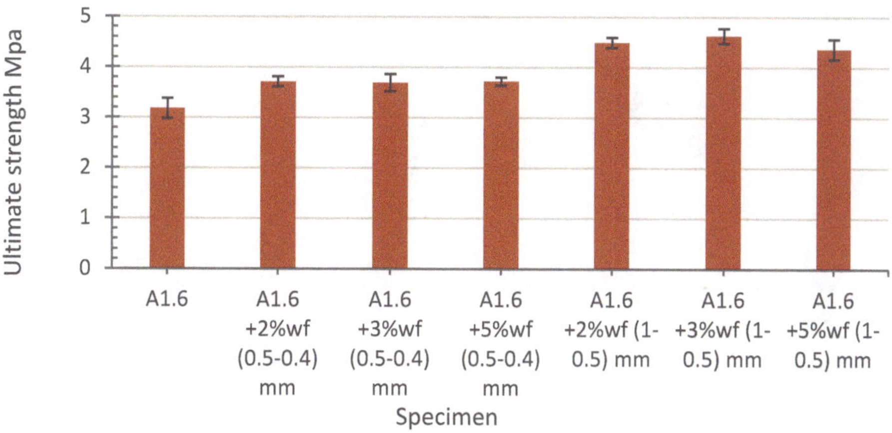

b

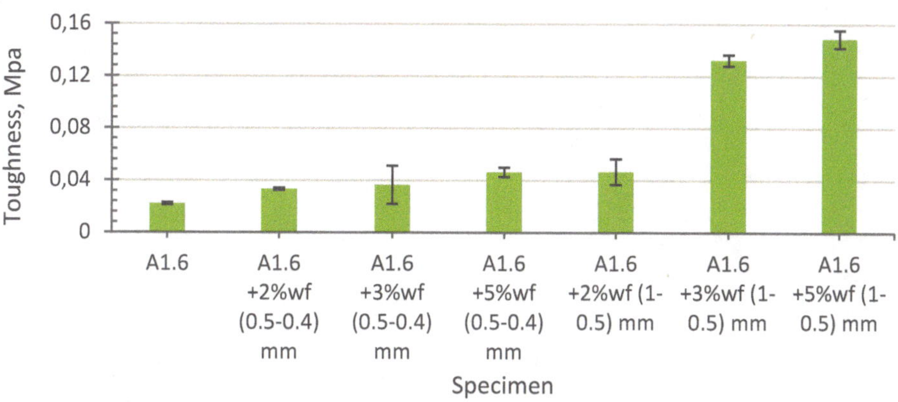

c

Figure 5.15 The results of tensile tests: secant Young's modulus (a) and ultimate tensile strength (b) and the toughness strength (c) for the specimens of different wood fibers percentages.

Fig. 5.17 illustrates that the better wood fibers percentage for the specimens with (0.5-1) mm wood fibers length is also the lowest percentage. The explanation is related to the bond strength provided by the components of the paste formulation. Here, the existing CMC content, which considers the adhesion agent between the formulation. If additional wood is added, the CMC would not be enough to cover all the surface areas of all the fibers (cellulose and the wood fibers). This means that the paste with a higher percentage of wood fibers has less adhesive bond and less slip resistance.

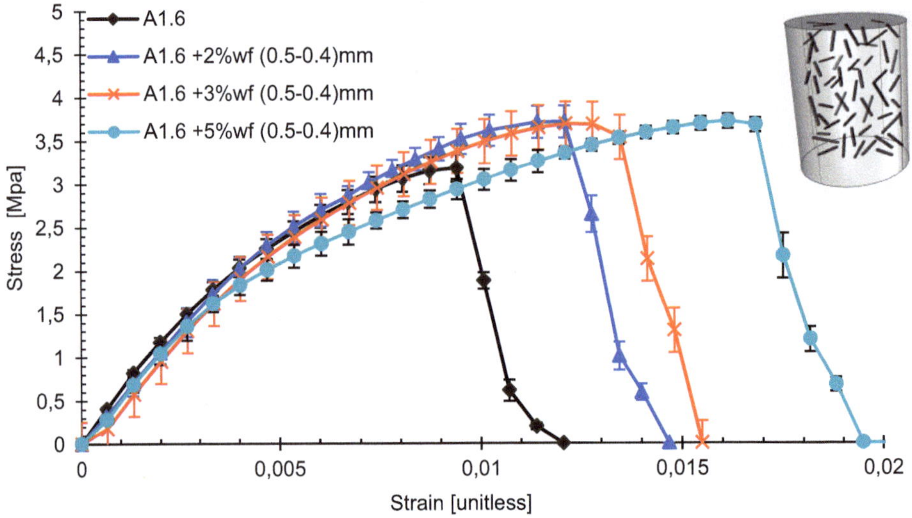

Figure 5.16 Stress-strain behavior using different ratios of wood fibers having length (0.5-0.4) mm.

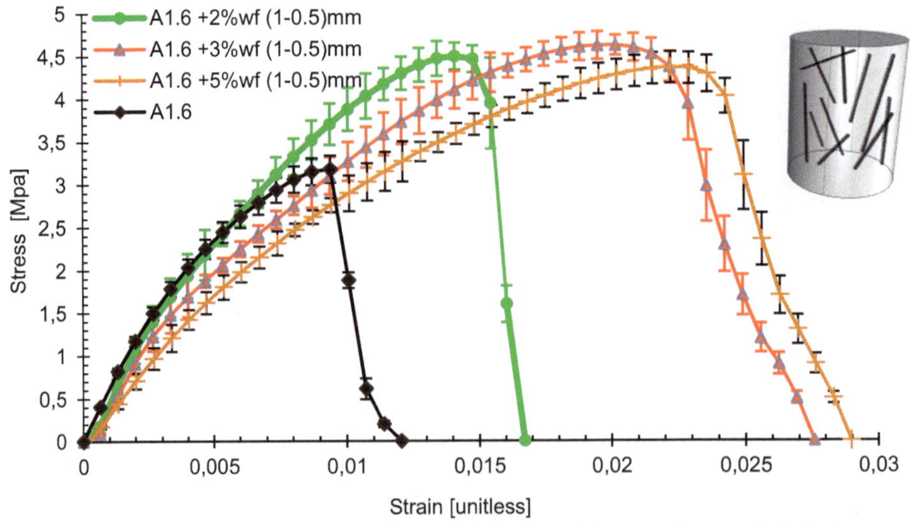

Figure 5.17 Stress-strain behavior using different ratios of wood fibers having length (1-0.5) mm.

Section 3.2.5 "Rheology: using wood fibers of different percentages and ranges" gives a more detailed explanation of this behavior and the comparison.
In terms of the higher modulus of elasticity, the specimen with 2% wood fibers of (0.5-1) mm (specimen A1.6 +2%wf (0.5-1) mm) is higher by 31.8% and 50.3% than the percentages 3% and 5% of the same wood fiber length; respectively. Meanwhile, the values between the fractions of (0.5-1) mm wood fibers do not show any significant difference in the ultimate strength. However, the specimen with 5% wood fibers ratio has more toughness stress than the specimens with 3% and 2% wood fibers of (0.5-1) mm fibers length, which absorbed energy higher by 12.6% and 218.25% respectively.
Fig. 5.18 shows the specimen behavior when comparing the use of different lengths of wood fibers and shows the improvements when using the wood fibers versus the same specimen of the same paste formulation but without wood fibers.
Figure 5.18 (a) shows the stress-strain behavior of the specimens with 2% wood fibers and the reference specimen without wood fibers. It's demonstrated that the first stage of the response has no significant difference in behavior. The modulus of elasticity remains approximately in the same range of values. Where, the addition of wood fibers doesn't improve the behavior. It's obvious that the specimen with wood fibers of (0.5-1) mm has higher tensile strength and toughness stress by 21% and 40% respectively than the specimen with wood fibers of (0.5-0.4) mm. Meanwhile, the behavior of the reference specimen (A1.6) in terms of tensile strength and toughness stress is improved by 16.67% and 51% respectively with the addition of 2% wood fibers of (0.5-0.4) mm and by 41.2% and 111.6% respectively with the addition of 2% wood fibers of (1-0.5) mm.
Fig. 5.15 (b) validates the stress-strain behavior of the 3% wood fiber specimens with the reference specimen. It's shown that the first stage of the response for the specimens have a different behavior. The Young's modulus of specimen (A1.6) is the highest because the addition of wood fibers didn't improve the Young's modulus. The specimen with wood fibers of (0.5-1) mm has higher tensile strength and toughness stress by 25% and 262% respectively than the specimen with wood fibers of (0.5-0.4) mm. However, the behavior of specimen (A1.6) is improved in terms of tensile strength and toughness stress by 16% and 65.3% respectively with the addition of 3% wood fiber of (0.5-0.4) mm and by 45% and 498.3% respectively with the addition of 3% wood fiber of (1-0.5) mm.

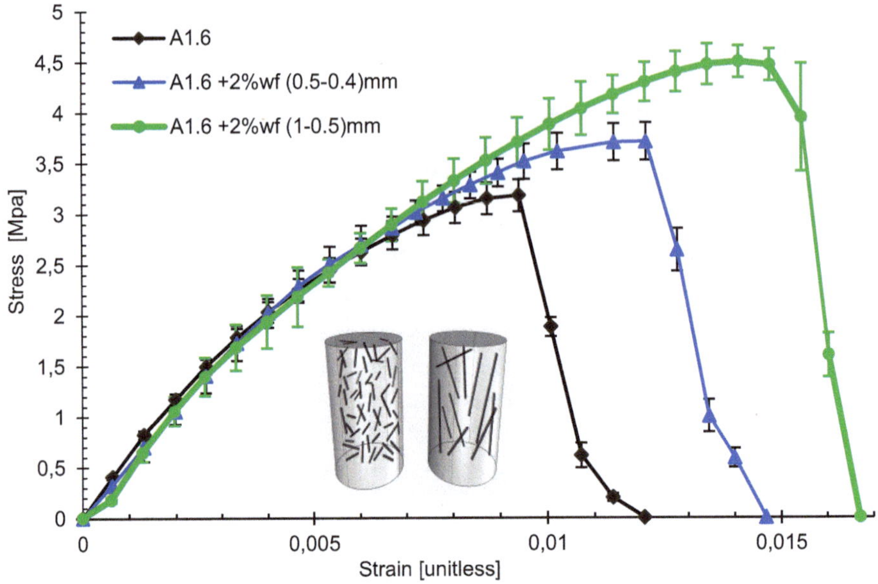

a

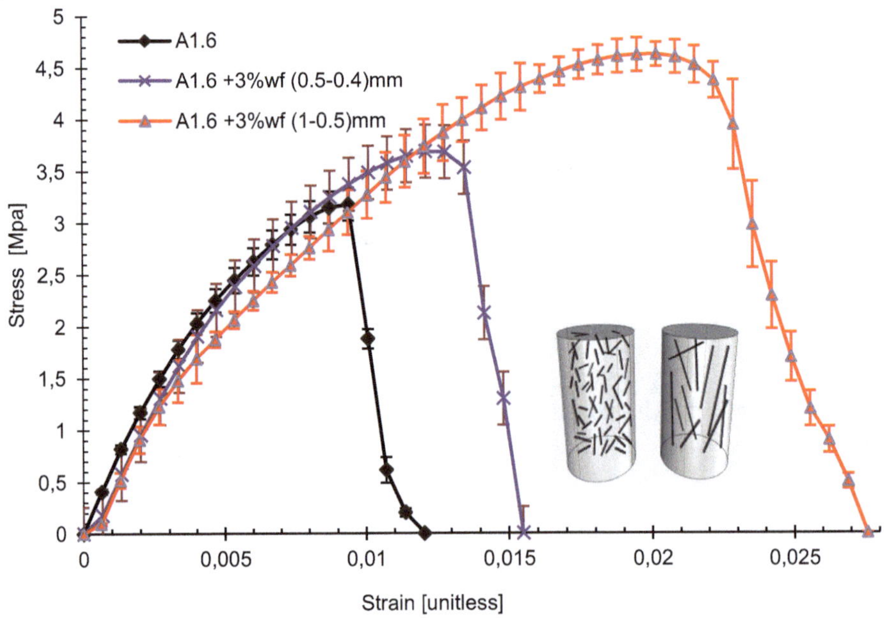

b

Figure 5.18 Stress-strain behavior comparing same ratios with different wood fibers length of (0.5-0.4) mm and (1-0.5) mm. for ratio 2Z (a) and 3%(b).

Figure 5.19 shows the stress-strain behavior of the 5% wood fiber specimens with the reference specimen. With the addition of wood fibers, it's obvious that the modulus of elasticity decreases. However, the behavior of specimen (A1.6) is improved in terms of tensile strength and toughness stress by 17% and 110.2% when 5% wood fibers of (0.5-0.4) mm are added and by 37.1% and 572.6% when 3% wood fibers of (1-0.5) mm are added, respectively.

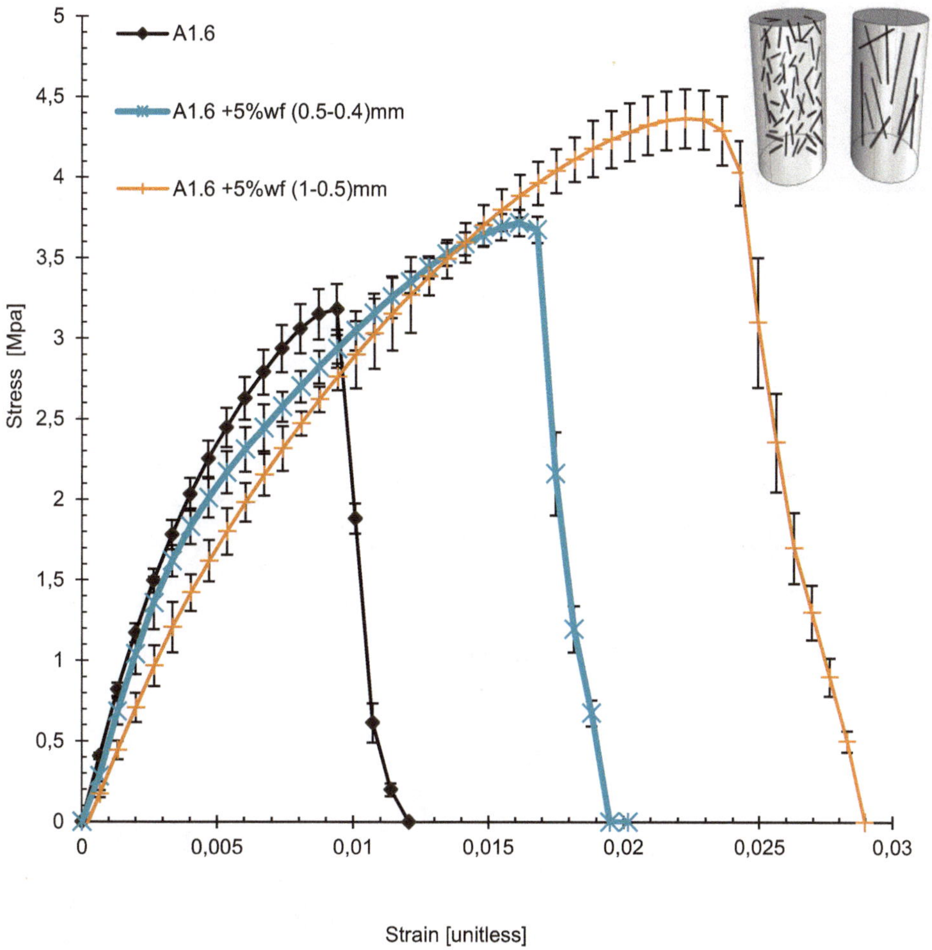

Figure 5.19 Stress-strain behavior comparing same ratios with different wood fibers length of (0.5-0.4) mm and (1-0.5) mm for ratio 5%.

3D printed paper models.

6 3D Models as Cases Study

This chapter examines various 3D printed models that specifically involve analyzing dimensional accuracy and integrating quality in detail to understand their outcomes in the case presented as a potential application. This will provide a glimpse or estimate of the results to illustrate the complexity and examine the efficiency, limitations, and practical implications of these models within specific case study contexts to design a qualified potential product which fits the functionality within acceptable quality.

These 3D models have been designed and printed for the façade courses at the TU Darmstadt as part of the tutorial assignments during the research program. The design technique and principles of the façade construction of these 3D models were based on the façade lectures /TU Darmstadt [Meijs, M. and Knaack, U. 2012], [Knaack, U. et al 2014], [Knaack, U. and Koenders, E. 2018]. Here, the final design is printed using a 3D printer with paper material. This material was designed and formulated in chapter three. Subsequently, a study of the printing results obtained would be examined in the context of the challenges of converting the design into reality as a 3D paper-printed product. High-resolution images within the Rhinoceros 3D software, as shown in Figure 6.1, were used to measure the dimensions. The results are then compared with the values from the use of equipment such as vernier / digital calipers, depth gauge, digital cross line laser, inclinometer protractor, digital protractor; as shown in Fig. 6.2.

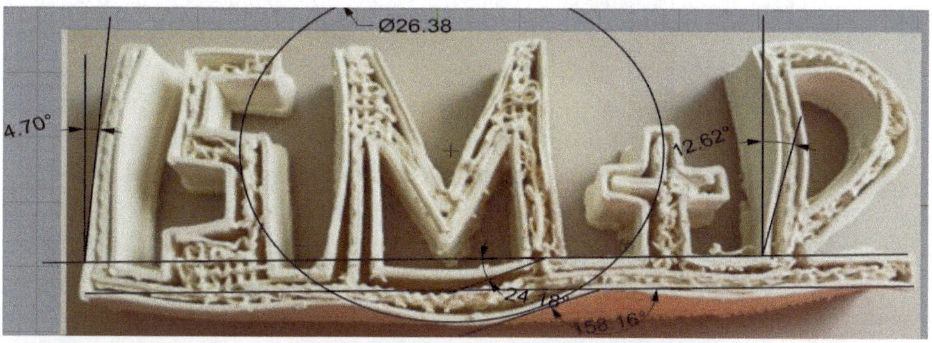

Figure 6.1 An example of the measurement technique by using the Rhinoceros 3D software.

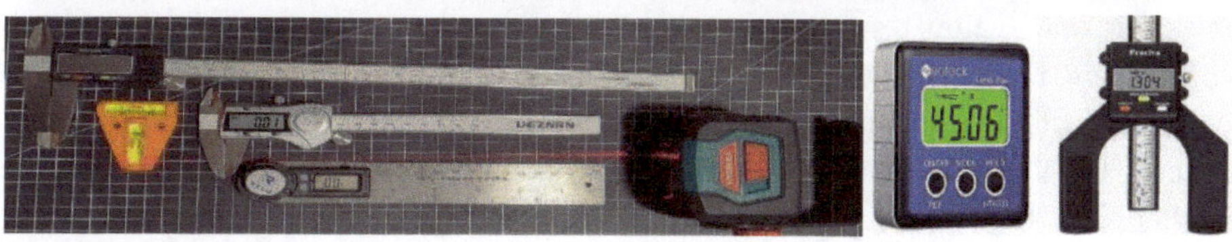

Figure 6.2 Equipment used to measure dimensional accuracy.

D. Abdullah Agha, *3D Paper Printing for the Built Environment*, Mechanik, Werkstoffe und Konstruktion im Bauwesen 73,
https://doi.org/10.1007/978-3-658-45076-2_6

Functionality refers to the set of features and capabilities that a product has that enable it to carry out a specific set of tasks or operations. It defines what the product is able to do and how it can be used to meet the requirements or objectives that are required.
It encompasses both the core functionality of a system and the additional features that are designed to provide users with a desired set of functionalities or to fulfil a particular purpose. It is important to ensure that the functionality is appropriate for the intended use and meets the needs and expectations of the users or stakeholders.
It includes basic functionality such as input/output operations, action manipulation and calculations, as well as more advanced features, customization, or integration capabilities. To prove that the functionality works as intended and meets the desired standards or criteria, such as quality integration and dimensional accuracy, testing and validation processes are often employed.
Here as applications for 3D paper printing and investment in functionality that is commensurate with its efficiency and effectiveness are panels for sun shading and for acoustic functionality that is the best fit for its material properties and avoidance of its drawbacks.

6.1 Acoustical Purposes & Interior Decoration

People spend most of their lives in buildings designed to provide them with safety and shelter. Therefore, the quality of these buildings plays an important role in providing them with peace and tranquility. This could be achieved with an aesthetic design and style that dazzles their eyes. It can be achieved by using a 3D paper printed model that has a complex geometry as sound absorbing and functioning the acoustic properties of the paper material. The material of the object and its geometric configuration greatly influence this functionality. Therefore, it's a prospective application that fits very well with the additive manufacturing technology and invests the material properties of the paper material as i) an insulator to be a sound absorber and create a complex geometry using AM to generate a geometry that could trap sound inside until total dissipation, or ii) as a reflector or sound defusing in a certain needed direction. This could be in the form of absorbing baffles or 3D printed paper wall coverings that reduce reflected sound and echoes. The acoustic parameters are acoustic properties and non-acoustic properties. Acoustic properties include absorption coefficient and damping properties. Non-acoustic properties include porosity, pore size, flow resistance and tortuosity within the designed model [Knaack, U. and Koenders, E. 2018]. These parameters should be considered when designing such 3D models. Acoustic Performance for 3D paper printed material experimentally is shown in Appendix G. A simulation for the acoustical behavior for each model was done using Grasshopper & Pachyderm; more detail is shown in Appendix G.1. Going into detail for each 3D model would help to understand the manufacturing process and the results of the manufacturing quality and accuracy.

6.1.1 Coral Panel

The inspiration behind the design came from the nature of the seamless organic lines from the coral pattern, as pictured in Fig. 6.3. The details of the design are pictured in Appendix G1.2.

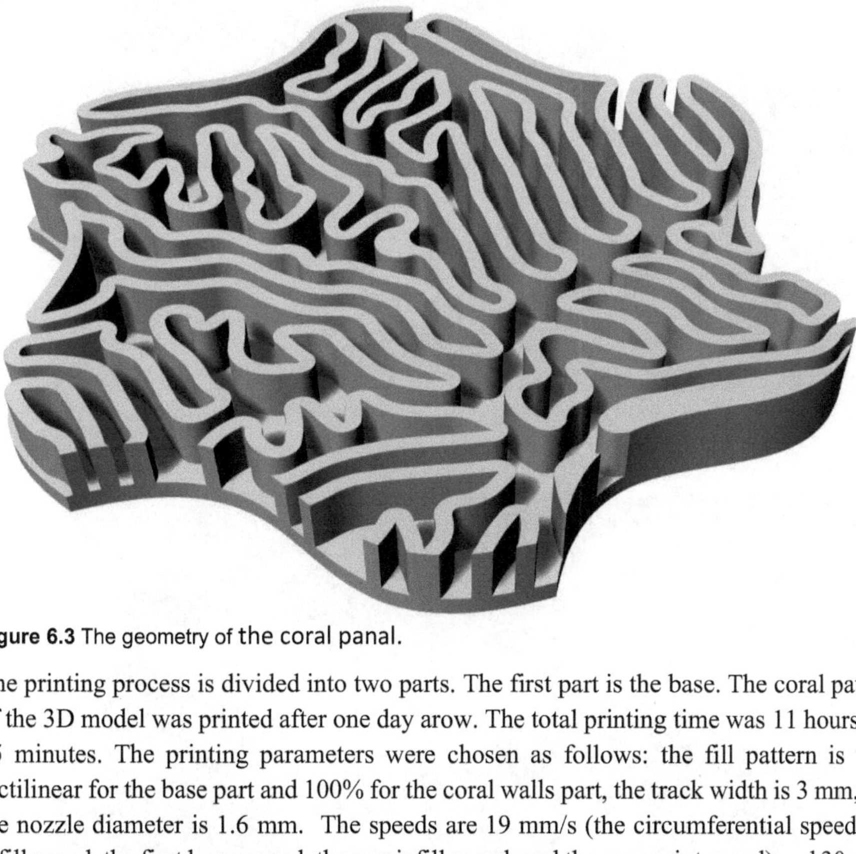

Figure 6.3 The geometry of the coral panal.

The printing process is divided into two parts. The first part is the base. The coral pattern of the 3D model was printed after one day arow. The total printing time was 11 hours and 15 minutes. The printing parameters were chosen as follows: the fill pattern is 95% rectilinear for the base part and 100% for the coral walls part, the track width is 3 mm, and the nozzle diameter is 1.6 mm. The speeds are 19 mm/s (the circumferential speed, the infill speed, the first layer speed, the gap infill speed, and the max print speed) and 30mm/s for the travel speed. The first layer height is 1.6 mm, and the layer height is equal to 1 mm. The extrusion width is 300%. The air pressure is 0.3 MPa and the extrusion multiplier is 0.8. The paste type is 13 from Tab. 3.9. Total used filament is 557.78 m. The samples were cured under the same conditions as mentioned in paragraph "3.4 Curing process". The slicing setting of 3D model is shown in Fig 6.4 and the volumetric flow rate is shown in Fig. 6.5. The printed model is shown in Fig. 6.6 and the dried sample in Fig. 6.7. A magnified image (Fig 6.8) shows a clean printed wall for the coral pattern. The values of the dimensional measurement in Tab. 6.1 are calculated as an average of at least 6 measurements distributed equally along the distance.

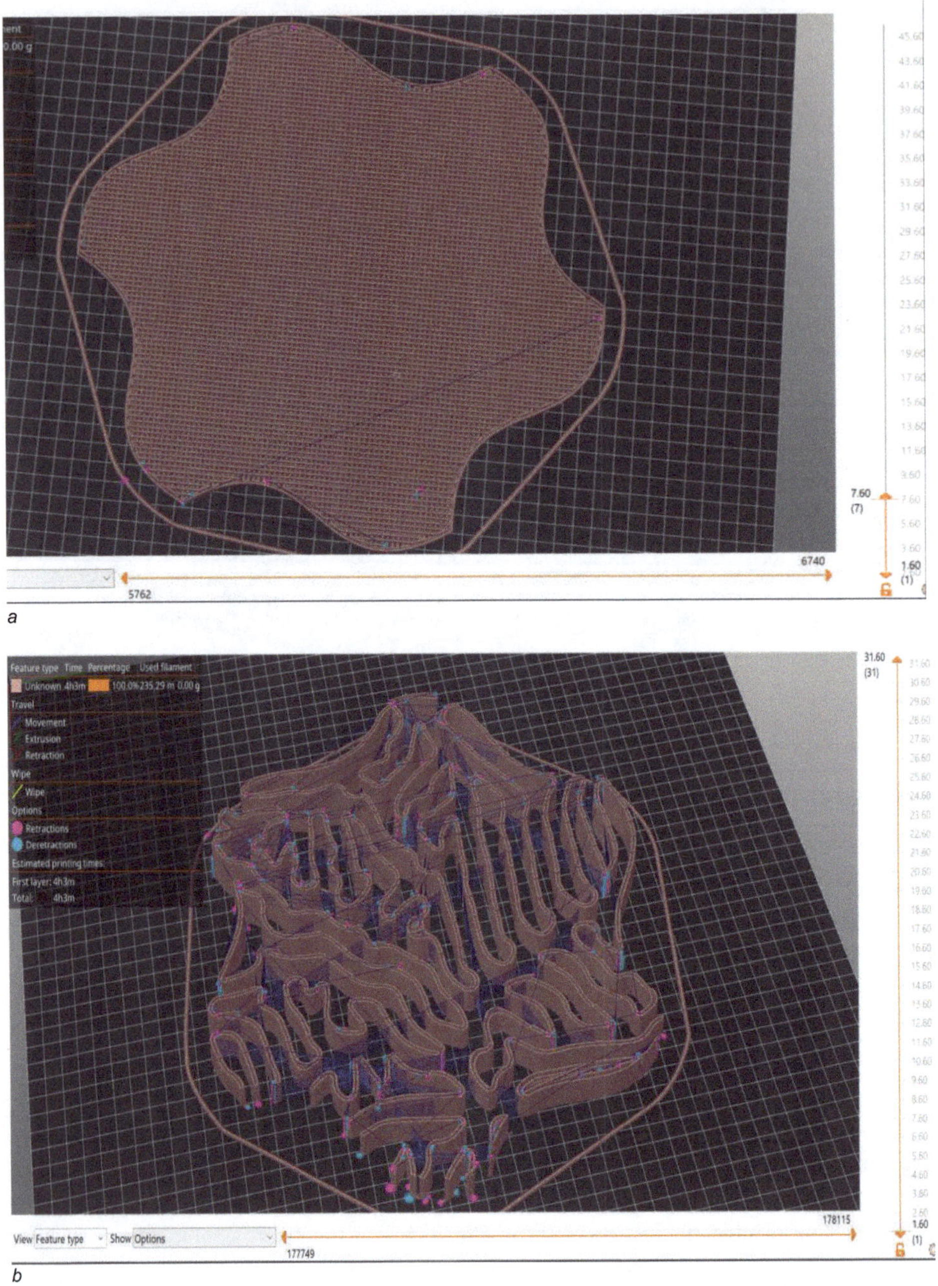

Figure 6.4 The slicing setting of the 3D coral panel model for the base part (a) and for the upper part (b).

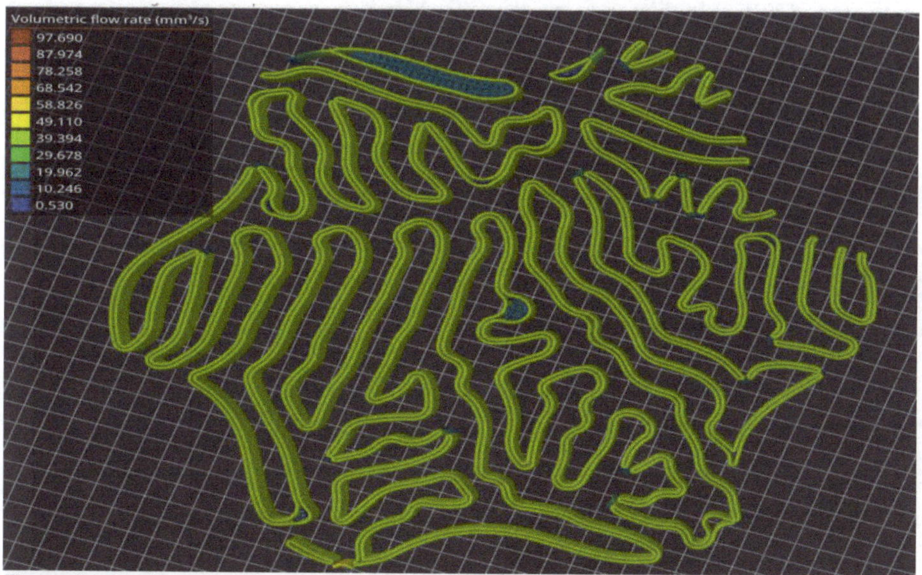

Figure 6.5 3D slice model for the coral panel according to volumatic flow rate.

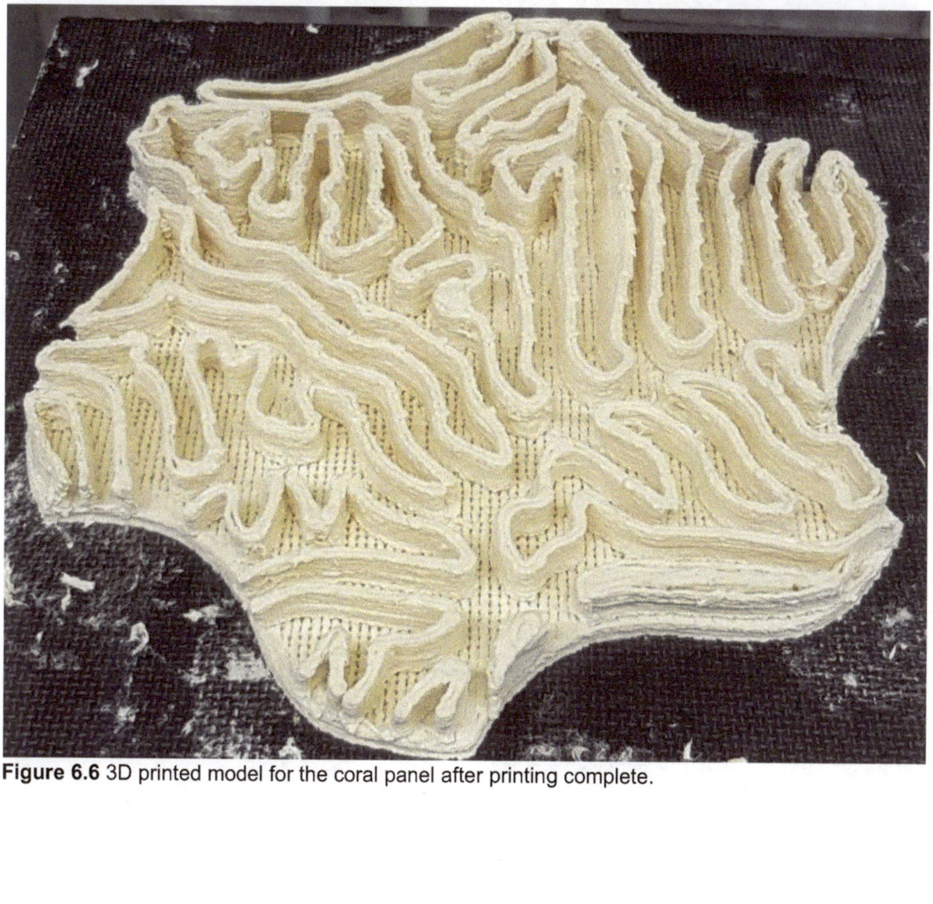

Figure 6.6 3D printed model for the coral panel after printing complete.

Figure 6.7 3D printed model for the coral panel after drying complete.

Figure 6.8 surface flatness without warping or buckling of the coral pattern.

As a result of shrinkage and the inapplicability of the curing technique, when the pause time exceeded hours, the coral pattern walls in some places became detached from the attached surface of the base, as pictured in Fig. 6.9. According to section 3.4 "Curing Process" this problem could have been avoided. The dimensions of the 3D model and the accuracy measurements are shown in Fig. 6.10 and Tab. 6.1. The surrounding environment is monitored as exemplified in Fig. 6.11.

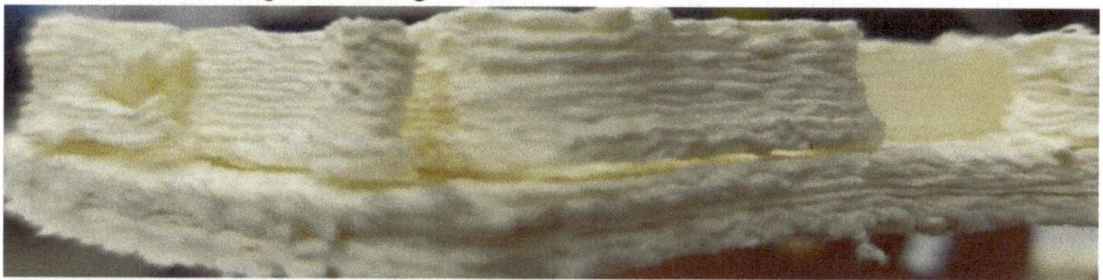

Figure 6.9 Debonding between the base of 3D model and the coral pattern.

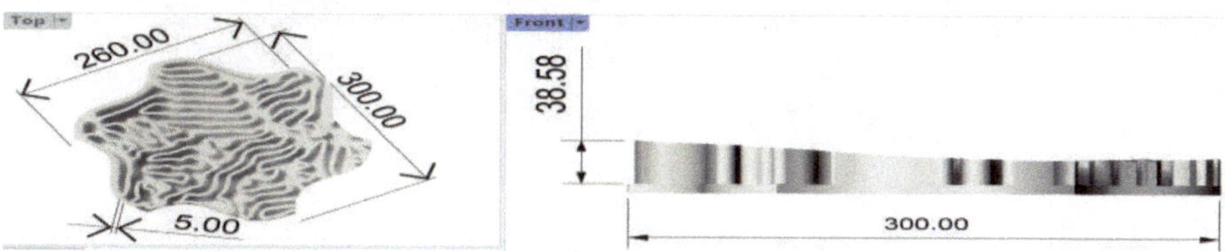

a *b*

Figure 6.10 Dimensions of the 3D model for the top view(a) and front view (b), the dimension in mm.

Table 6.1 The accuracy calculations for the 3D model of the coral panel sample.

The measurement	Dimensional difference %	Observation
The long aspect of the base	2.2 ± 0.05	The deformation after drying is uniform. No warping deformation is observed.
Short aspect of the base	2.3 ± 0.1	
Height of the base	32.5 ± 1.1	The effect of shrinkage and dimensional differences is increased by the permeability in the base of the 3D model caused by the print setting. The surfaces are nicely printed
Width of the coral body	23 ± 1.3	
Hight of the coral body	40.4 ± 3.6	
Corners angle	3.8 ± 0.2	The edges of the corners were pointed.
Curves' curvature	0.9 ± 0.05	Slight damage to the base underside caused by removing it from the print bed.
Flatness	1.4 ± 0.3	

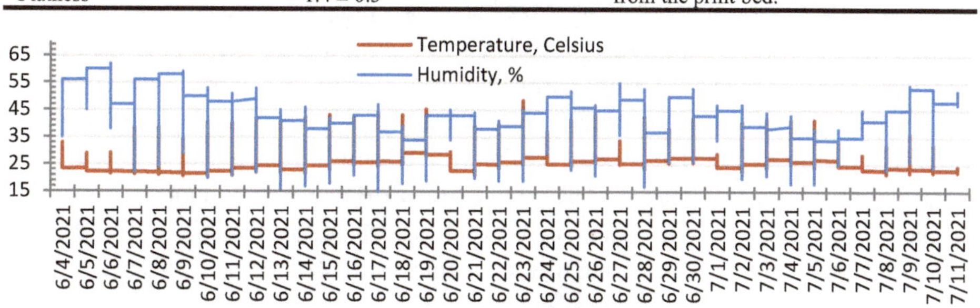

Figure 6.11 Monitoring of the sample room environment from printing day till complete drying.

6.1.2 Tides Panel (Ebb & Flow)

The organic structure of the ebb and flow ripple pattern was the inspiration for the design. This pattern is created by the movement of water and the deposition of sediment on the beach. Details of the design are given in Appendix G1.3. Fig. 6.12 shows the printed part of the geometry, which is the interface between the ebb and flow ripples.

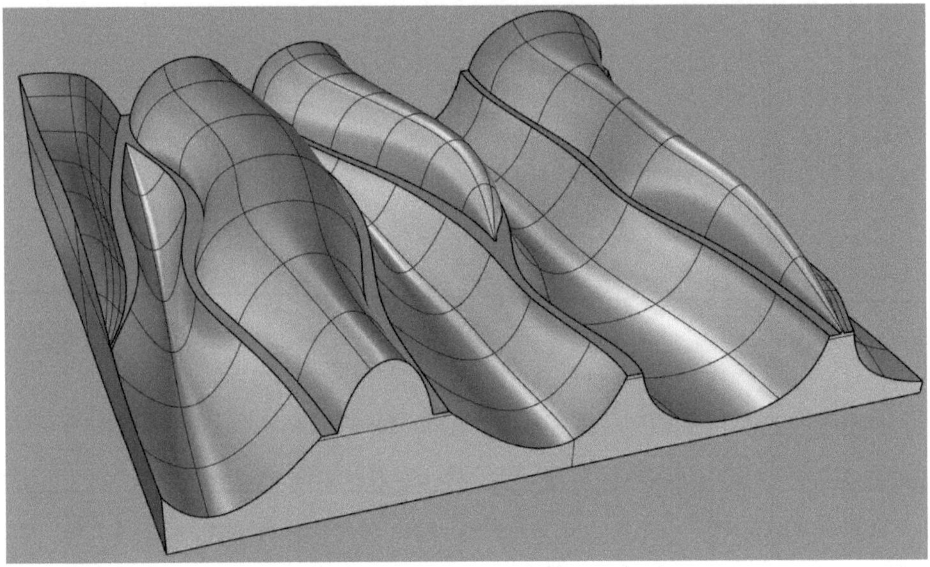

Figure 6.12 The geometry of 3D printed part for the tides panel.

The geometry was printed into two parts. The time taken to complete the print was 6 hours and 27 minutes. The print parameters were chosen as follows: the infill type is 100% rectilinear, the track width is 3 mm, and the nozzle diameter is 1.6 mm. The speeds are 22 mm/s (the circumferential speed, the infill speed, the first layer speed, the gap infill speed, and the max print speed), The layer height is 1 mm, the extrusion width is 300%, the air pressure is 0.33 MPa and the extrusion multiplier is 0.82. The paste type is 13 from Tab. 3.9. The slicing settings are displayed in Fig. 6.13. It's obvious to notice the deformation after drying, comparing the 3D model after complete printing (Fig. 6.14) and after drying (Fig. 6.15). The sample is not aesthetically pleasing due to the material behavior and printing parameters. The sample was dried under environmental conditions as shown in Fig. 6.16. The values of the dimensional measurement in Tab. 6.2 are calculated as an average of at least 6 measurements distributed along the distance.

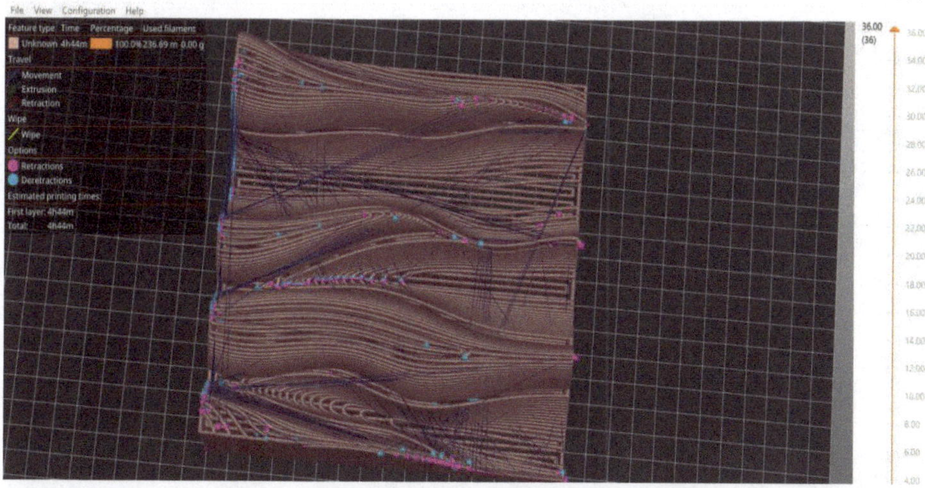

Figure 6.13 3D slice model for the second part of the printing for the tides panel.

Figure 6.14 3D tides model during printing.

Figure 6.15 3D printed tides model after drying.

Table 6.2 The accuracy calculations for the 3D printed model of the tides sample.

The measurement	Dimensional difference %	Observation
Side parallel to ebb flow	3.1 ± 0.1	The geomorphic configuration of the ebb stream contributed to constraining the shrinkage after drying only in the opposite direction. Therefore, the deformation after drying is not uniform on both sides of the square.
Side opposite to ebb flow	6.4 ± 3.5	
Height	37.5 ± 1.8	The surfaces of the specimens are not nicely done due to unbalance printing parameters
Corners angle	6.1 ± 1.7	The edges of the corners were rounded.
Curves' curvature and concave	4.1 ± 1.2	The sample is slightly damaged on the underside of the base when removed from the print bed.
Flatness	1.6 ± 0.8	

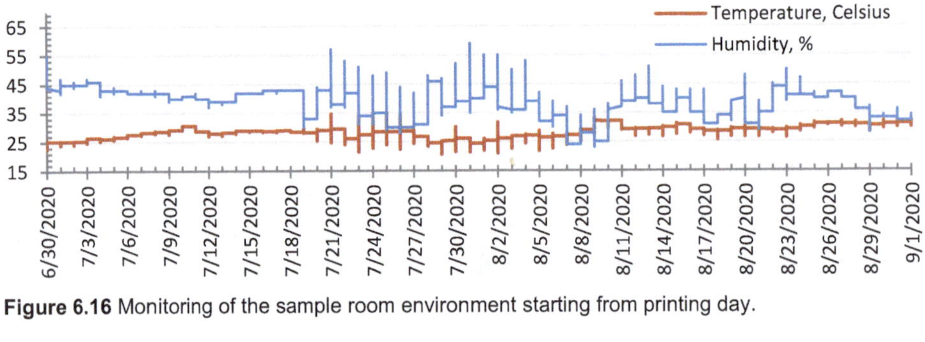

Figure 6.16 Monitoring of the sample room environment starting from printing day.

6.1.3 Gills 3D Model

The inspiration for this parametric design comes from the respiratory organ of fish. The basic idea is to take advantage of the deformation caused by shrinkage and distortion after the model dries. This means that the warping from shrinkage would not affect the aesthetics of the 3D model after drying because the geometry has designed gills that are randomly twisted, as illustrated in Fig. 6.17. The details of the design are pictured in Appendix G1.1.

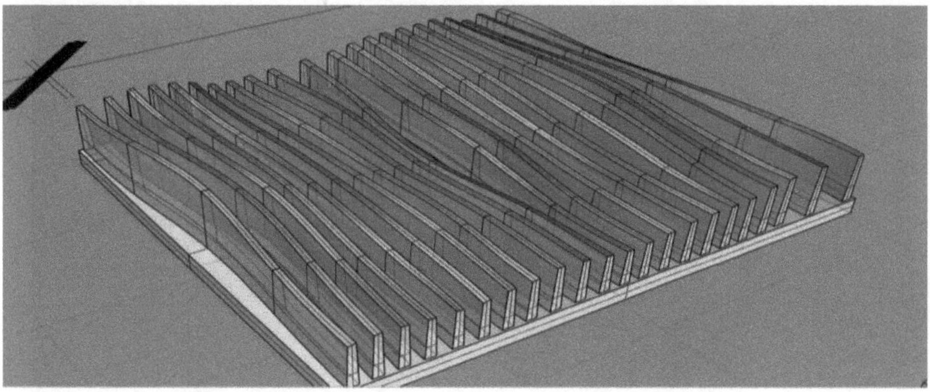

Figure 6.17 Geometry design for acoustic gills.

The printing process of the 3D model is divided into 3 parts. The first part is the base. The gills of the 3D model were printed after one day from printing the base part, in two stages due to refilling. The total printing time was 9 hours and 19 minutes. The printing parameters were chosen as follows: the fill type of the base is 99% concentric, the infill for the gills part is 99% Aligned rectilinear. The gills' part is resized by 90% from the height due to the material used is not reliable to be self-supported material with the available geometrical slenderness. The track width of the base is 3 mm but for the gills parts is 2 mm, and the nozzle diameter is 1.6 mm. For the base parts, the speeds are 25 mm/s (the circumferential speed, the infill speed, the first layer speed, the gap infill speed, and the max print speed), while for the gills of the 3D model, the speeds are 18 mm/s (the circumferential speed, the infill speed, the first layer speed, the gap infill speed, and the max print speed). The layer height is 1mm, the extrusion width is 300%, the air pressure is 0.3 MPa and the extrusion multiplier is 1. The paste type is 8 from Tab. 3.9. The samples were cured under the same conditions as mentioned in paragraph "3.4 Curing process".

Fig. 6.18 shows the 3D printed model during and immediately after complete printing. Fig. 6.19 shows the slicing setting of the 3D model for the parts. The zero level for part two is settled to be attached level to the top surface of the base. The base thickness is printed equal to 5 mm, the gills height is variable due to design variation.

However, the gills walls height seems to have approximately the same height due to material behavior and shrinkage.

a

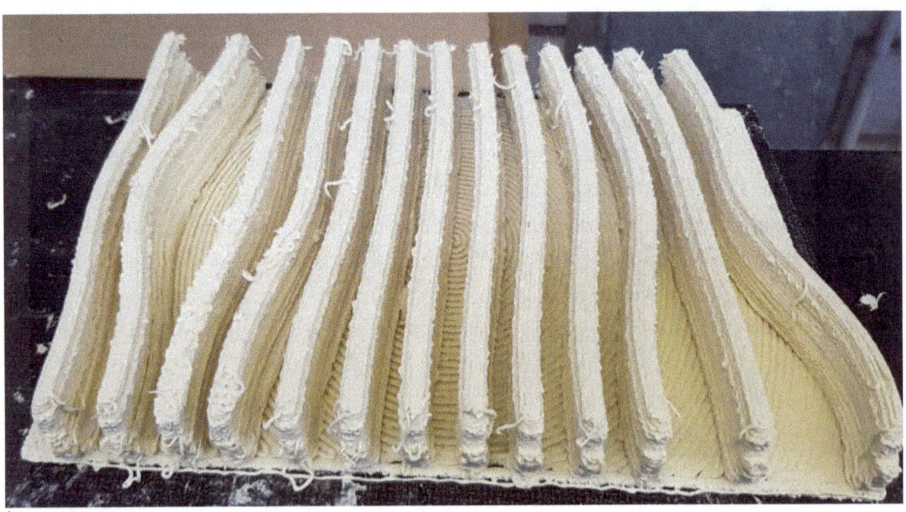

b

Figure 6.18 3D gill acoustic model during printing (a), and after finishing (b).

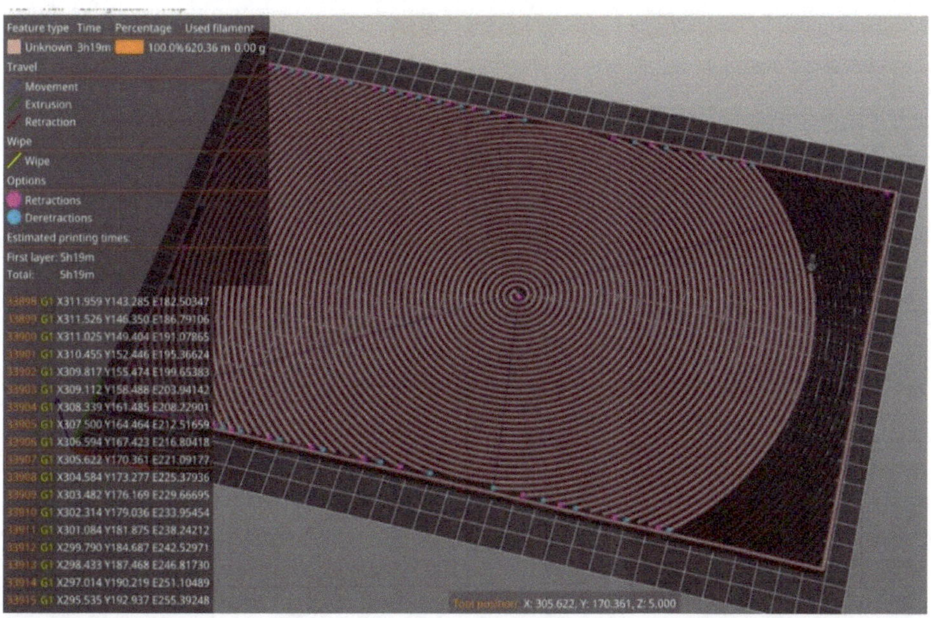

a

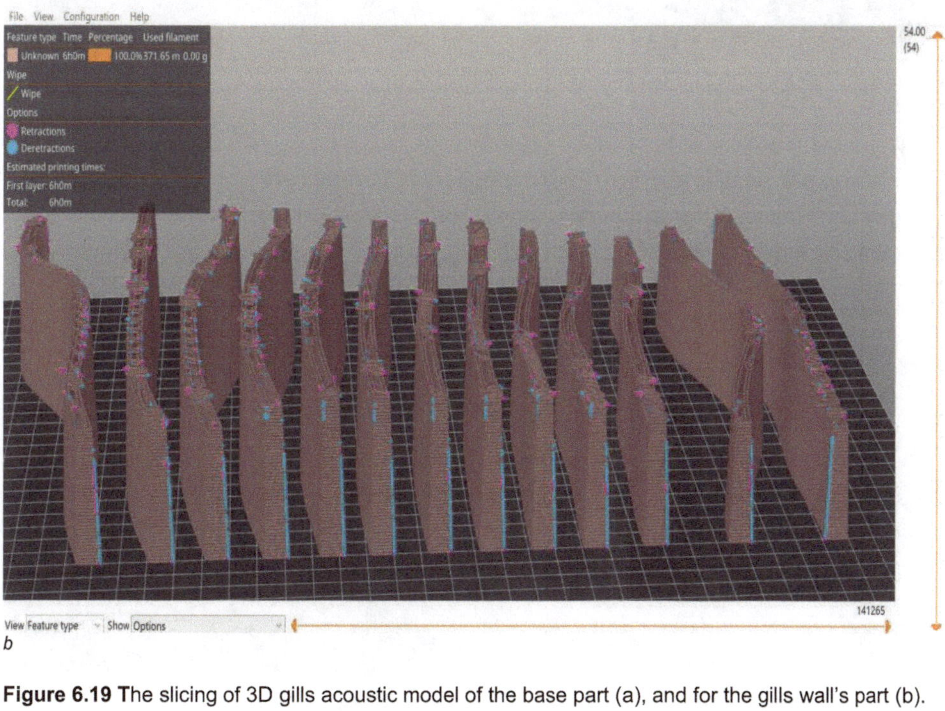

b

Figure 6.19 The slicing of 3D gills acoustic model of the base part (a), and for the gills wall's part (b).

Due to the thin base (5 mm) relative to the specimen volume and the surface area of the base, the model was damaged during removal from the print bed, as shown in Fig. 6.20. This problem has been overcome as shown in section "4.9 Removal from the print bed".

Figure 6.20 3D acoustic gill model after removal from the print bed.

The gill walls are separated and deboned from the attached surface between the gills and the base as shown in Fig. 6.21. This location is the interface between the two printed parts due to shrinkage and not using the curing technique when the break time exceeded hours caused the indicated separation. According to section 3.4 "Curing" this problem could have been avoided.

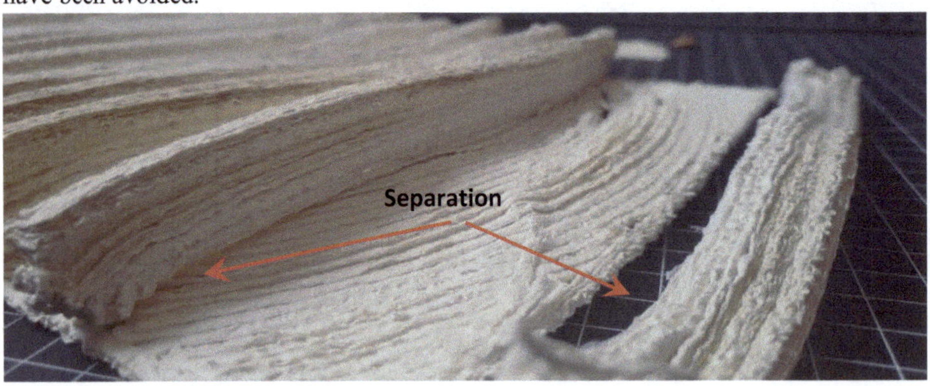

Figure 6.21 Debonding between the base part and gills part of the sample acoustic gills.

The buckling of the gills was caused due to the material characteristics according to the geometric configuration, which failed to support itself while drying, as shown in Fig. 6.22.

Figure 6.22 The buckling of the gills.

The values of the dimensional measurement in Tab. 6.3 are calculated as an average of 6 measurements distributed equally along the distance. The room environment for the specimen was monitored from printing day till complete drying, as shown in Fig. 6.23.

Table 6.3 The accuracy calculations for the 3D model of the acoustic gills sample.

The measurement	Dimensional difference %	Observation
Width of the base	1.1 ± 0.5	Surface topography also gives diverage in the accuracy.
Length of the base	1.32 ± 0.4	Difficult to measure accurately due to damage during detachment from the print bed
Height of the base	26.8 ± 3.6	
Width of the gills	14.1 ± 1	This difference is caused by the print resolution as well as the drying. Here the width dimension is variable.
Height of the gills	45.5 ± 4	Measurements are inaccurate due to buckling.
Length of the gills	1.5 ± 1	Overfilled filaments at the end edges due to time break printing result in more variation in accuracy.
Edges sharpness	5.2 ± 2	
Corners angle	1.2 ± 1	This issue is due to printing parameters
Gill curvatures	2 ± 1	Drying stresses variation according to geometrical configuration
Flatness	5.8 ± 1	Due to thin base and the sticking damaging of the detachment.

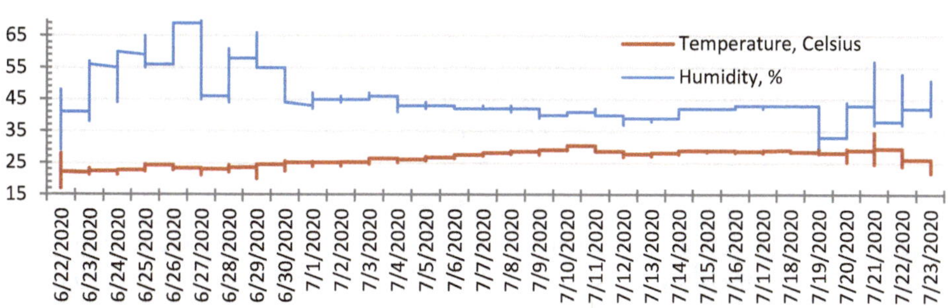

Figure 6.23 Monitoring of the sample room environment from printing day till complete drying.

6.1.4 3D Map Panel

The technical inspiration behind using 3D map-derived acoustic panels is their ability to provide both functional and aesthetic acoustic design benefits by dynamically adjusting acoustic properties based on real-time user preference data. The maps can create a visually appealing environment while also serving a functional purpose by depicting geographical features, cityscapes, or artistic patterns. The sample geometry is pictured in Fig. 6.24.

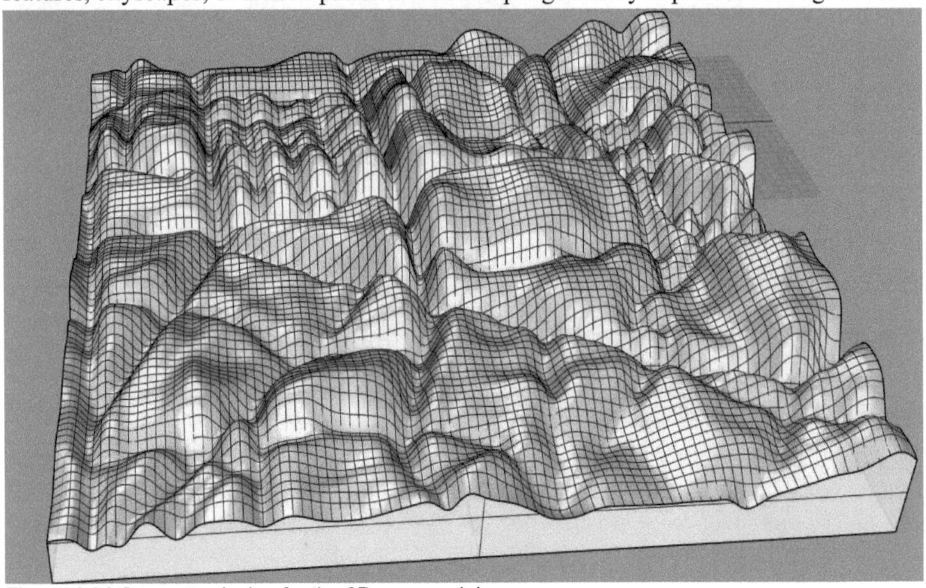

Figure 6.24 Geometry design for the 3D map model.

The geometry was printed completely with a pause. The time taken to complete the print was 6 hours and 24 minutes. The print parameters were selected as follows: the infill type is 100% aligned rectilinear, and the nozzle diameter is 1.6 mm, the speeds are 25 mm/s (the circumferential speed, the infill speed, the first layer speed, the gap infill speed, and the max print speed), while for the gills of the 3D model, the speeds are 18 mm/s (the circumferential speed, the infill speed, the first layer speed, the gap infill speed, and the max print speed). Travel speed is equal to 35mm/s. The layer height is 1 mm, the extrusion width is 300%, the air pressure is 0.32 MPa and the extrusion multiplier is 1.4. The paste type is 15 from Tab. 3.9. The dimension of the geometry is rescaled by the scale (60%, 60%, 100%) of the x, y, and z axis; respectively into (180 x 180 x 53.52) mm. The slicing and the printing movement are shown in Fig. 6.25 and the 3D printed models are pictured in Fig. 6.26 and the dried version in Fig. 6.27. The samples were dried under the surrounding room environment, as shown in Fig. 6.28. Tab. 6.4 displays the dimensional accuracy of the sample.

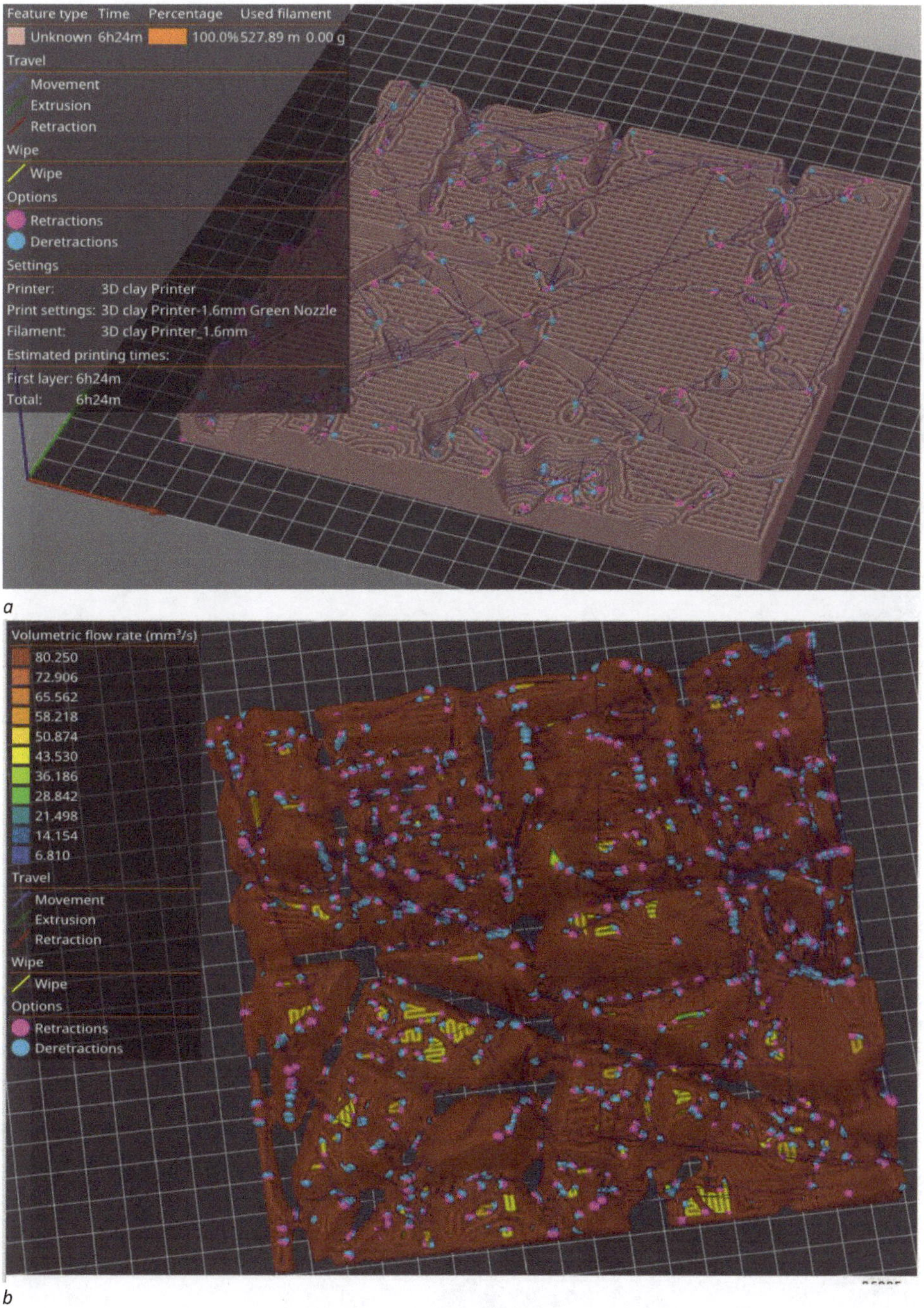

Figure 6.25 Slicing details of the specimen (a) and details of volumetric flow rate (b).

a

b

Figure 6.26 3D printed model during printing (a), and after complete (b).

Figure 6.27 3D printed model after drying.

Table 6.4 The accuracy calculations for the 3D map model.

The measurement	Dimensional difference %	Observation
Side of the square	6.2 ± 1.2	The print and travel speeds were inappropriate, resulting in a messy print. The choice of the nozzle was also the reason for this kind of result. Therefore, a finer nozzle could give more accurate detailing in higher resolution results.
Height	49.9 ± 1.4	
Curves' curvature	Inaccurate measurement in several positions	For the shrinkage results, it depends as well as on the materials used and the type of the paste used.
Corners angle	4.8 ± 0.5	The edges of the corners were rounded.
Layer shifting, as shown in Fig. 4.43	(73.4 ± 5 and 110.4 ± 3) degree	several damages to the underside of the base caused by removing the sample from the print bed.

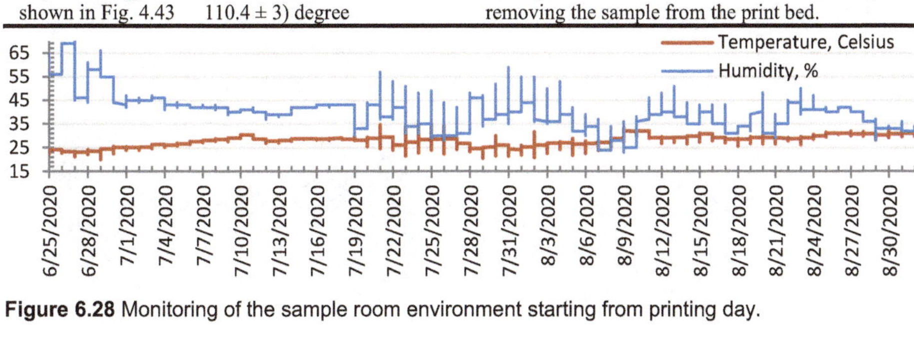

Figure 6.28 Monitoring of the sample room environment starting from printing day.

6.1.5 Triangular Seashell Panel

The insight of the design was from natural forms. The unique contours, patterns, and symmetries of leaves are the basis for creating this intriguing geometry. The geometry is displayed in Fig. 6.29 and the 3D model after printing is shown in Fig. 6.30.

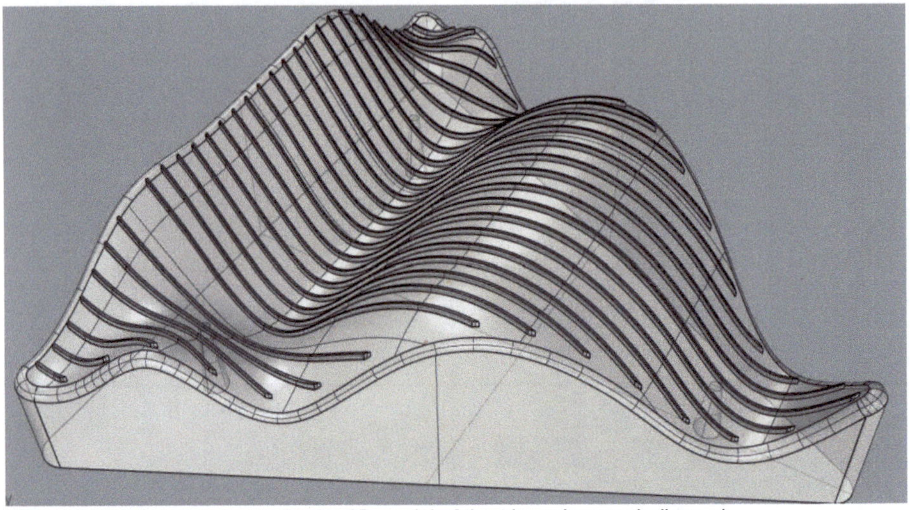

Figure 6.29 Geometry design for the 3D model of the triangular seashell panel.

Figure 6.30 3D printed model of the triangular seashell panel.

The geometry was printed completely on the same day. There was a pause to refill the material container. The print completion time was 8 hours and 27 minutes. The printing parameters were selected as follows: the type of filling is 99% concentric, the width of the track is 2 mm, and the diameter of the nozzle is 1.6 mm. The speeds are 17mm/s (the circumferential speed, the infill speed, the first layer speed, the gap infill speed and the maximum print speed) and the travel speed is 20mm/s. Thin wall detection is disabled, no external parameters. The layer height is 1 mm, the extrusion width is 200%, the air pressure is 0.8 MPa and the extrusion multiplier is 1.4. The paste type is 25 from Tab. 3.9. The samples were dried under ambient conditions, as mentioned in paragraph "3.4 Curing process"., the surrounding room environment is shown in Fig. 6.31. Because of this pause, an error occurred that caused the entire upper part to be displaced due to a mismatch in the origin of the two parts together, as described in Section 4.11.7 Layer shifting. This is shown in Fig. 6.32.

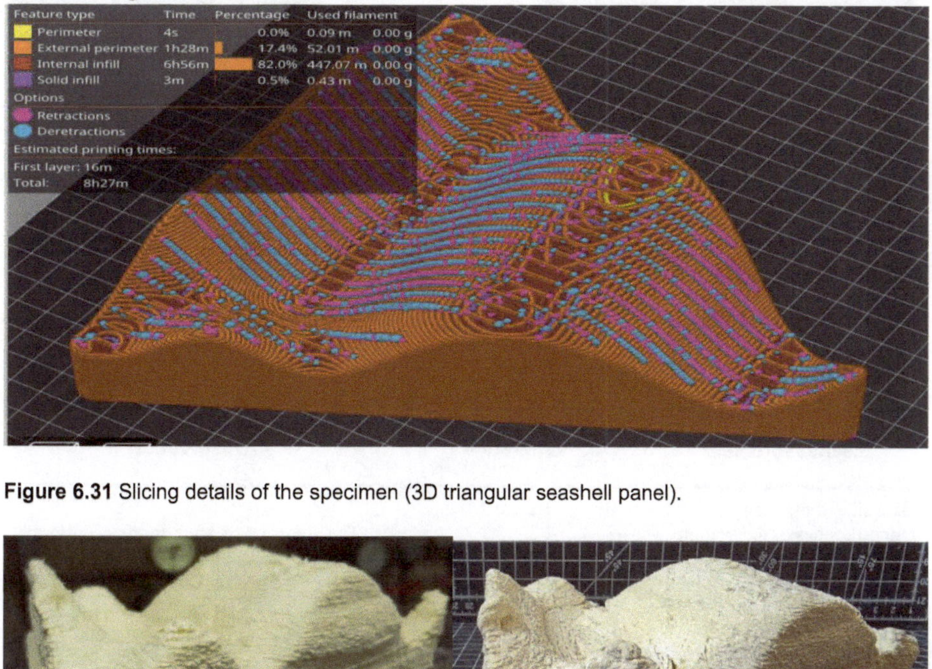

Figure 6.31 Slicing details of the specimen (3D triangular seashell panel).

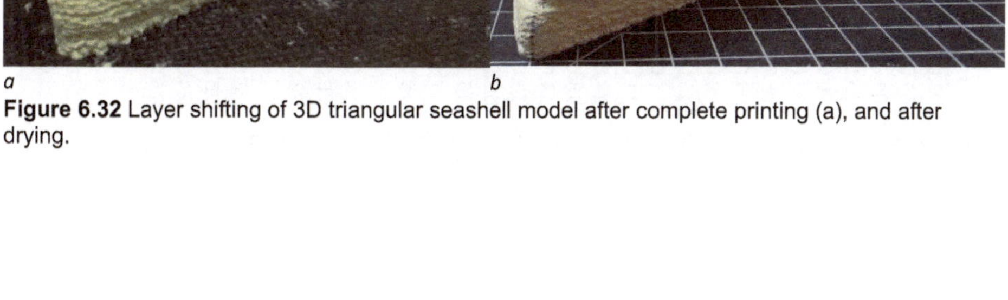

a *b*

Figure 6.32 Layer shifting of 3D triangular seashell model after complete printing (a), and after drying.

The dried sample is shown in Fig. 6.33, it shows that the 3D model is well printed with nice aesthetics and dried without warping or differential shrinkage. The sample has relatively accurate dimensions, as shown in Tab. 6.5. This can be attributed to mixture formulation, the properties of the material, the correct balance of the printing parameters and the slow drying caused by the cold climate. Fig. 6.34 displays the surrounding environment of the sample from printing time to drying.

Figure 6.33 3D triangular seashell shape model after drying.

Table 6.5 The accuracy calculations for the 3d model of the leaf shape panel sample.

The measurement	Dimensional difference %	Observation
Base side of the triangle	1.2 ± 0.2	
Opposite sides of the triangle	1.3 ± 0.3	The deformation after drying is uniform.
Height	18.5 ± 1.1	No warping deformation is observed.
Curves' curvature	0.9 ± 0.05	
Corners angle	3.8 ± 0.2	The edges of the corners were rounded.
Flatness	1.4 ± 0.3	The surfaces are nicely printed

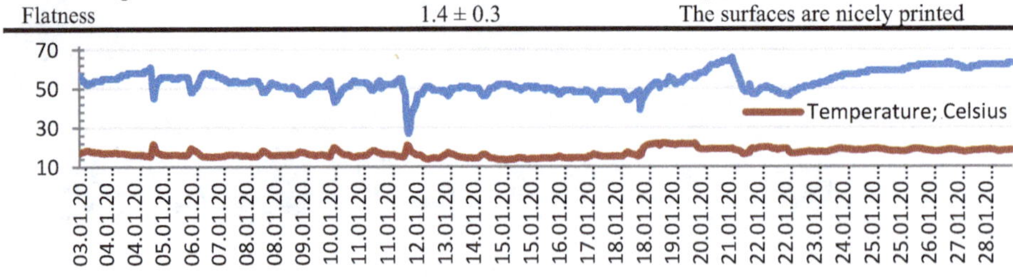

Figure 6.34 Monitoring of the sample room environment starting from printing day.

6.1.6 Bubbles Panel

The inspiration comes from the Nestle Aero Bubbles chocolate. The geometry is shown in Fig. 6.35.

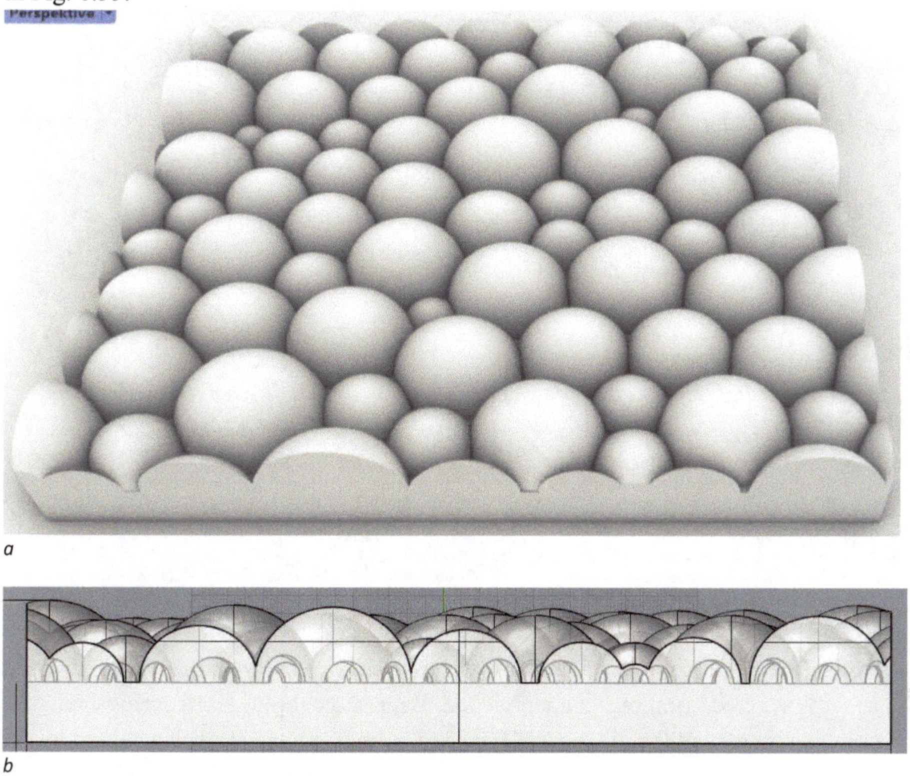

Figure 6.35 Geometry of the bubbles panel, perspective view(a), and for front view (b).

The geometry was printed completely on the same day. The print time was complete in 9 hours and 57 minutes. The base height is just 3 layers printed. The printing parameters were selected as follows: the type of filling is 100% concentric, the width of the track is 2 mm, and the diameter of the nozzle is 1.6 mm. The speeds are 18mm/s (the circumferential speed, the infill speed, the first layer speed, the gap infill speed, and the maximum print speed) and the travel speed is 22mm/s. The layer height is 1 mm, the extrusion width is 200%, the air pressure is 0.8 MPa and the extrusion multiplier is 1.4. The paste type is 27 from Tab. 3.9. The samples were dried under ambient conditions, as mentioned in paragraph "3.4 Curing process". The slicing setting of the bubble sample is shown in Fig. 6.36. The sample after complete printing is shown in Fig. 6.37.

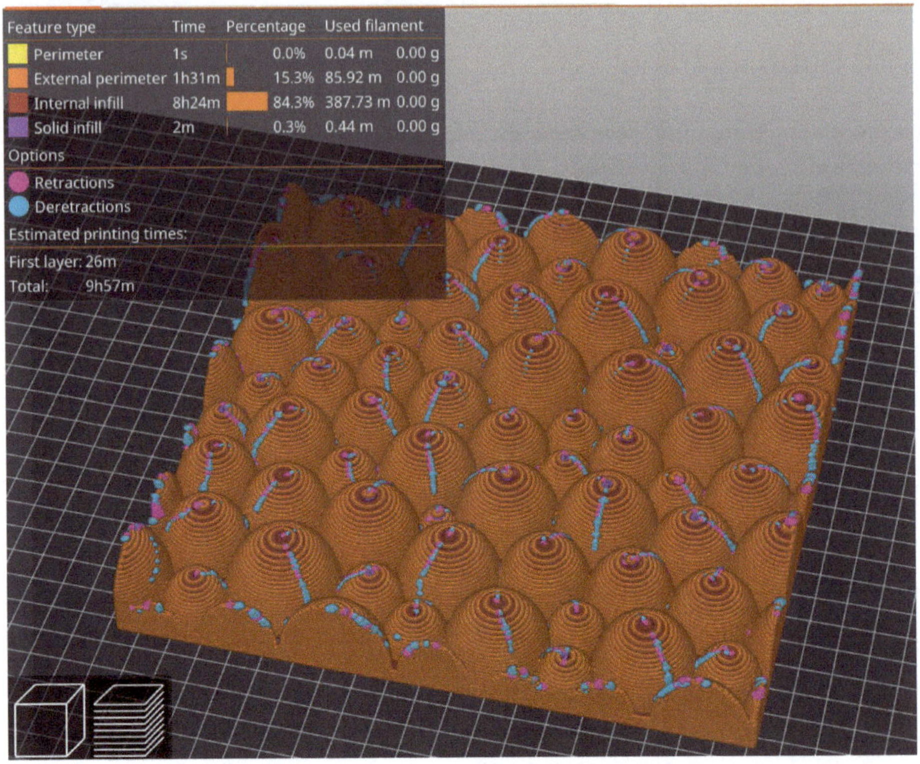

Figure 6.36 Slicing of the bubble sample for first part (a), and second part (b).

The base of the 3D model wasn't damaged when it was removed from the print bed due to the use of a screen-printing plate/plywood with a peel-ply fabric. However, the dimensional accuracy of the 3D model (Tab. 6.6) is deviant due to warping caused by the drying process under warm room curing conditions, as shown in Fig. 6.38. The used material and the type of the mixture is also a reason for this type of response. The sample after drying is displayed in Fig. 6.39.

Table 6.6 The accuracy calculations for the 3d model of bubbles panel sample.

The measurement	Dimensional difference %	Observation
Side square	7.07± 1.07	These percentages are increased when calculating the deformation after drying, due to unevenness of the base. The surfaces are nicely printed. The edges of the corners were rounded.
The other side square	6.96 ± 0.77	
Height	42.8 ± 1.2	
Diameter of the bubbles	5.2 ± 1.6	
Corners angle	1.07 ± 0.02	
Flatness	9.2 ± 4.8	

Figure 6.37 3D printed model of the bubbles panel immediately after complete printing.

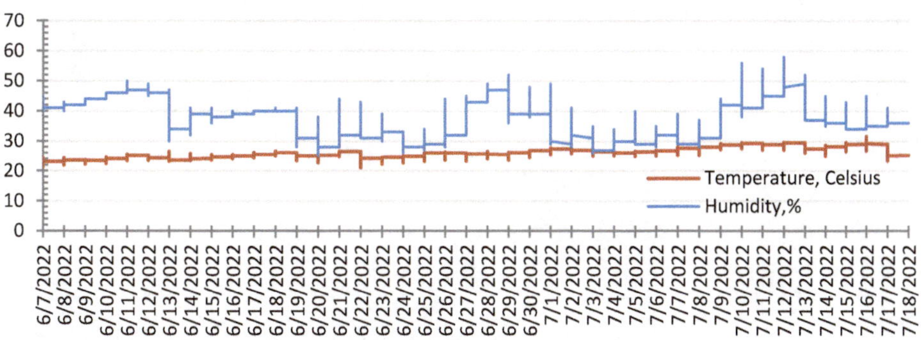

Figure 6.38 Monitoring of the sample room environment starting from printing day.

a

b

Figure 6.39 3D printed model of the bubbles panel after drying, side curvature (a), and top view (b).

6.1.7 Acoustic Reef

Combining functional sound control with visually appealing and nature-inspired designs is the technical inspiration behind the use of acoustic panels derived from natural reefs and Voronoi patterns. It creates a harmonious integration of technology, nature and mathematics in architecture and interiors. The geometry is displayed in Fig. 6.40.

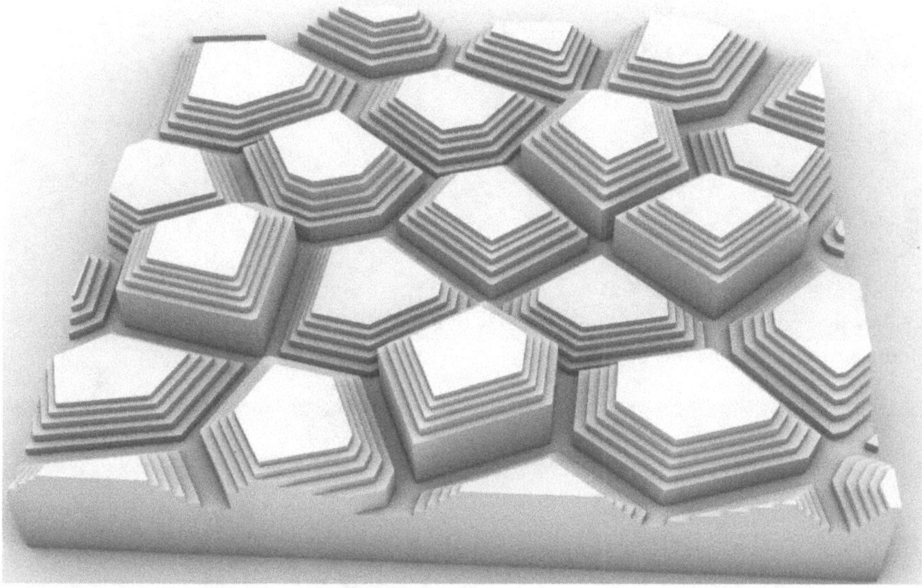

Figure 6.40 The geometry for the 3D acoustic reef panel.

The geometry was printed completely on the same day. The print time was complete in 8 hours and 54 minutes. The base height is just 3 layers printed. The printing parameters were selected as follows: the type of filling is 98% concentric, the width of the track is 2 mm, and the diameter of the nozzle is 1.6 mm. The speeds are 18mm/s (the circumferential speed, the infill speed, the first layer speed, the gap infill speed, and the maximum print speed) and the travel speed is 22mm/s. The layer height is 1 mm, the extrusion width is 200%, the air pressure is 0.8 MPa and the extrusion multiplier is 1.4. The paste type is 27 from Tab. 3.10. The samples were dried under ambient conditions, as mentioned in paragraph "3.4 Curing process". The slicing setting of the reef sample is shown in Fig. 6.41. The sample after complete printing is shown in Fig. 6.42. The dimensional accuracy of the 3D model was illustrated in Tab. 6.7. The sample after drying is displayed in Fig. 6.43 under the environmental conditions shown in Fig. 6.44.

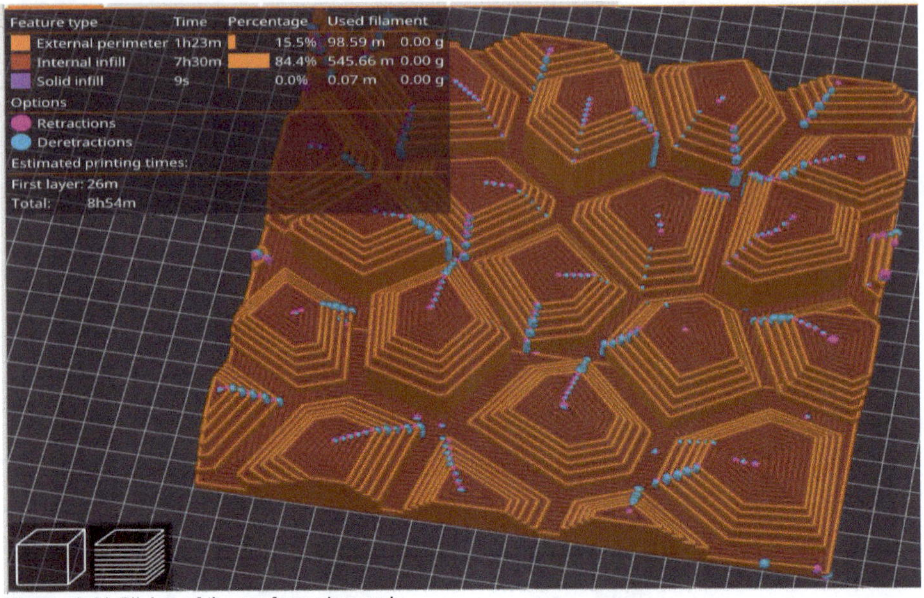

Figure 6.41 Slicing of the reef panel sample.

Figure 6.42 3D printed model of the reef panel immediately after complete printing.

Figure 6.43 3D printed model of the reef panel after drying.

Table 6.7 The accuracy calculations for the 3D model of reef panel sample.

The measurement	Dimensional difference %	Observation
Side square	5.33 ± 0.97	The surfaces are nicely printed. The material is stiff. No debonding is observed
The other side square	5.56 ± 1.7	
Height	32.52 ± 1.7	The reef body is shrinkage due to infill pattern and infill percentage.
Reef body corners	8.5 ± 2	
Corners angle	7.2 ±1.4	The edges of the corners are sharp and pointed.
Flatness of the base	0.7 ± 0.1	The base is easily removed from the print bed.

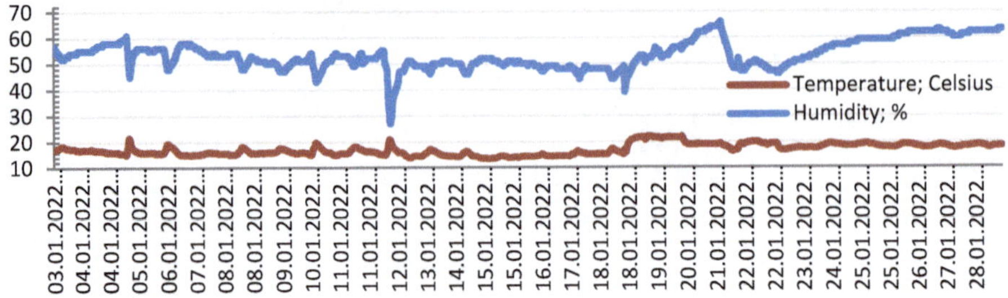

Figure 6.44 Monitoring of the sample room environment starting from printing day.

6.1.8 Inverted Crowned Buffel

This inspiration comes from using buffel as inverted crown. The geometry is displayed in Fig. 6.45.

Figure 6.45 The geometry for the 3D inverted crowned Buffel.

The geometry was printed in four stages in the four sequenced days. The print time was complete in 1 day 6 hours and 51 minutes. The printing parameters were selected as follows: the type of filling is 100% concentric, the width of the track is 2.5 mm, and the diameter of the nozzle is 1.6 mm. The speeds are 17mm/s (the circumferential speed, the infill speed, the first layer speed, the gap infill speed, and the maximum print speed) and the travel speed is 20 mm/s. The layer height is 1 mm, the extrusion width is 250%, the air pressure is 0.8 MPa and the extrusion multiplier is 1.4. The slicing setting of the sample is shown in Fig. 6.46. The paste type is 30 from Tab. 3.10. The sample during and after complete printing are shown in Fig. 6.47, 6.48. The dimensional accuracy of the sample indicates that this sample gives the best results, as demonstrated in Tab. 6.8. The sample was dried under surrounding room conditions, as mentioned in paragraph "3.4 Curing process", as shown in Fig. 6.49. The dried sample is shown in Fig 6.50.

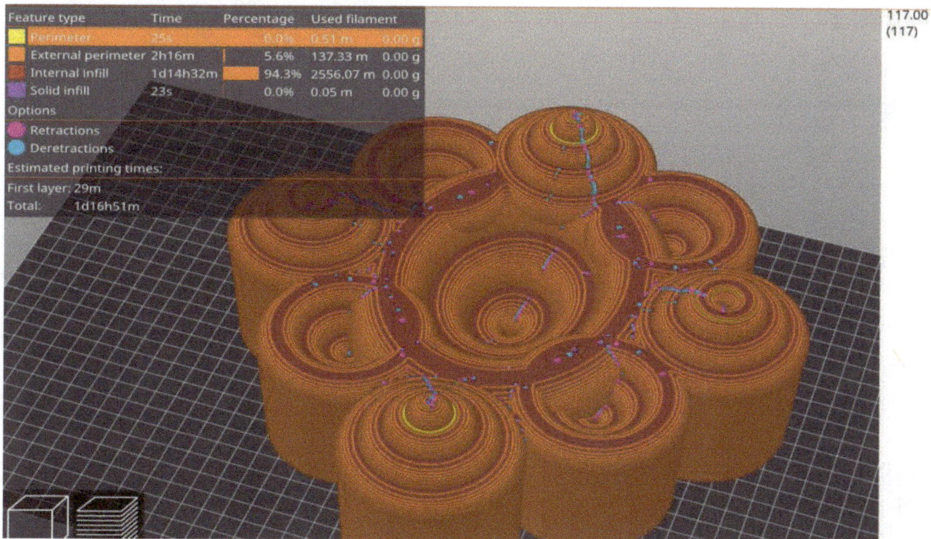

Figure 6.46 Slicing of the inverted crowned buffel sample

Figure 6.47 Balanced printing parameters for the 3D crowned buffel model.

a

b

Figure 6.48 3D printed model of the 3D crowned buffel while printing (a) and after complete(b).

Table 6.8 The accuracy calculations for the 3D crowned buffel model.

The measurement	Dimensional difference %	Observation
The long side	2.5 ± 0.3	The surfaces are nicely printed, and the material is stiff. No warping deformation is observed
The short side	2.6 ± 0.2	
The diagonal	1.2 ± 0.1	The shrinkage is limited and uniform
Height	14.8 ± 1.5	No cracks happened. No debonding occurs
Curves' curvature	0.5 ± 0.1	
Flatness of the base	0.5 ± 0.1	The base is a bit damaged due to sticking on the print bed.

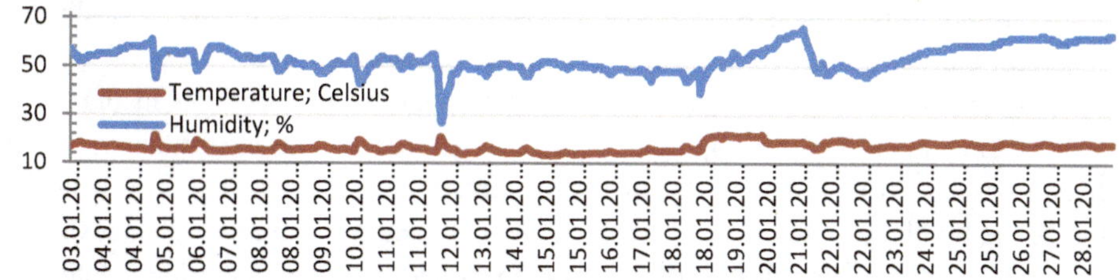

Figure 6.49 Monitoring of the sample room environment starting from printing day.

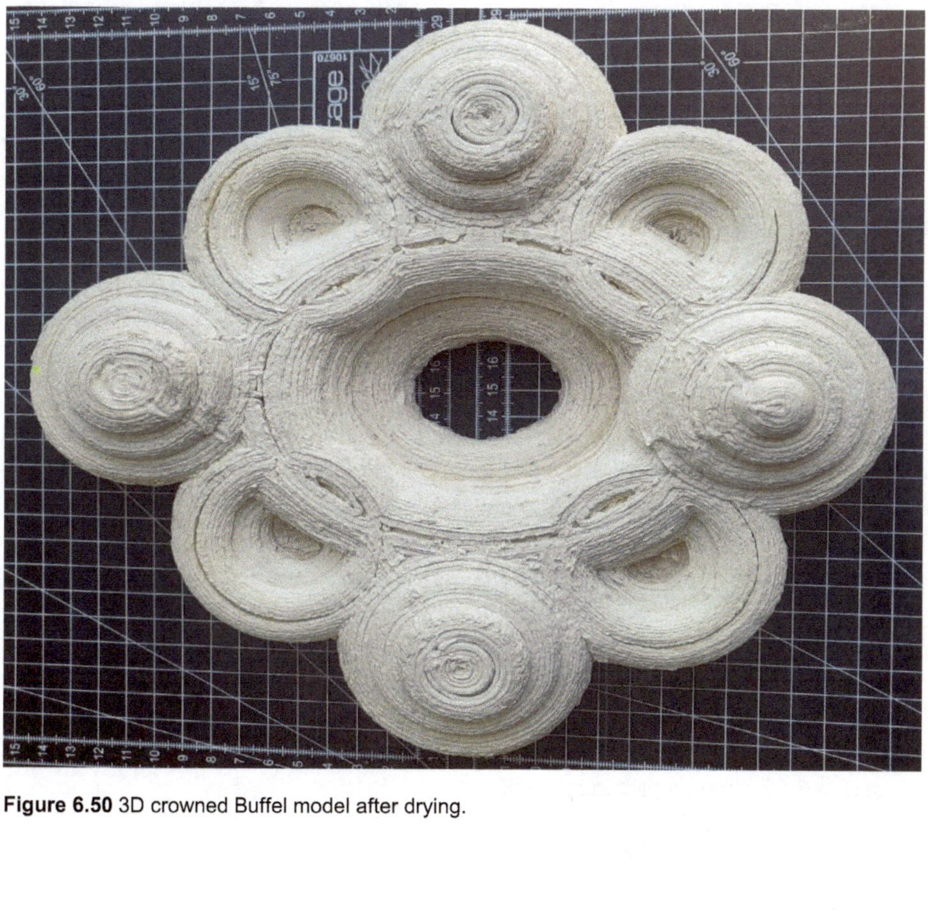

Figure 6.50 3D crowned Buffel model after drying.

6.2 Sunshade 3D Models

The material properties are essential purpose to define the benefit of adopting paper material as an application for solar shading, as it is an inert material for heat transfer. Specifically, the paper material is light, so adding it to the structure would not have an influential load. In this way, it's advantageous to use it as a solar shading panel inside the Insulated Glass Unit (IGU) windows. The insulation is also ideal for protecting the paper material from direct contact in the event of a fire. The final chosen design that was printed was optimised using Rhino & Grasshopper with Ladybug tools for solar shading by measuring the optimum view of the total radiant energy interference taking into consideration the optimum design to maximise the entrance of sunlight in the winter season and block it as much as possible in the summer season. This means that the design achieves passive solar design works; that is, benefits the most from sunlight in winter to increase the building's internal heat energy while reducing it in summer. This design has also achieved the creation to benefit the most from additive manufacturing technology. The calculations of radiation inside the building should only be in the critical part according to winter and summer sunlight. The details of the solar radiation analysis are illustrated in Appendix G.2, it was intinated as comparing the level of translucency, simulation for sun path and variation of the data of critical weather of (winter-summer) and daytime by using location specific weather data of Frankfurt am Main, Germany. These analyses were carried out to optimise the design of these shading panels.

6.2.1 Cell Shade Panel

The design is inspired from the cell form of the human. The geometry is shown in Fig. 6.51. The 3D model is printed twice from two different pastes. The printing parameters are different for each printed model. For both 3D models, the speeds are 20 mm/s the same (the circumferential speed, the infill speed, the first layer speed, the gap infill speed, and the max print speed), the fill pattern is 100% rectilinear for both models, and the nozzle diameter is 1.6 mm, layer height is equal to 1 mm and the extrusion width is 300%. The total printing time was 5 hours and 15 minutes for model with long fibers and 3 hours and 35 minutes for model with short fibers. The different between the printing parameter of the two samples are i) the air pressure is 0.4 MPa, the flow rate is 1.5, and the paste used is 11 for the model with long fibers, while ii) for the sample with short fibers, the air pressure is 0.3 MPa, the flow rate is 1, and the paste used is 12. The samples were cured under the same conditions as mentioned in paragraph 3.4 Curing process. The design geometry of the 3D model is shown in Fig 6.51. The slicing and printed models are shown in Fig. 6.52 and 6.53. The dried samples are pictured in Fig. 6.54 and 6.55. A magnified image (Fig 6.56) shows a printing defect.

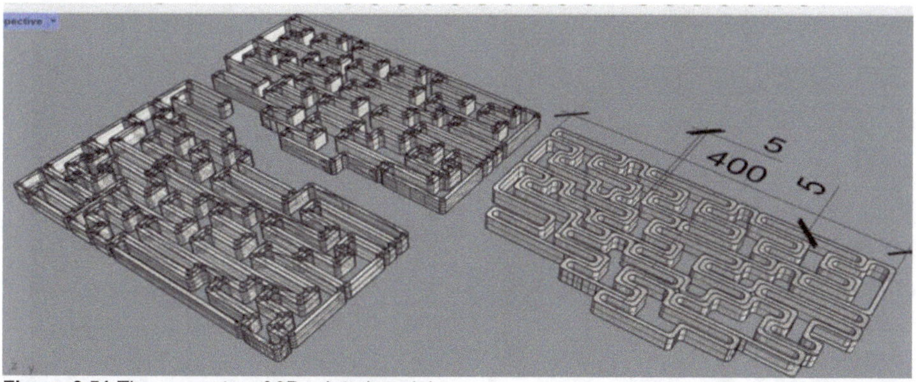

Figure 6.51 The geometry of 3D printed models.

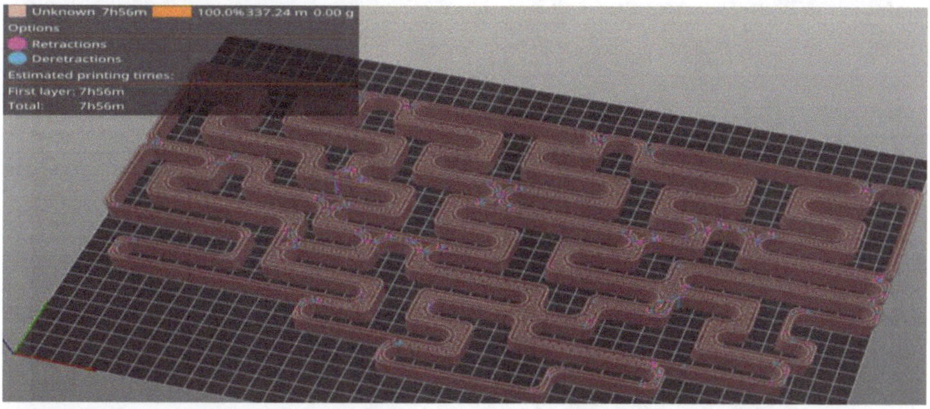

a

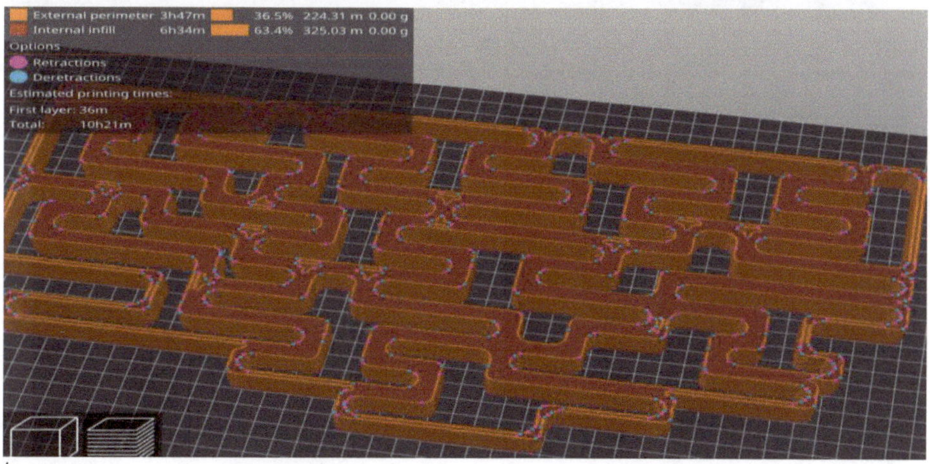

b

Figure 6.52 The slicing setting of the cell shade panel for short cellulose (a), and long cellulose (b).

a

b

Figure 6.53 The 3D printing of the cell shade panel for short fibers cellulose(a) and long fibers(b).

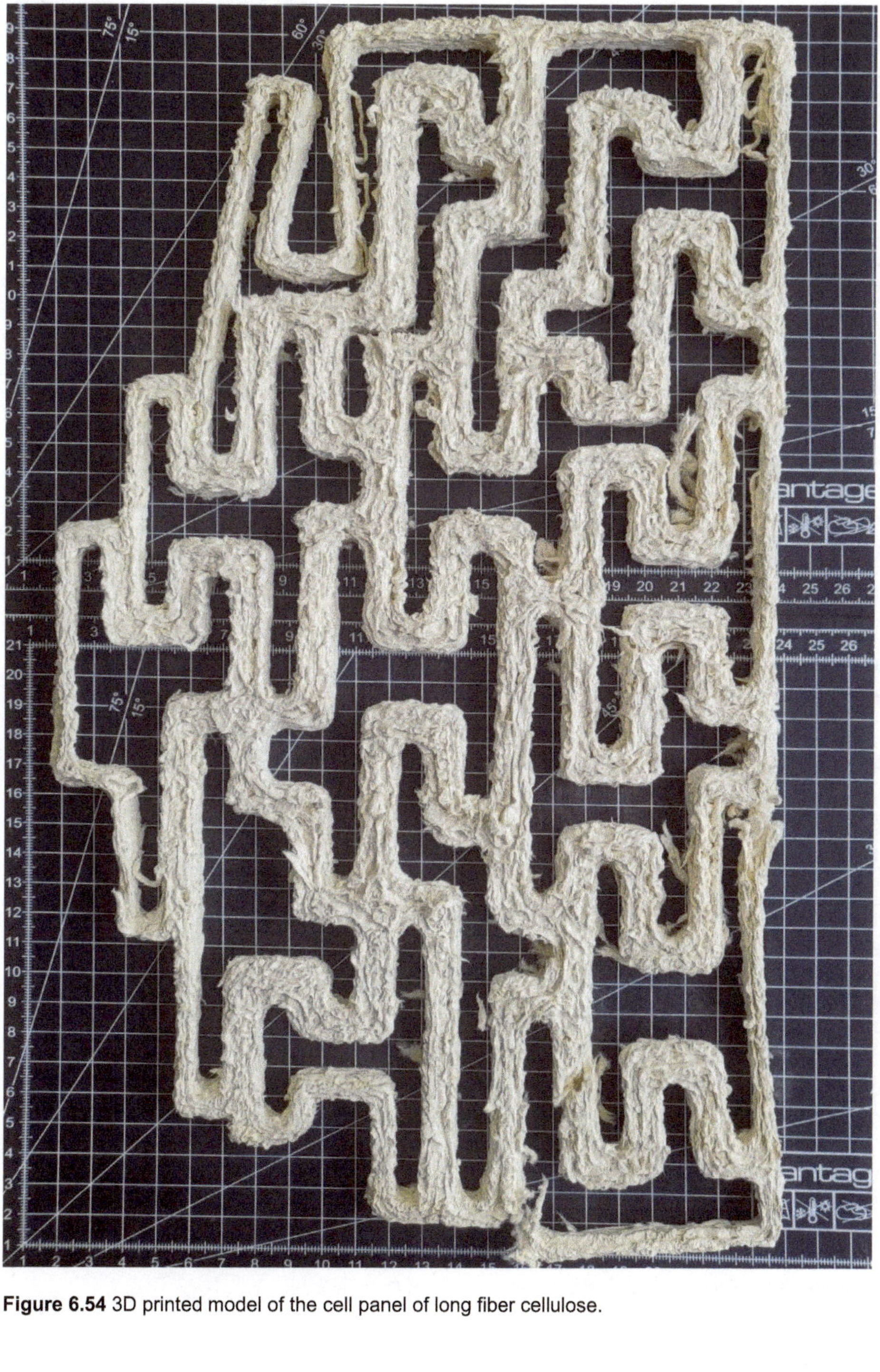

Figure 6.54 3D printed model of the cell panel of long fiber cellulose.

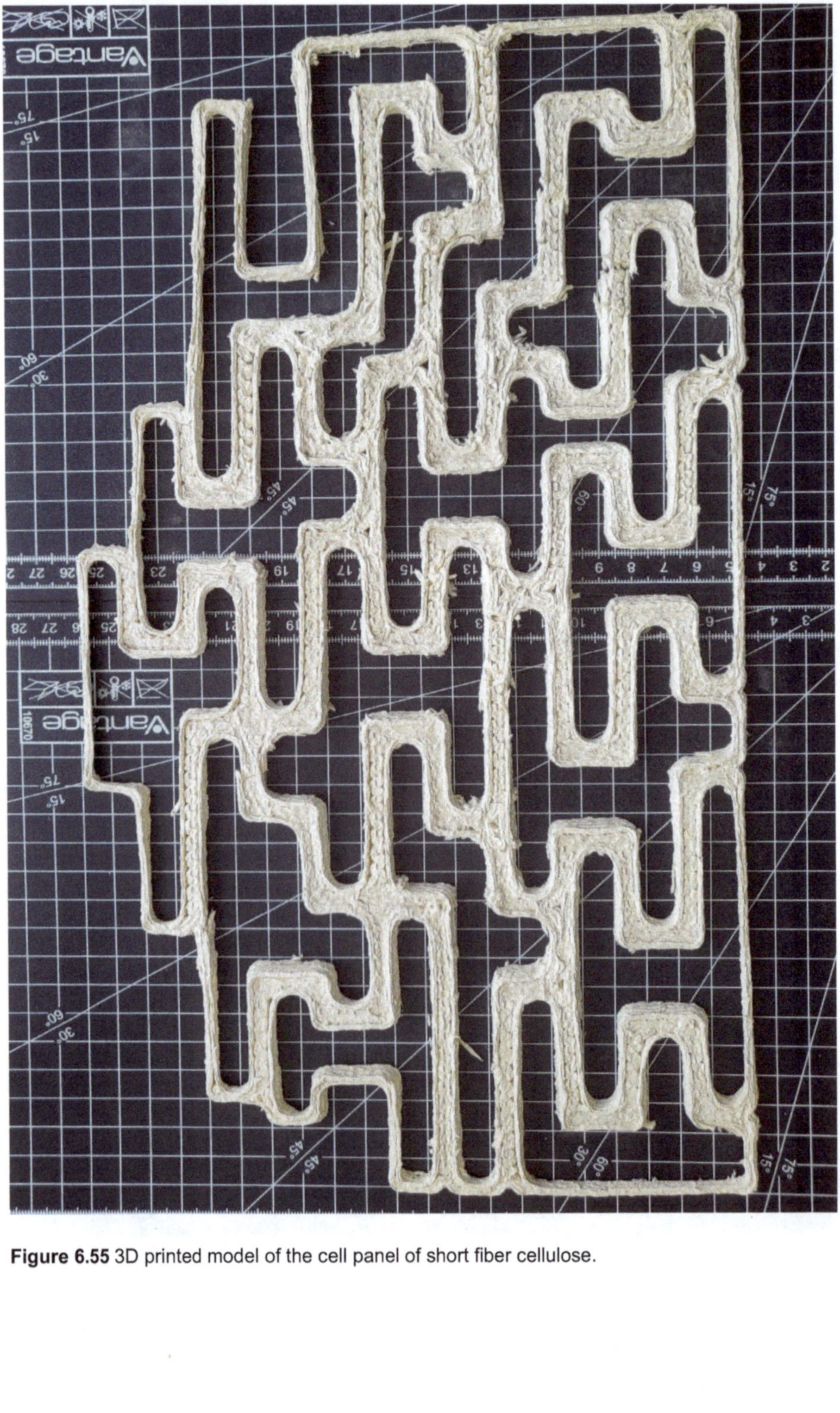

Figure 6.55 3D printed model of the cell panel of short fiber cellulose.

Tables 6.9 und 6.10 shows the dimensional accuracy for both samples. The sample with long cellulose fibers was printed at the beginning of June, from Fig. 6.57, and the sample with short cellulose fibers was printed on 10 June, from Fig. 6.57, which shows the surrounding environmental conditions of the samples.

Figure 6.56 Unbalanced printing parameters for the cell panel with long fibers (zik zak defect).

Table 6.9 The accuracy calculations for the 3d model of cell panel with long cellulose.

The measurement	Dimensional difference %	Observation
Length	1.1 ± 0.4	The printing parameters are not balanced, so the surface topography is rough and uneven.
Width	2.6 ± 0.8	
Height	21.5 ±0.9	The material is brittle. As a result, the sample was broken, and a part was torn in several places.
Corners angle	0.6 ± 0.1	Due to the unbalanced printing parameters, the corner and other dimensions are bulky and inaccurate to measure.
Flatness	0	

Table 6.10 The accuracy calculations for the 3d model of cell panel with short cellulose.

The measurement	Dimensional difference %	Observation
Length	1.2 ± 0.5	The printing parameters are well adjusted compared to the same geometry printed with long cellulose. Shrinkage was greater than the other sample. All corners were printed correctly and weren't affected by the shrinkage.
Width	4.5± 0.7	
Height	46.8 ± 1.7	
Corners angle	0	
Flatness	0	

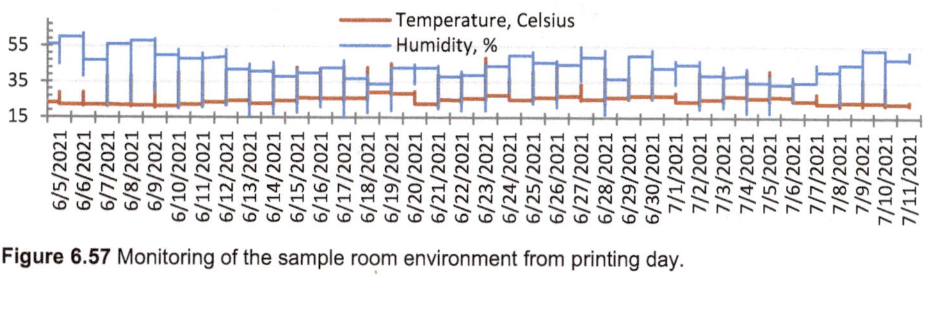

Figure 6.57 Monitoring of the sample room environment from printing day.

6.2.2 Shadow from Floral Leaf Buffel

The concept of the design is to combine sound acoustics and light shading for a special case where the roof is open to the sunlight, to create an integrated system that enhances both the auditory and visual experience while maintaining comfort and functionality. The details about the design and its functionality are listed in Appendix G2.6. The geometry of the model is shown in Fig. 6.58. The 3D model is divided into two parts that are printed separately. A connection lock is provided between the two parts.

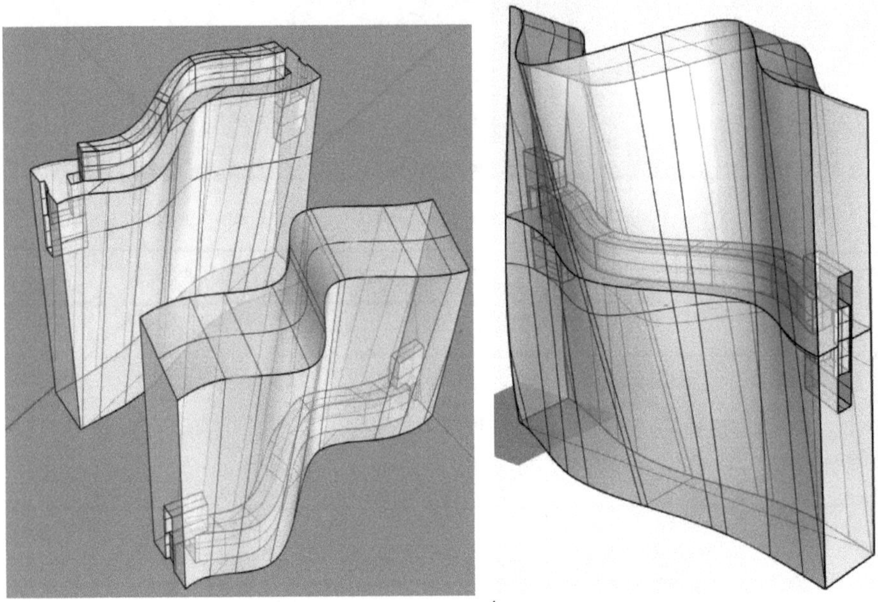

a *b*

Figure 6.58 The geometry of 3D printed models for the floral leaf Buffel (a), and when the two parts together linked by the connection lock (b).

The geometry of the two parts connecting is pictured in Fig. 6.58(b). The slicing and printing of the 3D model is shown in Figures 6.59 and 6.60. The dried samples are pictured in Fig. 6.61. Fig. 6.62 shows the details of interlocking of the dried sample. The printing parameters were selected as follows: the type of filling is 100% rectilinear, the width of the track is 2.5 mm, and the diameter of the nozzle is 1.6 mm. The speeds are 17mm/s (the circumferential speed, the infill speed, the first layer speed, the gap infill speed, and the maximum print speed) and the travel speed is 20 mm/s. The layer height is 1 mm, the extrusion width is 250%, the air pressure is 0.8 MPa and the extrusion multiplier is 1.4. The paste type is 27 from Tab. 3.10.

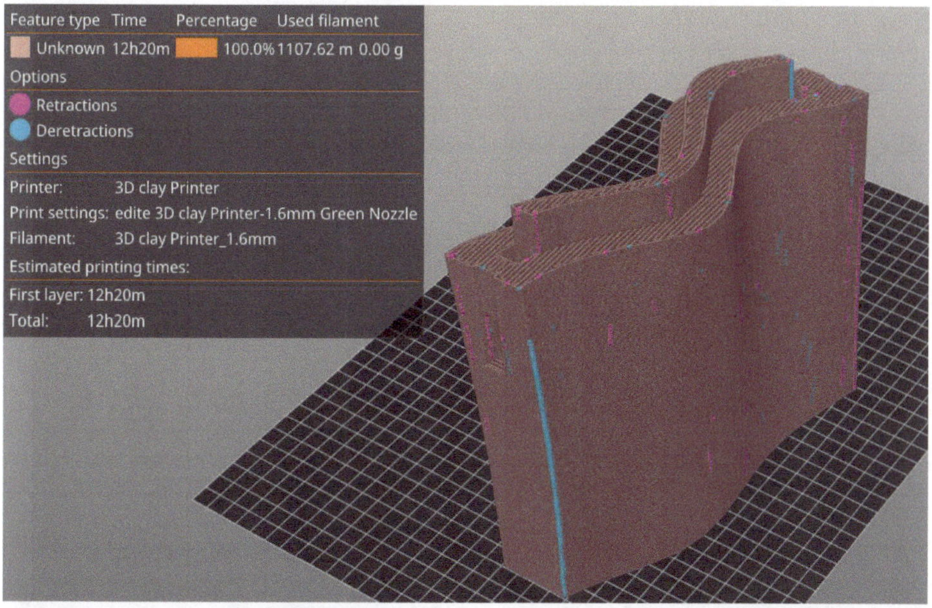

a

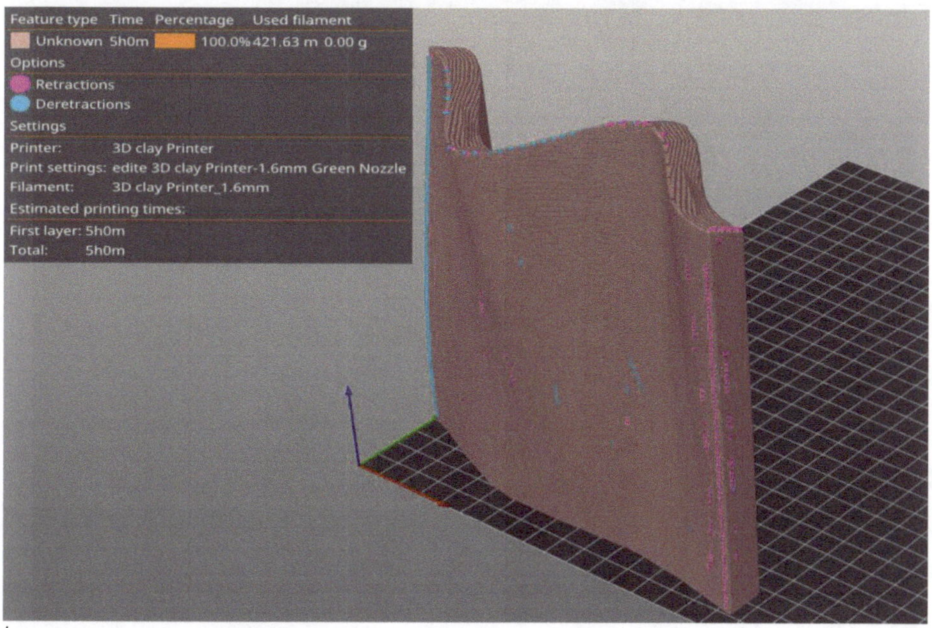

b

Figure 6.59 Slicing setting for the model for lower part(a) and for the upper part(b).

a

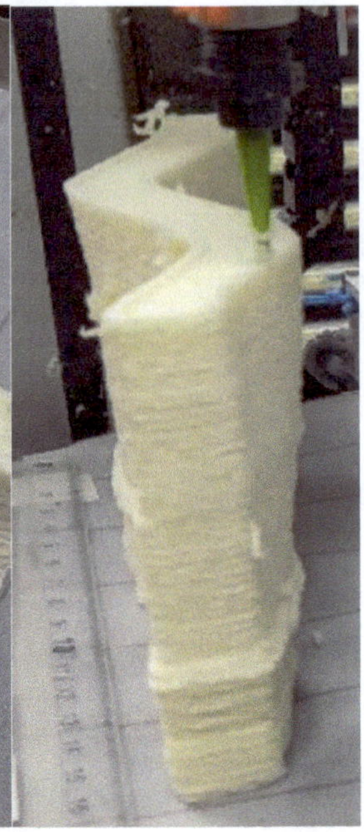

b

a

Figure 6.60 3D printed model while printing.

a

b

Figure 6.61 3D printed model after drying for the lower part (a) and for upper part (b).

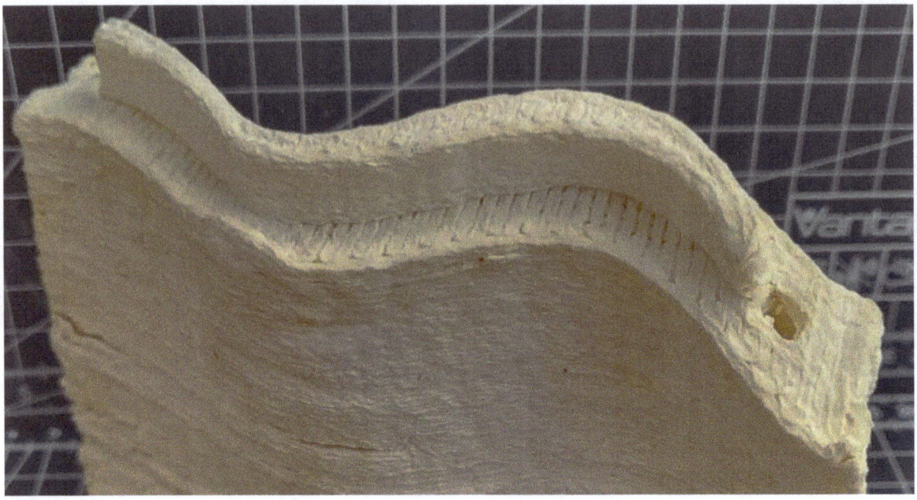

Figure 6.62 Top view of the lower part for the floral leaf model.

The dimensional accuracy of the sample was indicated in Tab. 6.11. The samples were dried under surrounding room conditions, as mentioned in paragraph 3.4 Curing process, as shown in Fig. 6.63.

Table 6.11 The accuracy calculations for the 3d model of floral leaf sample.

The measurement	Dimensional difference %	Observation
Length Project	1 ± 0.2	The surfaces are nicely done with good balanced printing parameters
width	1.1 ± 0.3	
Height	36.11 ± 1.5	There are creaks and debonding between the layer along the height due to shrinkage.
Corners angle	1.3 ± 0.2	The edges are sharped and pointed
Flatness	0	

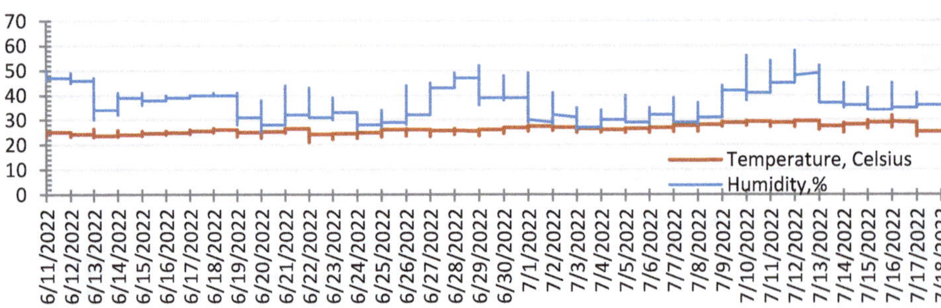

Figure 6.63 Monitoring of the sample room environment from printing day till complete drying.

6.2.3 Shading Tree Wall

The tree-inspired geometry with a Voronoi pattern for a shading wall is the principal idea of this design. The basic geometry for the 3D printed part is shown in Fig. 6.64.

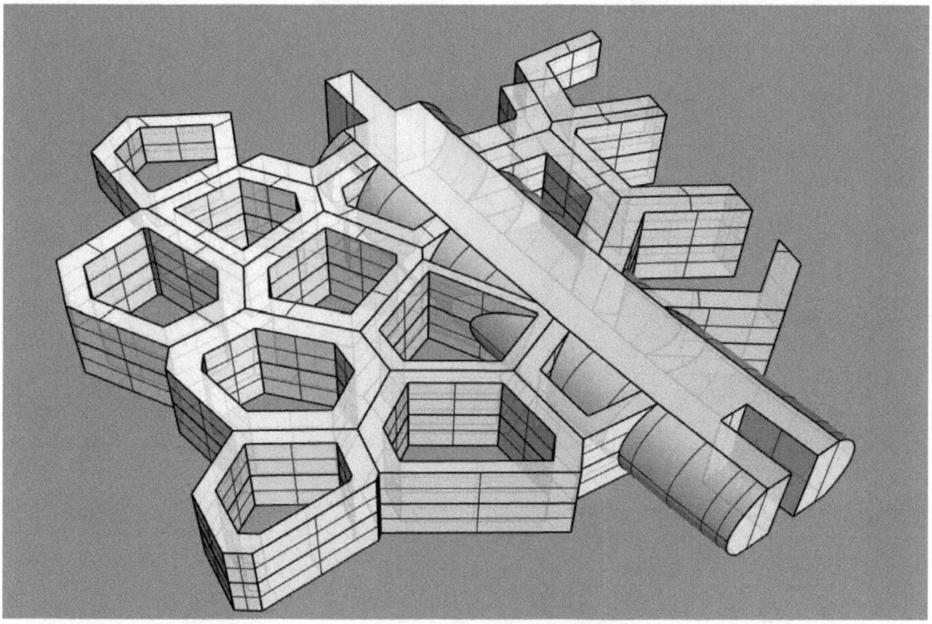

Figure 6.64 The geometry of 3D printed model for the tree wall sample.

The geometry was printed in four stages. The top part is printed differently from the other lower parts. The print time was complete in 15 hours and 51 minutes. The printing parameters for all parts except the top parts were selected as follows: the type of filling is 99% concentric, the width of the track is 3 mm, and the diameter of the nozzle is 1.6 mm. The speeds are 20mm/s (the circumferential speed, the infill speed, the first layer speed, the gap infill speed, and the maximum print speed) and the travel speed is 20 mm/s. The layer height is 1 mm, the extrusion width is 300%, the air pressure is 0.35 MPa and the extrusion multiplier is 1. The paste type is 13 from Tab. 3.9. For the top part only the type of infill and percentage is changed. The infill of the top part is 80% aligned rectilinear. The slicing setting of the sample is shown in Fig. 6.65. The sample during and after complete printing is shown in Figures 6.66, 6.67 and 6.68. The samples were dried under surrounding room conditions, as mentioned in paragraph "3.4 Curing process", as shown in Fig. 6.69. The dimensional accuracy of the sample is demonstrated in Tab. 6.12.

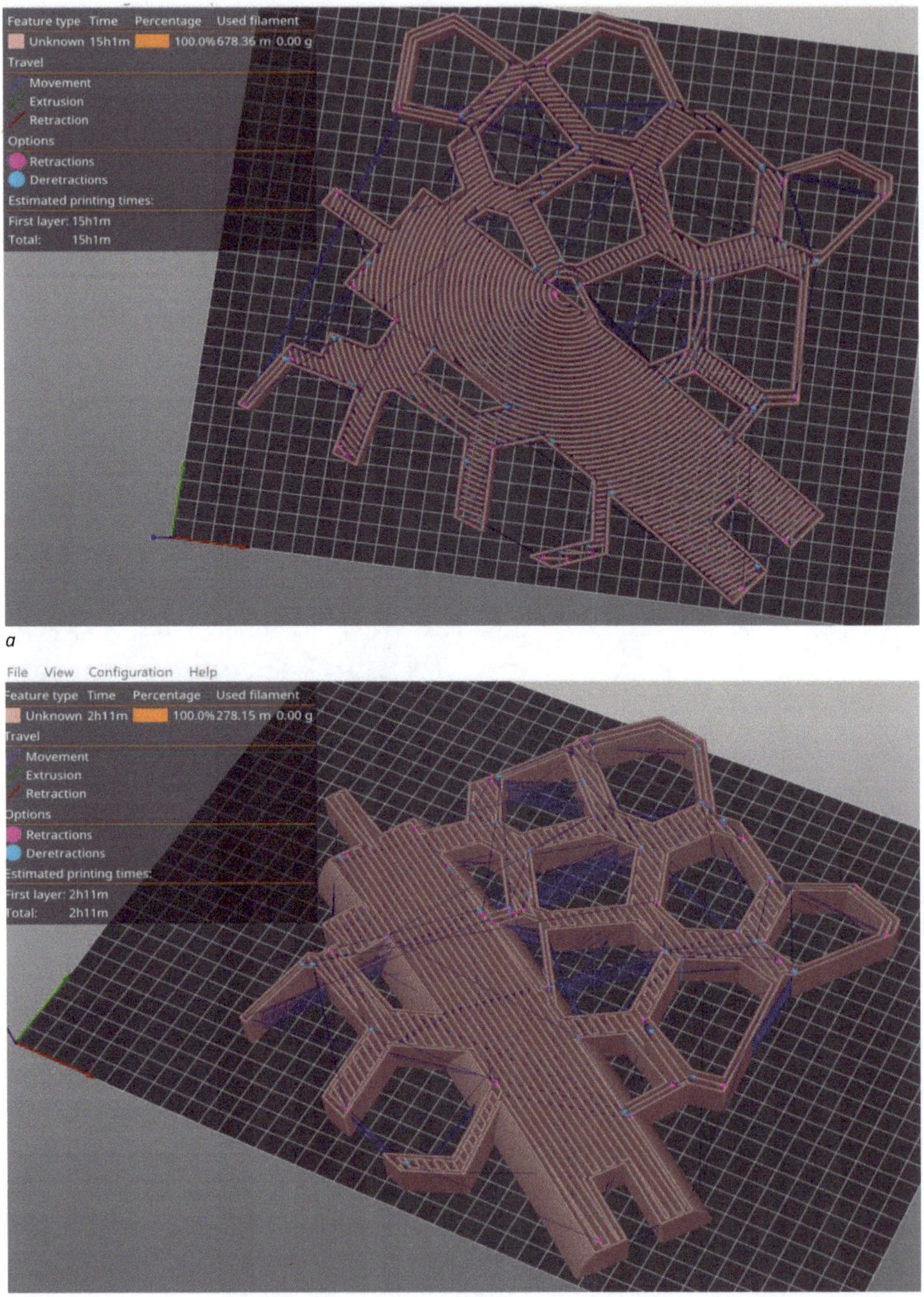

Figure 6.65 Slicing setting for lower part(a) and for the top part(b).

Figure 6.66 Shading tree wall sample during printing.

Figure 6.67 Side view for the tree wall sample after drying.

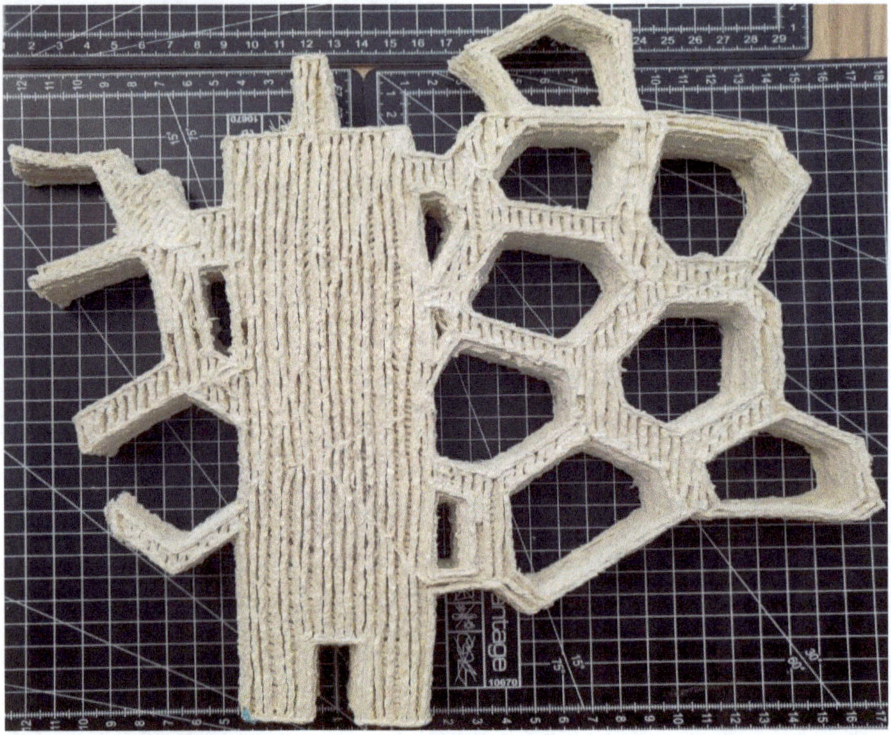

Figure 6.68 Top view for the tree wall sample after drying.

Table 6.12 The accuracy calculations for the 3D model of tree shade wall.

The measurement	Dimensional difference %	Observation
Tree trunk dimensions	6.7 ± 1.1	The geomorphic configuration of the 3D model helps to increase the shrinkage due to the 80% infill. The edges of the corners were rounded. The error in section 4.11.3 Zits & blobs causing dimensional deviation.
Tree branches	7.4 ± 3.5	
Height	40.5 ± 1.8	
Corners angle	6.1 ± 1.7	
Flatness	1.6 ± 0.8	

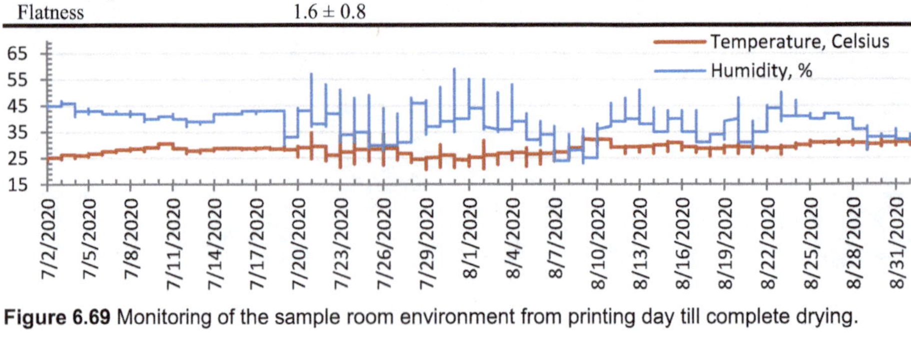

Figure 6.69 Monitoring of the sample room environment from printing day till complete drying.

6.2.4 Honeycomb Panel

The inspiration for the honeycomb shade design comes from the efficient and functional structure of honeycomb cells found in beehives. Honeycomb structures are made up of hexagonal cells that are interconnected to form a strong and lightweight framework. The geometry of the design is shown in Fig. 6.70.

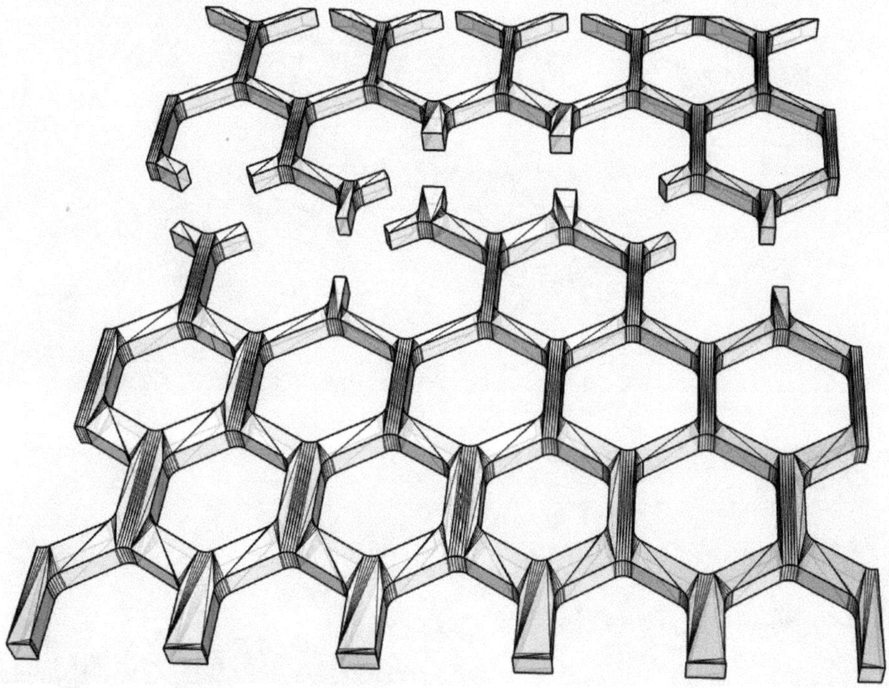

Figure 6.70 The geometry of the honeycomb sample.

The geometry was printed in 2 hours and 16 minutes. The printing parameters were selected as follows: the type of filling is 100% rectilinear, the width of the track is 3 mm, and the diameter of the nozzle is 1.6 mm. The speeds are 17mm/s (the circumferential speed, the infill speed, the first layer speed, the gap infill speed, and the maximum print speed) and the travel speed is 20 mm/s. The layer height is 1 mm, the extrusion width is 300%, the air pressure is 0.3 MPa and the extrusion multiplier is 0.8. The paste type is 12 from Tab. 3.10. The samples were dried under surrounding room conditions, as mentioned in paragraph "3.4 Curing process". The slicing setting of the sample is shown in Fig. 6.71. The sample during and after complete printing is shown in Fig. 6.72. and 6.73. The drying conditions were monitored as shown in Fig. 6.74. The dimensional accuracy of the sample is demonstrated in Tab. 6.13.

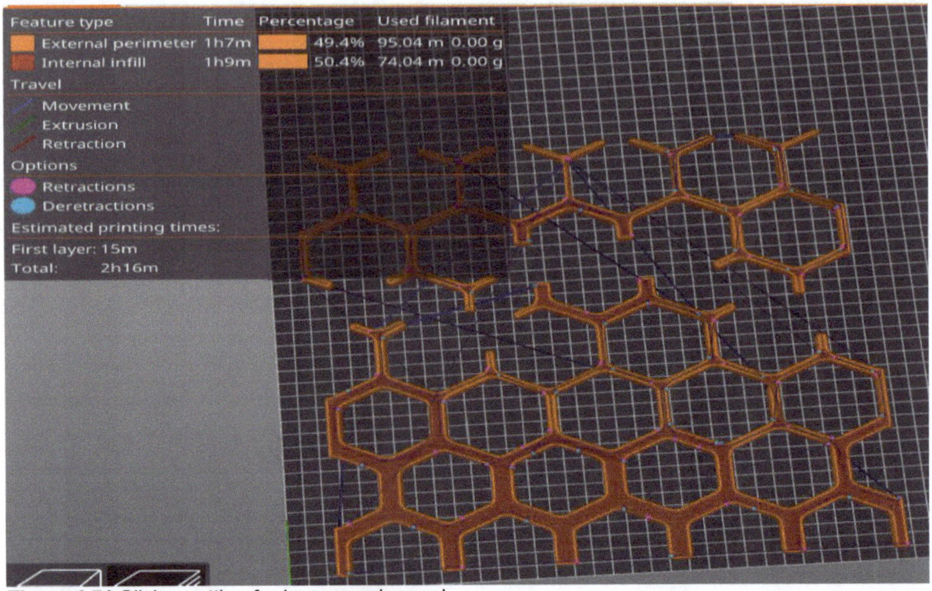

Figure 6.71 Slicing setting for honeycomb panel.

Figure 6.72 3D model for the honeycomb sample during printing.

Figure 6.73 3D model for the honeycomb sample after drying.

Table 6.13 The accuracy calculations for the 3D model of honeycomb panel.

The measurement	Dimensional difference %	Observation
Length Project	5.8 ± 1.2	The surfaces are nicely done with good balanced printing parameters.
width	6.1 ± 1.7	
Height	43.1 ± 2.5	There are creaks and debonding between the layer along the height due to shrinkage.
Corners angle	5.3 ± 1.2	
Flatness	4.2 ± 1.5	The edges is sharped and pointed

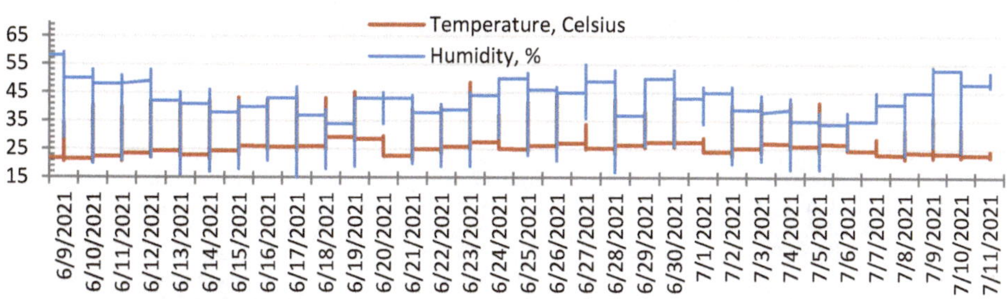

Figure 6.74 Monitoring of the sample room environment starting from printing day.

6.3 Summary

A comprehensive overview for the 3D printed paper model is shown in Tab. 6.14.

Table 6.14 The summary of the manufacturing configuration and critical outcomes for 3D models.

3D printed model	Cellulose typed	Materials formulation (Paste) % Wt.	Nozzle mm	Printing parameters	Drying condition & duration	Max. Dimensional difference %
Coral panel	Type2/ from Tab. 3.1, Aspen	Paste 13 15.5 cellulose, 74.6 Water, 3.8 CMC, 3.8 Lecithin, 2.3 Chalks	1.6	Print time 11:15 hr:min, track width 3 mm, speeds are 19 mm/s, Air pressure 0.3 MPa, the extrusion multiplier is 0.8, 95% infill for the base and 100% for coral walls	Room environment and drying condition in section 3.4. more than 12 days	40.4
Tides panel sample	Type2/ from Tab. 3.1, Aspen	Paste 13 15.5 cellulose, 74.6 Water, 3.8 CMC, 3.8 Lecithin, 2.3 Chalks	1.6	Print time 6:37 hr:min, track width 3 mm for base, speeds are 22 mm/s, Air pressure 0.33 MPa, the extrusion multiplier is 0.82, 99% aligned rectilinear infill	Room environment and drying condition in section 3.4. more than 15 days	37.5
Gills panel	Type2/ from Tab. 3.1, Aspen	Paste 8 14.5 cellulose, 77.4 Water, 3.34 CMC, 2.6 Lecithin, 2.23 Chalks	1.6	Print time 9:19 hr:min, track width 3 mm for base and 2mm for the gills part, speeds are 18 mm/s, Air pressure 0.3 MPa, the extrusion multiplier is 0.82, 99% concentric infill.	Room environment and drying condition in section 3.4. more than 14 days	45.5
3D map visualization	Type2/ from Tab. 3.1, Aspen	Paste 15 15.5 cellulose, 74.6 Water, 3.05 CMC, 2.7 Lecithin, 3.4 Starch	1.6	Print time 6:24 hr:min, track width 3 mm, Air pressure 0.32 MPa, the extrusion multiplier is 0.82, 100% aligned rectilinear infill print speed 25 m/s and travel speed 35mm/s.	Room environment and drying condition in section 3.4. more than 20 days	49.9
Triangular Seashell panel	Type2/ from Tab. 3.1, Aspen	Paste 25 16 cellulose, 74 Water, 4 CMC, 4 Lecithin, 3 Chalks	1.6	Print time 8:27 hr:min, track width 2 mm, Air pressure 0.8 MPa, the extrusion multiplier is 1.4, 99% concentric infill print speed 17 m/s and travel speed 20mm/s.	Room environment and drying condition in section 3.4. more than 21 days	18.5
Bubbles panel	Type2/ from Tab. 3.1, Aspen	Paste 27 16 cellulose, 74 Water, 4 CMC, 4 Lecithin, 3 Starch	1.6	Print time 9:57 hr:min, track width 2 mm, Air pressure 0.8 MPa, the extrusion multiplier is 1.4, 99% concentric infill print speed 18 m/s and travel speed 22mm/s.	Room environment and drying condition in section 3.4. more than 21 days	42.8

Table 6.14 continued.

3D printed model	Cellulose typed	Materials formulation (Paste) % Wt.	Nozzle mm	Printing parameters	Drying condition & duration	Max. Dimensional difference %
Acoustic reef	Type2/ from Tab. 3.1, Aspen	Paste 27 16 cellulose, 74 Water, 4 CMC, 4 Lecithin, 3 Starch	1.6	Print time 8:54 hr:min, track width 2 mm, Air pressure 0.8 MPa, the extrusion multiplier is 1.4, 98% concentric infill print speed 18 m/s and travel speed 22mm/s.	Room environment and drying condition in section 3.4. more than 21 days	32.5
Inverted crowned buffel	Type2/ from Tab. 3.1, Aspen	Paste 30 15 cellulose, 69.2 Water, 7.5 CMC, 2.8 Lecithin, 5.6 Chalks	1.6	Print time 1:6:51 day:hr:min, 3 parts, track width 2.5 mm, Air pressure 0.8 MPa, the extrusion multiplier is 1.4, 100% concentric infill print speed 17 m/s and travel speed 20mm/s.	Room environment and drying condition in section 3.4. more than 25 days	14.8
Cell shade panel_long fibers	Type3/ from Tab. 3.1, Pine	Paste 11 15.5 cellulose, 74.6 Water, 3.8 CMC, 3.8 Lecithin, 2.3 Starch	1.6	Print time 5:21 hr:min, track width 3 mm, Air pressure 0.4 MPa, the extrusion multiplier is 1.5, 100% rectangular infill print speed 20 m/s.	Room environment more than 4 days	21.5
Cell shade panel_Short fibers	Type2/ from Tab. 3.1, Aspen	Paste 12 15.5 cellulose, 74.6 Water, 3.8 CMC, 3.8 Lecithin, 2.3 Starch	1.6	Print time 3:35 hr:min, track width 3 mm, Air pressure 0.3 MPa, the extrusion multiplier is 1, 100% rectangular infill print speed 20 m/s.	Room environment. At least 3 days	46.8
Floral leaf buffel	Type2/ from Tab. 3.1, Aspen	Paste 27 16 cellulose, 74 Water, 4 CMC, 4 Lecithin, 3 Starch	1.6	Print time 12:20 hr:min, track width 2.5 mm, speeds are 17 mm/s, Air pressure 0.8 MPa, extrusion multiplier is 1.4, filling 100% rectilinear	Room environment and drying condition in section 3.4. At least 15 days	36.1
Tree wall	Type2/ from Tab. 3.1, Aspen	Paste 13 15.5 cellulose, 74.6 Water, 3.8 CMC, 3.8 Lecithin, 2.3 Chalks	1.6	Print time 15:51 hr:min, track width 3 mm, speeds are 20 mm/s, Air pressure 0.35 MPa, extrusion multiplier is 1, filling 99% concentric.	Room environment and drying condition in section 3.4. 20 days	40.5
Honycomb panel	Type2/ from Tab. 3.1, Aspen	Paste 12 15.5 cellulose, 74.6 Water, 3.8 CMC, 3.8 Lecithin, 2.3 Starch	1.6	Print time 2:16 hr:min, track width 3 mm, speeds are 17 mm/s, Air pressure 0.3 MPa, extrusion multiplier is 0.8, filling 100% rectilinear.	Room environment and drying condition in section 3.4. A least 3 days	43.1

The best result is 3D model "inverted crowned buffel", which max. dimensional difference reach to 14.8%.

3D Paper-printed model with poor resolution shows inappropriate pick for nozzle diameter.

7 Conclusion and Recommendation

This chapter offers an evaluation of the existing manufacturing process and identifies areas for improvement. It looks at the most valuable results that have been achieved throughout the research and shows which are the best and most effective for the purpose and requirements that need to be achieved.

7.1 Dimensional Accuracy And Geometrical Integration

Creating 3D paper models revolutionizes the way humankind thinks about paper, offering opportunities for innovation, customization, and complexity. However, to achieve a successful 3D printed paper model, it is crucial to prioritize quality, dimensional accuracy, and aesthetics. These factors are essential in ensuring the final product meets desired requirements and fulfills its intended purpose and customer satisfaction.
The approaches that answer the main research question (how to produce a 3D printed paper model and achieve the quality and dimensional accuracy) are summarized as follows:

7.1.1 Mixture Formulation

Choosing the right material to create a 3D printable part is critical. Different materials and proportions offer different properties that must consider factors such as extrudability, buildability, recyclability, strength, durability, flexibility, and the ability to achieve fine detail and other desired performance depending on preferences. Formulation involves understanding each ingredient's properties and interactions, determining appropriate ratios, and ensuring proper mixing techniques. Each contributes to the overall functionality and performance of the formulated mixture.
Cellulose is the main structure of the mix, acting as a skeleton for the paste. The cellulose particles or fibers are held together by an adhesive material, which in this case is carboxymethyl cellulose "CMC". CMC is a hydrogel material which when mixed with hot water provides the adhesive property between all the components of the mix.
Fillers are used to stabilize the gel formation of the CMC and to increase the number of interparticle contacts, which effectively improves buildability. Finally, lecithin is added to improve flowability and help the paste extrude easily through the die. Where lecithin is an organic material of fatty substances, its oily nature satisfies these needs.

D. Abdullah Agha, *3D Paper Printing for the Built Environment*, Mechanik, Werkstoffe und Konstruktion im Bauwesen 73,
https://doi.org/10.1007/978-3-658-45076-2_7

Long series of trial tests are carried out to find the suitable paste, which meets most of the requirements for 3D paper printing (extrudability, buildability, sufficient adhesive strength, limited shrinkage, and deformation outcomes after drying). The principle of the paste formulation is to reduce water content within the mixture. Water evaporation causes shrinkage problems. Reducing the amount of unnecessary water is therefore the objective to limit shrinkage and consequently improve quality and better workability. At the end of the experimental work, a mixture for the finest nozzles for 3D printers (LUTUM-V2.1) were found. The formulations that fit the selected nozzle are listed in Tab. 7.1.

Table 7.1 Mixture formulation suitable for nozzle diameters of 1.6 mm, 1.2 mm and 0.6 mm.

Nozzle used mm	**Cellulose** % by wt.	**Water** % by wt.	**CMC*** % by wt.	**Lecithin** % by wt.	**Filler** % by wt.
1.6	15, Aspen	69.2	7.5	2.8	5.6, Starch
1.2	15.5, Aspen	71.5	5.8	2.9	4.3, Starch
0.6	13.4, Aspen	78	3.6	2.6	2.3, chalk

7.1.2 Rheology Behavior

Rheological testing is a valuable tool for analyzing and understanding the flow and deformation properties of compounds. These tests provide important insight into the behavior of the material and help optimize the mixture formulation. This information is critical in achieving the desired consistency, e.g., providing a mixture with suitable extrudability for the specified fine nozzle, with as much stability as possible for buildability with appropriate adhesion strength. Rheological behavior is performed to evaluate viscosity, shear rate behavior, viscoelastic properties, and yield stress. The high shear yield stress correlates with higher buildability and better dimensional stability during production until drying. Conversely, viscosity is associated with ease of extrusion.

The rheological data obtained from these tests guide formulation decisions such as testing the addition of different ingredients and selecting appropriate additives (e.g., which filler is better for extrudability and buildability), adjusting component ratios, or modifying processing conditions to find an optimal mixture.

This is concluded as following:

7.1.3 Benefits of Lecithin and CMC Addition

Increasing lecithin improves extrudability. While increase CMC improves buildability.

✓ Adding 5.6% by weight of lecithin to a paste of (37.5% cellulose:50% water:12.5% CMC) improved flowability by 77% and 312.5%, respectively, in terms of reducing apparent viscosity and shear deformation at a load of 254 Pa. However, both pastes were unable to extrude.

- ✓ Increasing the lecithin content by 46% to a paste of (13% cellulose: 79.6% water: 2.5% CMC: 2.5% Lecithin) improves the flowability significantly; the yield stress and apparatus viscosity are reduced by 23.7% and 6.3% respectively.
- ✓ While the results are better when both the lecithin and CMC are increased by 40% to a paste of (13% cellulose: 79.6% water: 2.5% CMC: 2.5% Lecithin), the behavior is initially closer to reference paste, but it becomes stiffer after about 250 Pa. Both the yield stress and the apparatus viscosity increased by 31% and 16.6% respectively compared to reference paste.
- ✓ Conversely, increasing only the CMC content by 46% to paste (13% cellulose: 79.6% water: 2.5% CMC: 2.5% Lecithin) causes stiffer consistency. However, the yield stress of this paste decreases by 12.7% and the apparatus viscosity increases by 27% compared reference paste.
- ✓ Finally, increasing both lecithin and CMC give better results in term of buildability and flowability, compared to the pastes with increasing either lecithin or CMC alone. The yield stress of paste with increasing both CMC and lecithin increases by 50%, 71.7% and 31% compared to the paste of increased CMC, reference paste, and paste with only increased lecithin; respectively.

Advantages of Using Fillers

1. The viscoelastic behavior of the paste with starch is weaker in the early stages than that of the same paste but with chalk, due to the wettability behavior of the starch. Because of this behavior, the amount of water could be reduced where starch is insoluble in water. However, the starch filler paste is 12.6% higher in shear yield stress than the paste with chalk, giving a better buildability because the undissolved starch used here as an interparticle stabilizes the flow where the rearrangement of the microstructure of starch particles inside the fluid is hard by application of shear. Therefore, the apparent viscosity of the paste with starch is 15.8% higher than that of the same paste but with chalk.
2. The effect of increasing the filler content in the mixture have a positive aspect on the rheology behavior, increasing the starch to 46% increases the yield shear and apparent viscosity by 1.5% and 53.6% respectively, compared to the same mix but with CMC and Lecithin reduced by 20% and 29% by weight; respectively.

7.1.4 Cellulose Privilege

Cellulose is one of main factors that affects the quality and dimensional accuracy. The characteristics of cellulose vary according to the family of plant, age of the plants, location of the cellulose, and the manufacturing procedure, etc.

The cellulose that performs better in 3D paper printing is the cellulose from Mercer's pulp mills/ long fiber, NBSK_ECF, Pine STENDAL/ Germany. However, the short cellulose fibers (Aspen PHK-DMI/Canada) is the better choice for extrudability aspect for fine nozzles and the second better results in buildability and shrinkage results.

7.1.5 Printing Parameters

Printing parameters have a direct impact on the quality, accuracy, and overall success of the printed object. To achieve the desired results, it is essential to select and optimize these parameters, by carefully adjusting printing parameters (flow rate, print speed, air pressure, layer height, and infill density). Balanced printing parameters are determined through iterative testing and validation. Trials are conducted with different parameter combinations to evaluate the outcomes for dimensional accuracy, structural integrity, and overall aesthetics. The balanced printing parameters for the mixture fit the finest nozzle of the printer (0.6-, 1.2-, 1.6-) mm are shown in Tab. 7.2.

Table 7.2 Balanced printing parameters for the mixture fit the finest nozzle (0.6-, 1.2-, 1.6-) mm [Abdullah Agha and Knaack, 2022].

The used paste for the nozzle mm	**Printing speed** mm/s	**Flow rate**	**Layer height** mm	**Air pressure** MPa	**Path width,** mm
1.6	18	1	1	0.8	2
1.2	18	1.2	1	0.8	2
0.6	15	1.4	0.25	0.8	0.5

7.1.6 Shrinkage Outcomes

It is a major challenge when it comes to 3D printing of semi-liquid materials due to water evaporation. Shrinkage creates internal stresses, potentially deforming the printed object and changing dimensions. This primarily affects the aesthetics of the printed product.

Several factors contribute to shrinkage, including material properties, curing mechanisms, process parameters and geometry configuration. Optimizing the materials formulation and adjusting the curing parameters are the main ways to control shrinkage. Different materials exhibit different levels of shrinkage due to their inherent chemical and physical properties. Manipulating the mix formulation, selecting different materials, trying different proportions, and reducing the water content as much as possible are the approaches followed in the study to achieve better performance for the 3D printed part in terms of maintaining quality and dimensional accuracy.

Effect of Cellulose Type

Using long fiber cellulose instead short cellulose reduced shrinkage by 28.9% when comparing cellulose (SÖDRA Golden Sweden) to (Short cellulose fiber SÖDRA- Blue Sweden) within the mixture (13% cellulose: 79.6% water: 2.5% CMC: 2.5% Lecithin). While using (Long cellulose fiber, NBSK_ECF, Kiefer STENDAL Germany) reduces shrinkage by 60.4% and by 36% in comparison to paste with powder cellulose and paste with (Short cellulose fiber, Aspen PHK, DMI Canada).

Effect of Filler (Magnesium sulfate, Starch, chalk)

1. Magnesium sulphate is not a suitable filler for 3D paper printing. Despite its lower shrinkage, it is irregular due to its crystallization in water, which gives acceptable measurements in some places but suffers from higher shrinkage in others. As a result, the shrinkage results are scattered and have a higher standard deviation. In addition, Mgso4 affects the recyclability of paper material.
2. Starch is the most acceptable filler among the other fillers investigated. Due to the possibility to reduce the water content when using indissoluble starch. Then it is possible to manipulate the other components to provide sufficient extrudability by increasing Lecithin and sufficient buildability and adhesion strength by increasing CMC.

The minimum shrinkage results are listed in Tab.7.3 for the printable pastes for the nozzles (0.6-, 1.2-, 1.6-) mm diameter.

Table 7.3 Shrinkages measurements after drying for filaments extruded from the pastes.

The used paste for the nozzle mm	**Diagonal shrinkage of the diameter** %	**Longitudinal shrinkage of the filament** %
1.6	17.4 ± 0.7	1 ± 0.3
1.2	15.7 ± 0.9	1.1 ± 0.5
0.6	38.6 ± 1.2	2.8 ± 0.9

7.2 Design Considerations and Limitations

It is essential to understand and address the specific considerations and limitations associated with this manufacturing approach. The subsequent process can then be optimized to avoid identified problems, overcome potential challenges, and create a completable, functional approach. Some of these keys are:

- Dimensional compensation strategy (Rescale Z-, X-, Y- axis)

To achieve the target dimensions, it is helpful to edit (scale) the 3D CAD by the same shrinkage ratio before cutting and printing. In this way, the target dimensions can be reached after the drying process. For the 3D printing of paper, there are several aspects to consider for the shrinkage to calculate the dimensional compensation, namely:

1. The shrinkage in direction of the layer height is much higher than in the other two directions (the layer plane).
2. Larger volume specimens have less shrinkage due to their lower ratio of exposed surface area to specimen volume. It preserves a lesser water evaporation rate within the core.
3. The cylindrical shape has higher shrinkage results by a range (17.9%, 18.7% and 20.5%) than the cubic shape with the same surface to volume ratio as (0.15, 0.2, 0.3)1/mm respectively.

The explanation lies in the geometric shape and height of the sample. The cylindrical shape has a greater height than the cubic shape and (Y axial shrinkage) shrinks more than the width (Z or X axial shrinkage). In addition, the cubic shape has corners which act as a cross-connection, confining the shrinkage compared to the cylindrical shape which has in the radial strain distribution and no confinement.

4. The type of infill determines the percentage of shrinkage. If the layer has more extruded material, the volumetric shrinkage will be lower. This means that more material within the duplicate layers fills the space and compacts the area. It also depends on the type of pattern that has knots or cross contacts.

When comparing the concentric and rectangular fills, the rectilinear fill is preferred in terms of improving the shrinkage results by 10.4%, 41.43% and 13.43% in terms of Y-axis shrinkage, X/Z-axis shrinkage and volumetric shrinkage respectively, for a cubic of dimension (50x50x50) mm The rectilinear infill provides intersecting nodes and junctions that reinforce the confinement and constraint.

- G-code optimization: when using the printing parameters listed in Tab. 7.2, it is better to follow the additional settings mentioned in section 3.3.7.

- Paste validation

The paste rotted over time. Experimental observations have shown that even if the paste is kept in a refrigerator, it is no longer a good material for printing approximately six days after mixing. It is therefore preferable to use the paste immediately after mixing.

- Printing Orientation

Parallel Filament Orientation: When the filaments are oriented parallel to the applied load or force direction, the tensile strength equal to 3.18 ±0.2 MPa.
Perpendicular Filament Orientation: When the filaments are oriented perpendicular to the applied load or force direction, the tensile strength is equal to 0.53±0.2 MPa.
This means, that Adhesive strength between the filaments was 16.67% of unidirectional tensile strength of 90-degree filaments orientation to force direction without frame (using 1.6mm nozzle). This indicates that this formulated mix has acceptable bond strength and coalescence.

- Strengthening with wood fibers

5. The most acceptable wood fiber content is 2% which passes through the (1mm) sieve and remains on the (0.5mm) sieve.
6. The effect of the wood fiber is seen in the final stages of loading where it increases the ultimate load and toughness by 41.2% and 111.6% respectively, but it has no effect on the modulus of elasticity.

- Resolution of the print

This is depended on nozzle diameter and layer height. High-resolution products give stronger rigid prints, more consistent, and smaller and more closely spaced infill lines, thus increasing the layering effect, contact area, and stiffness of the print.
The ultimate tensile strength and modulus of elasticity of the smaller nozzle (1.2 mm nozzle) are 11.9% and 7.3% higher respectively than those of the larger nozzle (1.6 mm nozzle). For 45 degrees filament aligned to the direction of the force, are 5% and 6.9% higher, respectively. For adhesion strength between the filament, are also higher by 24.5% and 56.7% higher, respectively.

- Accuracy calibration

Samples (floral leaf shape of triangles and inverted crowned Buffel) have the best dimensional accuracy, reaching 18.5% and 14.8% respectively, in terms of shrinkage results in layer height direction.

7.3 Recommendation & Prospects

It is important to note that these recommendations and perspectives are general and may vary according to specific contexts, goals and advances in the field of 3D paper printing with cellulose-based materials. Here are suggestions to facilitate a new approach as complementary prospects to improve the production of paper-based 3D printing.

1. Optimization of the mixture formulation (Using MgO as filler & for fire resistance)

MgO (Magnesium Oxide) seems to be an interesting material for engineering applications, for the printing of 3D paper with a cellulose based material. It is commonly used as a filler material in various applications due to its unique properties, known for its excellent fire-retardant properties and used as construction products.
It's interesting to use it as a filler for the mix formulation and investigate the production performance, especially it improves the paper properties for fire resistance. Fig. 7.1 shows the behavior of wettability of the starch, MgO and chalk. MgO seems to solidify faster, that maybe causing extrusion issues. Despite starch having better performance for 3D paper printing, in comparing to chalk, MgO seems to have less shrinkage results than chalk.

b

Figure 7.1 Wettability behavior of the 46% by weigh starch, MgO or chalk with 56% water after one day of mixing (a), and after drying(b).

2. Drying rate

Experimental work has shown that drying during the winter months gives better performance than drying during the warmer months. In other words, dry slowly. This could be a study approach to improve the drying process for 3D paper printing production.

3. Characterizing the material behavior

From the undocumented experimental work, it was investigated that the paste becomes stiffer with time. So, this could be an investigation for characterization of material behavior and deep into material properties.

4. Efficiency of the product's shading requirements for 3D paper printed models.

Exposing the material to sunlight for a long period of time and then investigating the efficiency of the material properties to have enough durability to function as a shading product, as shown in Fig.7.2.

Fig. 7.3 shows the material was exposed to the sun light for over one year, it was observed that the material color didn't change into yellow like to normal A4 paper sheets.

a b c d

Figure 7.2 The location of designed 3D paper printed model for shading functionality(a) Visualization of the 3D model (b), during printing of the sample (c), and the final view.

Figure 7.3 Exposing the I-bone and cube specimens to sunlight to test the mechanical properties and determine the deformation to assess the capability of the material for this potential purpose.

5. Efficiency of 3D paper printed models as acoustics functionality.

An interesting direction of research is to explore the development of novel sound absorbing materials using 3D paper printing techniques. To achieve improved sound absorption properties over specific frequency ranges, the material composition, porosity and surface patterns need to be investigated.

As a starting point, the acoustic properties of the 3D paper printing material are being tested at the Department of System Reliability, Adaptive Structures and Machine Acoustics SAM/TU Darmstadt, see Appendix E. This application requires further research and experimentation to fully understand the potential of 3D paper printing in the field of acoustics.

6. Efficiency of 3D printed models to resist humidity.

Some experiments were conducted to investigate the moisture resistance of the paper material. It was found that Paste 25 of Tab. 3.10 has good resistance to humidity after exposing the sample to intensive humidity environment, which it was placed in the bathroom for about 6 months, as shown in Fig. 7.4. Further investigation is required to identify the material likely to be used in this application.

Figure 7.4 Exposing a sample to intensive humidity environment (bathroom).

7. Seek potential for application.

While 3D printing with cellulose-based semi-liquid paper material is an innovative approach for products that are still under research. It is attracting attention and is being studied and developed in various aspects to realize its full potential and find new applications. The growth and development of this approach needs to be supported by further research and collaboration between researchers, industry professionals and innovators. Tab. 7.4 exhibits some suggestions.

Table 7.4 Possible application for 3D paper models.

Potential Application for	Related categories	Description of the Research Field
3D isolation models and impacting effect models	Investigation of different paper material formulations, patterns, and techniques to find the best way to represent the insulation within the 3D models, taking into account the overall accuracy and purpose of the rendering. Investigation of the thickness requirement for the 3D paper model to have full insulation capability. Thicker layers of paper models can provide better insulation by creating a greater barrier for heat to pass through. Investigation of the possibility of air trapping within the design and printing process, which will help to slow down the transfer of heat through the material.	The creation of the design, the internal fill and patterns are key parameters for this application. Studying the environmental impact compared to conventional methods. It has a certain insulating capacity due to its structure and composition. To increase the potential of using 3D paper models as an insulating structure, several factors need to be considered.
Welding materials for paper material (cardboard or cartoon, etc.)	Investigating the best mixture to perform as adhesion material. Investigating the technique and printing process to apply the welding to other material. Investigation of the challenge of the application of 3D paper printing and the mechanical capability to produce intricate and detailed models of welded components, joints and assemblies.	Welding material for paper products such as board is an innovative application of a high strength adhesive. This is where the strength of the bond is reliable enough to hold the parts together. Adhesive properties are an important measure of a compound's ability to bond to other surfaces as well as to itself, which is a critical factor in determining its potential applications. As a starting point, the adhesion of the compound to other surfaces was tested. See Appendix C for further details.
Educational tools and 3D mock-up models Rapid Prototyping	Investigation the different nozzles to achieve wide possible resolution for the printed mock-ups.	Where physical representations aid understanding, such as in architecture, engineering, and in exhibition. Recyclability plays an important role here.
Short term usage: Packaging	Leverage the complexity of additive manufacturing to create custom packaging solutions with optimized shapes for better protection and space utilization. To enhance visual appeal, information delivery and consumer engagement, paper models can be incorporated into packaging.	For example, packaging a product can include a small 3D paper model that represents the product itself or what it contains. This interacting element not only provides a unique unboxing experience, but also communicates product features and how to use the product in a creative way.

Table 7.4 Continued

Potential Application	Related categories	Description of the Research Field c
Art and Sculpture Decoration purposes	The design and Print process are the trick for this application. Investigating the possibility of accompanying the paper material with other materials to formulate a piece part of art.	In the field of architecture and building physics, incorporating art and sculpture into designs can have both functional and aesthetic benefits.
Short term usage: support structure for other additive manufacturing techniques	Investigation of the challenges of paper's material properties, potential interference with the final object's integrity, and ensuring easy removal without damaging the printed part to use 3D paper modes as support structure. Balancing structural support and easy removal is crucial for successful implementation.	The use of 3D paper models as temporary support structures in additive manufacturing involves the creation of paper structures that help complex parts retain their shape during printing. This reduces the need for extensive post-processing.
Sandwich panel applications	Investigation the possible pattern and the printing process	The use of 3D paper printing for sandwich panel applications can be an innovative approach for lightweight structures that have specific aesthetic or temporary requirements.

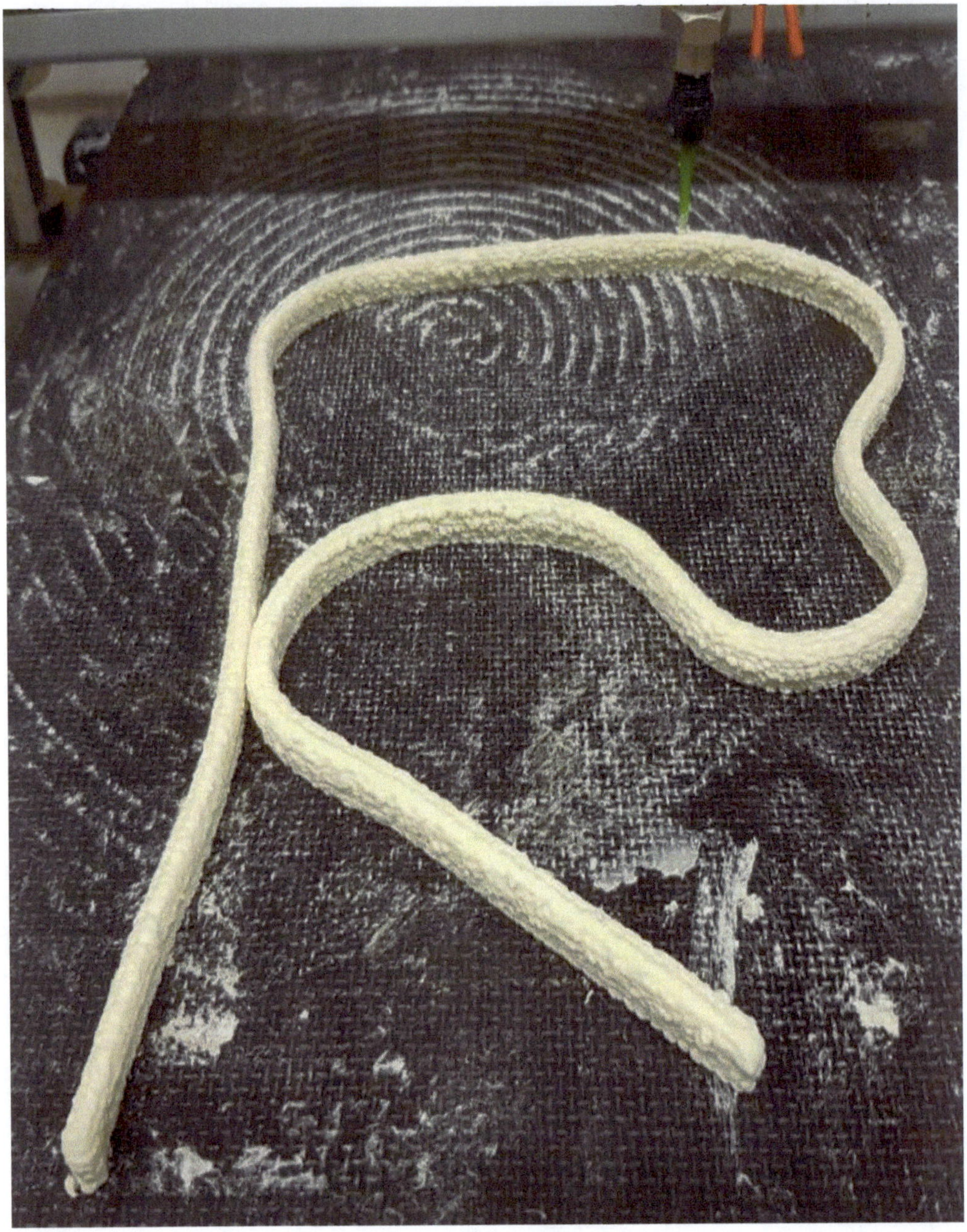

3D paper printed model to demonstrate buildability and stability while printing till drying.

Reference

21 CFR 111.70 – "What specifications must you establish?" Part 111 - Current Good Manufacturing Practice In Manufacturing, Packaging, Labeling, Or Holding Operations For Dietary Supplements,Subpart E - Requirement to Establish a Production and Process Control System, Section 111.70. GPO U.S. Government Publishing Office, April 1, 2011. https://www.govinfo.gov/content/pkg/CFR-2011-title21-vol2/pdf/CFR-2011-title21-vol2-sec111-70.pdf

Abdullah Agha, D. and Knaack, U. 2022. Additively manufactured paper products from cellulose-pulped fibers and its quality according to rheology and 3D printing performance. Book: Structures and Architecture a Viable Urban Perspective? Taylor & Francis Group Logo, 1st Edition, First Published2022, Imprinter Presse Book ISBN9781003023555.

Abdullah Agha, D. and Knaack, U. 2022. Additive Manufacturing Technologies for the Build Environment, 3D Printed Acoustic Absorbers, BEAM booklet.

Abdullah Agha, D. and Knaack, U. (2021). Additive Manufacturing Technologies for the Build Environment (3D Paper Printing), BEAM booklet.

Abolhasani, M. M., Salari, M., Amanatfard, F., & Madani, S. Y. (2019). Mechanical and physical properties of 3D-printed paper-based composites. Cellulose, 26(12), 7345-7359.

Allen, S.T., Sprenger, M., Bowen, G.J., Brooks, J.R. (2022). Spatial and Temporal Variations in Plant Source Water: O and H Isotope Ratios from Precipitation to Xylem Water. In: Siegwolf, R.T.W., Brooks, J.R., Roden, J., Saurer, M. (eds) Stable Isotopes in Tree Rings. Tree Physiology, vol 8. Springer, Cham. https://doi.org/10.1007/978-3-030-92698-4_18

Asdrubali, F; Schiavoni, S; Horoshenkov, KV. (2012). A Review of Sustainable Materials for Acoustic Applications. Building Acoustics.;19(4):283-311. https://doi.org/10.1260/1351-010X.19.4.283

ASTM Designation: D695 – 15 Standard Test Method for Compressive Properties of Rigid Plastics.

ASTM F2971- 13, Standard Practice for Reporting Data for Test Specimens Prepared by Additive Manufacturing1.

D. Abdullah Agha, *3D Paper Printing for the Built Environment*, Mechanik, Werkstoffe und Konstruktion im Bauwesen 73,
https://doi.org/10.1007/978-3-658-45076-2

Azimi, B.; Maleki, H.; Gigante, V.; Bagherzadeh, R.; Mezzetta , A.; Milazzo, M.; Guazzelli, L.; Cinelli, P.a; Lazzeri, A.; Danti. Se. (2022). Cellulose-based fber spinning processes using ionic liquids, Cellulose 29:3079–3129 https://doi.org/10.1007/s10570-022-04473-1

Baksa, T.; Ngenea, I. S.; Jeroen J.G.van Soestb, Anja E.M.Janssena, and Remko M.Boom (2007). Comparison of methods to determine the degree of gelatinisation for both high and low starch concentrations. Carbohydrate Polymers Volume 67, Issue 4, 19 February 2007, Pages 481-490. https://doi.org/10.1016/j.carbpol.2006.06.016

Baranwal, J.; Barse, B.; Fais, A.; Delogu, G.L.; Kumar, A. (2022). Biopolymer: A Sustainable Material for Food and Medical Applications. Polymers 2022, 14, 983.https://doi.org/10.3390/polym14050983

BE-AM 2021 Symposium and Exhibition booklet: Additive manufacturing for building environment, pages 127-128, 3D Printed Acoustic Absorbers.https://beam1photo.files.wordpress.com/2021/12/be-am_2021-1.pdf

Brenken, B.; Barocio, E.; Favaloro, A.; Kunc, V. (2018). Pipes, R.B. Fused filament fabrication of fiber-reinforced polymers: A review. Addition Manufacturing, 21, 1–16.https://doi.org/10.1016/j.addma.2018.01.002

Brethauer, S., Shahab, R.L. & Studer, M.H. (2020). Impacts of biofilms on the conversion of cellulose. Appl Microbiol Biotechnol 104, 5201–5212 https://doi.org/10.1007/s00253-020-10595-y

Briffaza, A.; Bohuonb, P.; M.Méota, J.; Dornierb M.; and Mestres C., (2014). Modelling of water transport and swelling associated with starch gelatinization during rice cooking. Journal of Food Engineering, ELSEVIER. Volume 121, Pages 143-151https://doi.org/10.1016/j.jfoodeng.2013.06.013

Campos, T.; Cruz, P.J.S.; and Figueiredo, B. (2022) Exploration of natural materials in additive manufacturing in architecture: Use of cellulose-based pulps. Structures and Architecture (ICSA 2022) – Hvejsel & Cruz (eds) © 2022 Copyright the Author(s), ISBN: 978-0-367-90281-0

Chatterjee, S., and Madras, G. (2016). Carboxymethyl cellulose: properties and applications. In Carbohydrate Polymers: Properties and Applications (pp. 271-288). CRC Press.

Chen, T; Xie, Y; Wie, Q; Wang, X A.; Hagman, O; Karlsson, O; and Liu, J. (2016). Effect of refining on physical properties and paper strength of Pinus massoniana and china fir cellulose fibers BioResources 11 7839–48 Paper F 2012 paper & Physics Full Paper 17 583–90

Chen, W., Yu, H., Liu, Y., Chen, P., Zhang, M., & Hai, Y. (2011). Individualization of cellulose nanofibers from wood using high-intensity ultrasonication combined with chemical pretreatments. Carbohydrate Polymers, 83(4), 1804–1811. https://doi.org/10.1016/j.carbpol.2010.10.040

Chen, Y. and Wang, X. (2019). 3D Printing in Architecture: Recent Advances and Future Directions. Architectural Engineering and Design Management, 15(4), 284-301.

Daniele, R. (2019). Polymer composites for sustainable 3D printing materials. Doctoral Thesis in Materials, Mechatronics and Systems Engineering Published in Trento (Italy) – by University of Trento.

De Schutter, G; Lesage, K; Mechtcherine, V; Nerella, VN; Habert, G; Agusti-Juan I. (2018). Vision of 3D printing with concrete - Technical, economic, and environmental potentials. Cem Concr Res; 112:25–36.

Devicharan R. and Garg R. (2019). Optimization of the Print Quality by Controlling the Process Parameters on 3D Printing Machine, Printing and Media Engineering Department, Manipal Institute of Technology, Manipal University, Manipal, India, ©Springer Nature Singapore Pte Ltd. https://doi.org/10.1007/978-981-13-0305-0_16

El-Sakhawy, M., and Hassan, M. L. (2007). Physical and mechanical properties of carboxymethyl cellulose films. Carbohydrate Polymers, 67(4), 1-10.

EN 1991-1-1 (2002) (English): Eurocode 1: Actions on structures - Part 1-1: General actions - Densities, self-weight, imposed loads for buildings [Authority: The European Union Per Regulation 305/2011, Directive 98/34/EC, Directive 2004/18/EC].

Fabricius, I. L. (2007). Chalk: composition, diagenesis and physical properties. Bulletin of the Geological Society of Denmark, Vol. 55, pp. 97–128 by Geological Society of Denmark. ISSN 0011–6297.

Ferretti, P.; Leon-Cardenas, C.; Ciotti, E.; Santi, G.; Donnici; G.; and Frizziero; L. (2021). FDM 3D Printing Parameters Optimization: The Key Role of Line Width. © IEOM Society International. Proceedings of the International Conference on Industrial Engineering and Operations Management Bangalore, India, August 16-18, 2021 https://www.ieomsociety.org/proceedings/2021india/257.pdf

Fuwen, Hu.; Mikolajczyk, T.; Pimenov D. Y.; and Gupta M. K. (2021). Extrusion-Based 3D Printing of Ceramic Pastes: Mathematical Modeling and In Situ Shaping Retention Approach, Materials 2021, 14, 1137. https://doi.org/10.3390/ma14051137. https://www.mdpi.com/journal/materials

Gao, H.; Rao, J.; Guan, Y; Li, W.; Zhang, M.; Shu, T.; Lv, Z. (2018). Investigation of the Thermo-Mechanical Properties of Blend Films Based on Hemicelluloses and Cellulose. Research article. International Journal of Polymer Science. 2018. https://doi.org/10.1155/2018/9620346

Gauss, Ch.; Pickering Kim L.,and Muthe, L. P. (2021). The use of cellulose in bio-derived formulations for 3D/4D printing: A review, Composites Part C: Open Access, Volume 4, 100113, ISSN 2666-6820, https://doi.org/10.1016/j.jcomc.2021.100113

Gibson RF (2007) Principles of composite material mechanics. CRC Press, New York. https://doi.org/10.1201/b19626

Giubilini, A.; Siqueira, G.; Clemens, FJ.; Sciancalepore, C.; Messori, M.; Nystrom,G. (2020). 3D printing nanocellulose-poly(3-hydroxybutyrate-co-3-hydroxyhexanoate)biodegradable composites by fused deposition modeling, ACS Sustain. Chem. Eng., https://doi.org/10.1021/acssuschemeng.0c03385

Glatthor, A.; Schweizer, D. (1994). Rheological Lab Testing of Building Formulations (Hercules GmbH, Düsseldorf). Printed in: ConChem International Exhibition & Conference Proceedings 1994. Verlag Ziolkowsky, Augsburg. https://construction-chemistry.com/wp-content/uploads/2019/03/Rheological-Lab-Testing-of-Building-Formulations-1994.pdf

Gordeev, E. G.; Galushko, A. S.; and Ananikov, V. P. (2018). Improvement of quality of 3D printed objects by elimination of microscopic structural defects in fused deposition modeling, N.D. Zelinsky Institute of Organic Chemistry, Russian Academy of Sciences, Moscow, RussiaPublished: June 7, 2018. https://doi.org/10.1371/journal.pone.0198370

Habbal, O.; Ayoub, G.; and Pannierl Ch. (2021). Variable Extrusion Width for Interlocking Features in Fused Filament Fabrication 3d Printing. Solid Freeform Fabrication 2021: Proceedings of the 32nd Annual International Solid Freeform Fabrication Symposium – An Additive Manufacturing Conference. https://utw10945.utweb.utexas.edu/sites/default/files/2021/126%20Variable%20Extrusion%20Width%20for%20Interlocking%20Features.pdf

Harini, K.; and Mohan, C.C, (2020). Isolation, and characterization of micro and nanocrystalline cellulose fibers from the walnut shell, corncob and sugarcane bagasse. Int. J. Biol. Macromol, 163, 1375–1383.

Harini, K.; and Mohan, C.C. (2020). Isolation and characterization of micro and nanocrystalline cellulose fibers from the walnut shell, corncob and sugarcane bagasse. Int. J. Biol. Macromol, 163, 1375–1383. https://doi.org/10.1016/j.ijbiomac.2020.07.239

Hong, S.; Sanchez, C.; Du, H.; and Kim, N. (2015). Fabrication of 3D Printed Metal Structures by Use of HighViscosity Cu Paste and a Screw Extruder, Journal of ELECTRONIC MATERIALS, Vol. 44, No. 3, 2015 The Minerals, Metals & Materials Society. https://doi.org/10.1007/s11664-014-3601-8

Hopewell, J; Dvorak, R; Kosior, E. (2009). Plastics recycling: challenges and opportunities. Philos Trans R Soc Lond B Biol Sci.27;364(1526):2115-26. https://doi.org/10.1098/rstb.2008.0311 PMID: 19528059; PMCID: PMC2873020.

Hu, Q., Chen, C., Zhang, W., & Wang, S. (2017). Effects of the paper properties on the 3D printing quality. TAPPI Journal, 16(4), 217-222.

Immonen, K; Willberg-Keyriläinen, P; Ropponen J; Nurmela A; Metsä-Kortelainen S, Kaukoniemi OV, Kangas H. (2021). Thermoplastic Cellulose-Based Compound for Additive Manufacturing. Molecules. 18;26 (6):1701 PMID: 33803734; PMCID: PMC8003161. https://doi.org/10.3390/molecules26061701 ..

ISO 527 Plastics: Determination of tensile properties

ISO/ASTM 52900, 2015(E), Standard Terminology for Additive Manufacturing, General Principles, Terminology1,2.

ISO/ASTM 52910, 2018(E), Additive manufacturing, Design, Requirements, guidelines, and recommendations.

ISO2811-2Second edition2011-03-01Paints and varnishes — Determination of density — Part 2: Immersed body (plummet) method.

Journal of Chemical Education **2022** *99* (2), 553-560

Kazemian, A.; Yuan, X.; Meier, R.; Khoshnevis, B. (2019). Chapter 2 - Performance-Based Testing of Portland Cement Concrete for Construction-Scale 3D Printing. 3D Concrete Printing Technology, Butterworth-Heinemann, Pages 13-35. ISBN 9780128154816. https://doi.org/10.1016/B978-0-12-815481-6.00002-6. https://www.sciencedirect.com/science/article/pii/B9780128154816000026.

Khorasani, S. N., Mohseni, S. M., & Shariati, P. (2019). A review on the 3D printing of paper-based materials. Cellulose, 26(15), 8465-8481.

Khordagui, E.L. (2010). Carboxymethyl cellulose: Uses and applications. Cellulose-Structure and Properties, Derivatives and Industrial Uses, 173-206.

Knaack, U. and Koenders, E. (2018) Building Physics of the Envelope: Principles of Construction, Berlin, Boston: Birkhäuser, 2018. https://doi.org/10.1515/9783035609493

Knaack, U; Klein, T; Bilow, M ; Auer, T. (2014) Facades: Principles of construction. Basel, Switzerland: Birkhäuser Verlag, 2014. 135 p.

Kotnik, T., & Dolšak, B. (2019). Analysis of the accuracy of 3D printing on the FDM printer. Tehnički Vjesnik, 26(6), 1619-1625.

Krieger, K. J.; Bertollo, N. Dangol M., Sheridan, J. T. Lowery M. M., and Cearbhaill E. D. (2019). Simple and customizable method for fabrication of high-aspect ratio microneedle molds using low-cost 3D printing, Krieger et al. Microsystems & Nanoengineering 5:42. https://doi.org/10.1038/s41378-019-0088-8

Kuder, K. G., & Shah, S. P. (2010). Processing of high-performance fiber-reinforced cement-based composites. Construction and Building Materials, 24(2), 181-186. https://doi.org/10.1016/j.conbuildmat.2007.06.018

Laith, A. A. and Al-Hashimi, A. G. (2019). Mechanical Properties of Carboxymethyl Cellulose Edible Films. Basrah Journal of AgriculturalSciences 32 (1): 68-78. ISSN 1814 – 5868. https://doi.org/10.21276/basjas

Lamberti, F.M., Román-Ramírez, L.A. & Wood, J. Recycling of Bioplastics: Routes and Benefits. J Polym Environ 28, 2551–2571 (2020). https://doi.org/10.1007/s10924-020-01795-8

Larysa Okhrimenko, Loïc Favergeon, Kevyn Johannes, Frederic Kuznik, Michèle Pijolat (2017). Thermodynamic study of MgSO4 – H2O system dehydration at low pressure in view of heat storage. Thermochimica Acta, Elsevier, 2017, 656, pp.135 à 143. https://hal.archives-ouvertes.fr/hal-01585099

Lei Pu, Peng Xu, Mingbiao Xu, Jianjian Song, and Miao (2021). Effect of Fiber on Rheological Properties and Flow Behavior of Polymer Completion Fluids. ACS Omega 2021, 6, 27, 17136–17148. Publication Date:July 1, 2021. Published by American Chemical Society. https://doi.org/10.1021/acsomega.0c05346

Li, L Chen, Y Yu, T Wang, N Wang, C. Wang, H (2019). Preparation of polylactic acid /TEMPO-oxidized bacterial cellulose nanocomposites for 3D printing via Pickering emulsion approach, Compos. Commun. 16 (2019) 162–167,https://doi.org/10.1016/j.coco.2019.10.004

Li, Y., Li, S., & Zhang, Z. (2020). Effect of fiber length and concentration on the mechanical properties of 3D-printed paper-based composites. Journal of Materials Science, 55(17), 7331-7344.

List. G.R. (2015). Soybean Lecithin: Food, Industrial Uses, and Other Applications, Editor(s): Moghis U. Ahmad, Xuebing Xu, Polar Lipids. Elsevie, pages 1-33, ISBN 9781630670443. https://doi.org/10.1016/B978-1-63067-044-3.50005-4.

Liu, Q.; Donner, E. ; Tarn, R.; Singh, J.; and Chung, H. (2009). Advances in Potato Chemistry and Technology, 8.6.3. ISBN 978-0-12-374349-7Elsevier Inc. https://doi.org/10.1016/B978-0-12-374349-7.X0001-3

Mahmood A, Akram T, Chen H, and Chen S. (2022). On the Evolution of Additive Manufacturing (3D/4D Printing) Technologies: Materials, Applications, and Challenges. Polymers (Basel). 2022 Nov 3;14(21):4698. https://doi.org/10.3390/polym14214698 PMID: 36365695; PMCID: PMC9656270.

Mahmood, A; Akram, T.; Chen, H, Chen, S. (2022). On the Evolution of Additive Manufacturing (3D/4D Printing) Technologies: Materials, Applications, and Challenges. Polymers (Basel). 2022 Nov 3;14(21):4698. https://doi.org/10.3390/polym14214698 PMID: 36365695; PMCID: PMC9656270.

Marwah, O.; Yahaya, N.; Darsani, A.; Mohamad, E.; Haq, R.; Johar, M. and Othman, M. (2019). Investigation for Shrinkage Deformation in the Desktop 3D Printer Process By Using DOE Approach Of The ABS Materials, International Conference on Mechanical and Manufacturing Engineering (ICME2018) IOP Conf. Series: Journal of Physics: Conf. Series 1150 012038 IOP Publishing.https://doi.org/10.1088/1742-6596/1150/1/012038

Meijs, M. & Knaack, U. (2012). Components and Connections: Principles of Construction. Berlin, Boston: Birkhäuser. https://doi.org/10.1515/9783034610636

Moon, R. J., Martini, A., Nairn, J., Simonsen, J., & Youngblood, J. (2011). Cellulose nanomaterials review: Structure, properties and nanocomposites. Chemical Society Reviews, 40(7), 3941. https://doi.org/10.1039/C0CS00108B

Naboni, R. and Jakica, N. (2022). Chapter 17 - Additive manufacturing in skin systems: trends and future perspectives. Rethinking Building Skins/Transformative Technologies and Research Trajectories Woodhead Publishing Series in Civil and Structural Engineering, Pages 425-451. https://doi.org/10.1016/B978-0-12-822477-9.00004-8

NIST NCSTAR1-3D 2005 Federal Building and Fire Safety Investigation of the World Trade Center Disaster. Mechanical Properties of Structural Steels. William E. Luecke J. David McColskey Christopher N. McCowan Stephen W. Banovic^ Richard J. Fields Timothy Foecke Thomas A. Siewert Frank W. Gayle

Ostadfar A. (2016). Fluid Mechanics and Biofluids Principles. PhD, in Biofluid Mechanics, https://doi.org/10.1016/C2014-0-01583-3

Pandey J.K. (2015) Handbook of Polymer Nanocomposites. Processing, Performance and Application – Volume C: Polymer Nanocomposites of Cellulose Nanoparticles, Springer-Verlag Berlin Heidelberg 2015 https://doi.org/10.1007/978-3-642-45232-1_72

Paoletti, P., Farina, I., Scalise, L., and Rossi, M. (2020). Investigation on the Accuracy of Consumer 3D Printers for the Manufacturing of Scale Models for Structural Analysis. Materials, 13(16), 3664.

Paul, C.P., and Jinoop, A.N. (2023). Additive Manufacturing: Bringing a Paradigm Shift. In: Shrivastava, A., Arora, A., Srivastava, C., Dhawan, N., Shekhar Singh, S. (eds) New Horizons in Metallurgy, Materials and Manufacturing. Indian Institute of Metals Series. Springer, Singapore. https://doi.org/10.1007/978-981-19-5570-9_3

Plaza, E. G.; López, P. N.; Torija, M. Á.; and Muñoz, J. M. (2019). Analysis of PLA Geometric Properties Processed by FFF Additive Manufacturing: Effects of Process Parameters and Plate-Extruder Precision Motion. Polymers 2019, 11, 1581 https://doi.org/10.3390/polym11101581 www.mdpi.com/journal/polymers.

Polyzos, E.; and Pyl, L. 2022. Effect of Printing Direction on the Elastic Properties of 3D-Printed Nylon Materials. Phys. Sci. Forum. https://doi.org/10.3390/ psf2022004021

Polyzos, E.; Katalagarianakis, A.; Polyzos, D.; Van Hemelrijck, D.; and Pyl, L. (2020). A multi-scale analytical methodology for the prediction of mechanical properties of 3D-printed materials with continuous fibres. Addit. Manuf. 2020, 36, 101394.https://doi.org/10.1016/j.addma.2020.101394

Prusa3D_ help und Guide.

Rajabi, H., and Fathi, M. (2017). Applications of carboxymethyl cellulose (CMC) in pharmacy and pharmaceutical industry. Cellulose, 24(6), 2629-2640.

Redwood, B., Schöffer, F. and Garrent, G. (2018). The 3D printing Handbook: technologies, Design and applications. Amsterdam, 3D hubs B.V.

Redwood, B.; Schöffer F.; and Garret, B. (2017). The 3D Printing handbook: Technologies, Design and Applications. Coers & Roest © 3D Hubs B.V. ISBN, 9082748509, 9789082748505.

Rigotti, D; Dorigato, A; Cataldi, A; Fambri, L; and Pegoretti, A. (2020). Nanocellulose as reinforcing agent for biodegradable polymers in 3D printing fused deposition modeling, in:ECCM 2018 - 18th Eur. Conf. Compos. Mater., 2020, pp. 24–28.https://doi.org/10.1016/j.mtcomm.2018.02.007

Saiful S.; Hasima, S.; Kamila, N.; and Rahmi S. (2022). Cellulose acetate from palm oil bunch waste for forward osmosis membrane in desalination of brackish water. Results Eng. 2022, 15, 100611.

Schirmer, J.; Roudenko, J.; Reichenberger, M.; Neermann, S.; Franke, J. (2018) Print Quality Assessment by Image Processing Methods for Printed Electronics Applications. 41st International Spring Seminar on Electronics Technology (ISSE). https://doi.org/10.1109/ISSE.2018.8443617

Schirmer, M.; Jekle, M. and Becker, T. (2015). Starch gelatinization and its complexity for analysis. 2014 WILEY-VCH Verlag GmbH & Co. KGaA, Weinheim. Starch/Stärke 2015, 67, 30–41. https://doi.org/10.1002/star.201400071

Schmidt, J. Huber, T. and Müssig, J. (2022). Improving Material Property Understanding with Virtual Experiments: A New Approach to Teach About Mechanical Testing of Materials Using an Interactive Desktop App. https://doi.org/10.1021/acs.jchemed.1c01087

Shamsuri, A. A., Abdan, K. and Jamil, S. N. (2021). Properties and applications of cellulose regenerated from cellulose/imidazolium-based ionic liquid/co-solvent solutions: A short review. e-Polymers, vol. 21, no. 1, 2021, pp. 869-880. https://doi.org/10.1515/epoly-2021-0086

Siqueira, G., Bras, J., & Dufresne, A. (2010). Cellulosic bionanocomposites: a review of preparation, properties and applications. Polymers, 2(4), 728-765.

Slotte, S. (2021). Production process of carboxymethyl cellulose University of Oulu, Bachelor's thesis of Process Engineering at the university: Markku Ohenoja D.Sc.

Song, J., Su, X. and Liu, X. (2019), "Study on shape retention properties of filament/staple fiber composite yarns and fabrics", International Journal of Clothing Science and Technology, Vol. 31 No. 6, pp. 790-801. https://doi.org/10.1108/IJCST-12-2018-0164

Tan, X., Liu, S., Cao, Y., & Li, W. (2021). Design and manufacturing of paper-based actuators by 3D printing. Journal of Manufacturing Processes, 61, 786-796.

Tekinalp; HL; Meng, X. Lu, V Kunc, LJ Love, WH Peter, (2019). High modulus biocomposites via additive manufacturing: cellulose nanofibril networks as "microsponges, Compos. Part B Eng. 173 106817, https://doi.org/10.1016/j.compositesb.2019.05.028

Thakker, M.; Shukla, P.; and Shah D. O. (2015). Surface and colloidal properties of chalks: A novel approach usingsurfactants to convert normal chalks into dustless chalks. 0927-7757/© 2015 Elsevier B.V. All rights reserved: Colloids and Surfaces A: Physicochemical and Engineering Aspects 480 (2015) 236–244 http://dx.doi.org/10.1016/j.colsurfa.2015.01.054

Thibaut, C. (2020). Development of fibrous cellulosic materials for the production of bio-based 3D printed objects by extrusion. Material chemistry. Université Grenoble Alpes English.2020GRALI005ff. fftel-02570560f. https://tel.archives-ouvertes.fr/tel-02570560

Torres, F. G., Commeaux, S., & Troncoso, O. P. (2012). Biocompatibility of bacterial cellulose based biomaterials. Journal of Functional Biomaterials, 3(4), 864–878. https://doi.org/10.3390/jfb3040864

Tu, K.; Ding, Y.; Keplinger, T. (2022). Review on design strategies and applications of metal-organic framework-cellulose composites, Carbohydrate Polymers, Volume 291, 119539, ISSN 0144-8617, https://doi.org/10.1016/j.carbpol.2022.119539

Utz, J.; Zubizarreta, J.; Geis, N.; Immonen, K.; Kangas, H.; Ruckdäschel, H.(2022). 3D Printed Cellulose-Based Filaments— Processing and Mechanical Properties. Materials 2022, 15, 6582. https://doi.org/10.3390/ma15196582

Voisin, H. P., K Gordeyeva, G Siqueira, M K. Hausmann, A R. Studart, and Lennart Bergström. (2018). 3D Printing of Strong Lightweight Cellular Structures Using Polysaccharide-Based Composite Foams ACS Sustainable Chemistry & Engineering, 17160-17167https://doi.org/10.1021/acssuschemeng.8b04549

Waghmare, A. and Sahoo, P. (2017). Adhesive Friction Based on Accurate Elastic-Plastic Finite Element Analysis and n-Point Asperity Concept. International Frontier Science Letters. 11. 1-28. 10.18052/www.scipress.com/IFSL.11.1.

Wang Q., C Ji, L Sun, J Sun, J. Liu, (2020). Cellulose nanofibrils filled poly (lactic acid) biocomposite filament for FDM 3D printing, Molecules 25https://doi.org/10.3390/molecules25102319

Wang, Z.; and Liu, F.; (2021). Industrial Ventilation Design Guidebook. Volume 2: Engineering Design and Applications. 2nd Edition . Editors: Howard Goodfellow, Yi Wang. Paperback ISBN: 9780128166734. eBook ISBN: 9780128167793. Chapter 7 - Environmental assessment tools. Pages 435-448.

Xia, M.; Nematollahi, B.; and Sanjayan J. G. (2019). Chapter 11- Development of Powder-Based 3D Concrete Printing Using Geopolymers. Editor(s): Jay G. Sanjayan, Ali Nazari, Behzad Nematollahi. 3D Concrete Printing Technology, Butterworth-Heinemann, Pages 223-240, ISBN 9780128154816. https://doi.org/10.1016/B978-0-12-815481-6.00011-7.

Xiao, S.; Gao, R.; Gao, L.; Li, J. Poly(vinyl alcohol) films reinforced with nanofibrillated cellulose (NFC) isolated from corn husk by high intensity ultrasonication. Carbohydr. Polym. 2016, 136, 1027–1034.

Xiao, S.; Gao, R.; Gao, L.; Li, J. Poly(vinyl alcohol) films reinforced with nanofibrillated cellulose (NFC) isolated from corn husk by high intensity ultrasonication. Carbohydr. Polym. 2016, 136, 1027–1034.

Xiaoyu Bi, Runzhou Huang (2022). 3D printing of natural fiber and composites: A state-of-the-art review, Materials & Design, Volume 222, 111065, ISSN 0264-1275. https://doi.org/10.1016/j.matdes.2022.111065

Xin, Q. (2013). Diesel Engine System Design Chapter 10: Friction and lubrication in diesel engine system design in Diesel Engine System Design, 20132013, Pages 651-758https://doi.org/10.1533/9780857090836.3.651

Xu X, and Lin P. Parameter identification of sound absorption model of porous materials based on modified particle swarm optimization algorithm. PLoS One. 2021 May 4;16(5): e0250950. PMID: 33945538; PMCID: PMC8096118. https://doi.org/10.1371/journal.pone.0250950.

Xu, F.; Zhang, L.; Liu, W.; Liu, Q.; Wang, F.; Zhan, H.; Honghai; and Blecker, Ch. (2021) Physicochemical and Structural Characterization of Potato Starch with Different Degrees of Gelatinization. Foods 2021, 10, 1104. https://doi.org/10.3390/foods10051104

Xu, Q ; Nakajima, M; Liu, Z; and Shiina, T. (2011). Soybean-based Surfactants and Their Applications, Soybean - Applications and Technology, Prof. Tzi-Bun Ng (Ed.), ISBN: 978-953-307-207-4.

Youssef, A. M., El-Sayed, H., & El-Zeany, B. A. (2018). Applications of carboxymethyl cellulose (CMC) in food industry: A review. Food Hydrocolloids, 81, 1-11.

Yu Liu, Bo Li, Wanbo Mao, Wen Hu, Gang Chen, Yingyao Liu, and Zhiqiang. Fang Strong Cellulose-Based Materials by Coupling Sodium Hydroxide–Anthraquinone (NaOH–AQ) Pulping with Hot Pressing from Wood. ACS Omega 2019 4 (4), 7861-7865 https://doi.org/10.1021/acsomega.9b00411

Yüce, C., Willenbacher, N. Challenges in Rheological Characterization of Highly Concentrated Suspensions — A Case Study for Screenprinting Silver Pastes. J. Vis. Exp. (122), e55377, doi:10.3791/55377 (2017) https://doi.org/10.3791/55377

Zhang, Y., Jia, J., Guo, J., & Han, G. (2020). Study on 3D printing technology of paper pulp. Journal of Porous Materials, 27(1), 17-27

3D paper printed modelusing 1.2 mm nozzle diameter.

Appendix A: Material properties and preparations

A.1 Cellulose

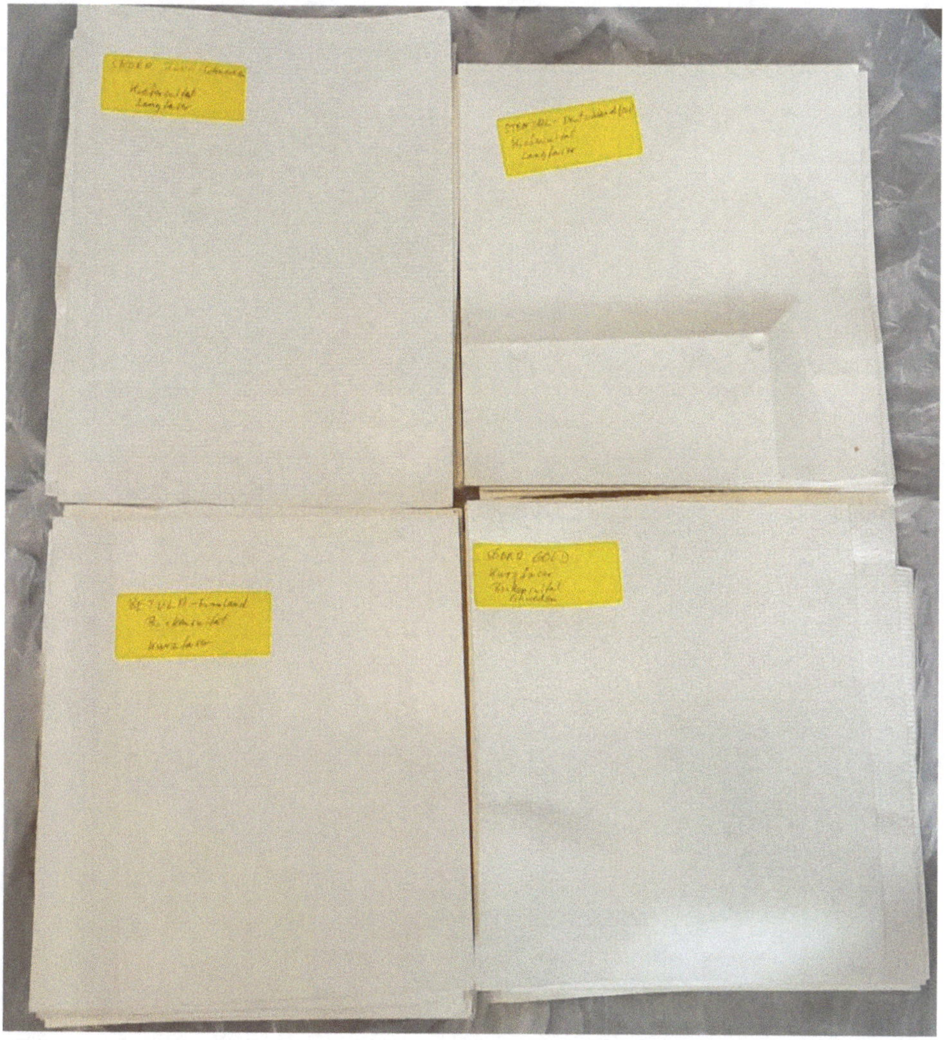

Figure A.1 Form of cellulose fibers that used for 3D paper printing.

D. Abdullah Agha, *3D Paper Printing for the Built Environment*, Mechanik, Werkstoffe und Konstruktion im Bauwesen 73,
https://doi.org/10.1007/978-3-658-45076-2

A.1.4 Pine STENDAL Deutschland.

MERCER PULP

ZELLSTOFF STENDAL

Stendal ECF-Typical properties

NBSK made of	
Pine	40 - 70%
Spruce	30 - 60%

PFI-mill at 23°C and 50% relative humidity

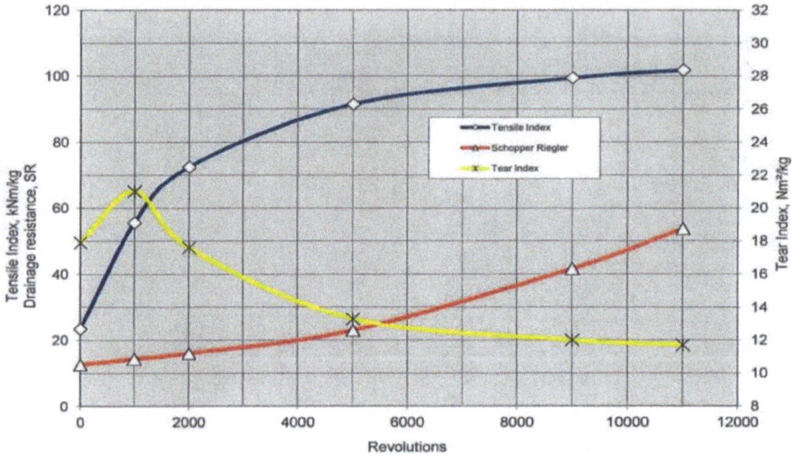

Beating revolutions		0	1000	2000	5000	9000	11000	6800	3000
Density	kg/m³	581	646	689	751	791	803	771	713
Drainage resistance*	SR	12,6	14,3	16,0	23,0	41,7	53,8	30	18,0
Breaking length	m	2386	5671	7395	9424	10138	10567	9792	8240
Tensile index	kNm/kg	23,4	55,6	72,5	91,5	99,4	101,8	96,0	80
Tear Index	Nm²/kg	17,9	21,0	17,6	13,3	12,0	11,7	12,8	15,9
Elongation	%	2,4	3,8	4,0	4,2	4,0	3,9	4,0	4,0
Burst Index	MN/kg	1,5	4,0	5,2	7,0	7,4	7,6	7,2	6,1
Gurley air permeability	s / 100 ml	1,7	3,4	6,1	26,8	117	237	62	9,1
Light scattering coefficient	m²/kg	32,4	26,3	23,2	20,2	18,4	17,6	19,0	22,2
Opacity	%	76,4	72,0	70,0	66,5	64,3	62,7	66,0	68,3
Quantity of analysis		160	27	27	1211	69	27	Interpolated	

*Analysed using tap water
Sheet former (Rapid Köthen)

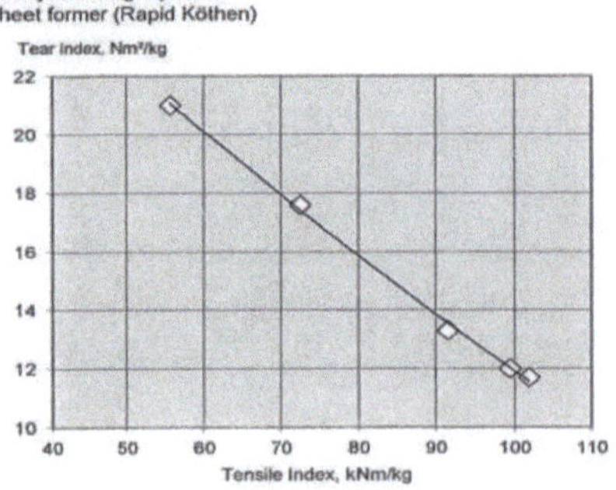

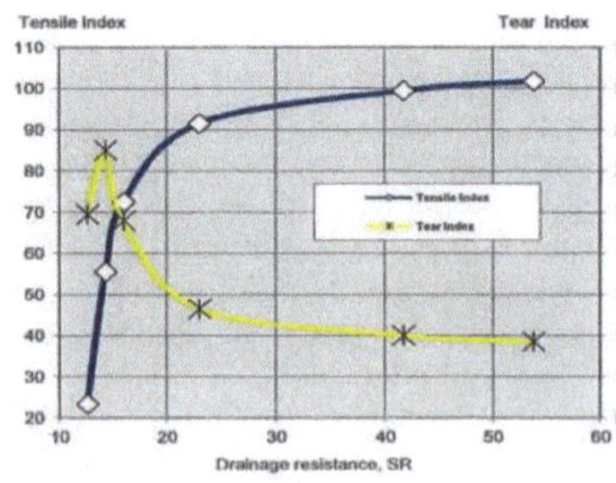

Date: 09.03.2017

Figure A.2 Data sheet of Pine STENDAL Deutschland

Mercer Pulp Stendal ECF

General Properties	Typical values	Guarantee value	Unit	Method SCAN	Method EN ISO	Frequency
ISO Brightness	89,4	> 88,5	% ISO	-	3688	1 Day
Sheet brightness	88,7	-	%	-	2470	24 Day
Brightness reversion	3,7	-	%	-	5630-1*	1 Day
Dirt count	1,0	< 3	mm^2/kg	-	5350-2	48 Day
Dirt count	1,0	< 3	ppm	-	5350-2	48 Day
Intrinsic Viscosity	734	> 600	ml / g	-	5351	6 Day
Extractives	0,06	< 0,10	%	CM 49:03	-	4 Month
Ash	0,20	< 0,30	%	-	1762	4 Month
pH	5,3	> 5,00		P 14:65**	-	4 Month
Organic chlorine in pulp	61	< 100	mg/kg	CM 52:94	-	12 Month
AOX removable by washing	3	< 5	mg/kg	CM 44:97	-	4 Month

Fibre dimensions, Pulp Eye	Typical values	Unit	Range ± 2 SD***	CoV****	Method SCAN	Method EN ISO	Frequency
Fibre length	2,10	mm	0,09	2,1	-	-	Continous
Fibre width	28,2	µm	0,7	1,2	-	-	Continous

Preparation of test samples	Target	Variation	Unit	Method SCAN	Method EN ISO	
Temperature	23	±1	°C	-	187	Continous
Relative humidity	50	±2	% RH	-	187	Continous
Preparation of laboratory sheets				-	5269-2	

Physical Properties	Typical value	Unit	Range ± 2 SD***	CoV**** %	Method SCAN	Method EN ISO	Frequency
PFI beating revolutions	0				-	5264-2	
Water Retention Value	1,10	g H_2O/g pulp	0,06	2,5	C 62:00	-	1 Day
Density	581	kg/m^3	640	55,1	-	534	1 Day
Drainage resistance	13	SR	5	18,2	-	5267-1	1 Day
Breaking length	2386	m	201	4,2	-		1 Day
Tensile index	23,4	kNm/kg	2,0	4,2		1924-3	1 Day
Tear index	17,9	Nm^2/kg	10,7	29,9	-	1974	1 Day
Burst index	1,5	MN/kg	0,2	5,8	-	2758	1 Day
PFI beating revolutions	2000				-	5264-2	
Density	689	kg/m^3	15	1,1	-	534	4 Month
Drainage resistance	16	SR	1	3,7	-	5267-1	4 Month
Breaking length	7395	m	398	2,7	-		4 Month
Tensile index	72,5	kNm/kg	3,9	2,7		1924-3	4 Month
Tear index	17,6	Nm^2/kg	1,6	4,5	-	1974	4 Month
Burst index	5,2	MN/kg	0,4	3,4	-	2758	4 Month
PFI beating revolutions	5000				-	5264-2	
Density	751	kg/m^3	20	1,3	-	534	6 Day
Drainage resistance	23	SR	2	4,4	-	5267-1	6 Day
Breaking length	9424	m	472	2,5	-		6 Day
Tensile index	91,5	kNm/kg	4,6	2,5		1924-3	6 Day
Tear index	13,3	Nm^2/kg	1,9	7,2	-	1974	6 Day
Burst index	7,0	MN/kg	0,5	3,4	-	2758	6 Day
PFI beating revolutions	9000				-	5264-2	
Density	791	kg/m^3	32	2,0	-	534	8 Month
Drainage resistance	42	SR	8	9,4	-	5267-1	8 Month
Breaking length	10138	m	578	2,9	-		8 Month
Tensile index	99,4	kNm/kg	5,7	2,9	-	1924-3	8 Month
Tear index	12,0	Nm^2/kg	1,1	4,4	-	1974	8 Month
Burst index	7,4	MN/kg	0,5	3,6	-	2758	8 Month

* time 24 h
** use of KCl instead of deion. Water
*** standard deviation
**** coefficient of variation, SD*100/average

Approval:

A. Koppensteiner
Managing Director

i. V. Dr. Martin Zenker
Production Manager

Date: 09.03.2017

Figure A.2 continued

A.1.5 Short cellulose fiber, Aspen PHK, DMI Canada.

Peace River Pulp Division

ASPEN HARDWOOD KRAFT PULP

Life begins in the forest and that is why DMI manages this precious resource with responsible practices.

DMI is dedicated to the sustainable use and responsible stewardship of the forest resources entrusted to our care. In keeping with this commitment, DMI has implemented a very deliberate and comprehensive certification strategy for it Forests and Mill Operations positioning the organization as a world leader. DMI has chosen to pursue the Sustainable Forestry Initiative for its Forest Management Agreement area with an Environmental Management System, based on ISO 14001:2004, providing the foundation and direction. Our mill operations are also certified with the ISO 14001:2015 and 9001:2015 Environmental & Quality Management Systems. Our certification achievements confirm that we are meeting the highest environmental standards and ultimately, results in higher levels of environmental performance, providing a framework for continued improvement.

DAISHOWA-MARUBENI INTERNATIONAL LTD.

700, 510 Burrard Street
Vancouver, BC, Canada
V6C 3A8

Tel: 604-684-4326
Fax: 604-681-8659

Exclusive sales agent:
Marubeni Corporation Pulp Department

Tokyo Nihombashi Tower
2-7-1 Nihonbashi, Chuo-ku,
Tokyo, 103-6060 Japan

Tel: 03-3282-4467
Fax: 03-3282-9657

SPECIES

Trembling Aspen (85%)
(Populous tremuloides)

Balsam Poplar (15%)
(Populus balsamifera)

Figure A.3 Data sheet of Aspen PHK, DMI Canada.

DAISHOWA-MARUBENI INTERNATIONAL LTD.

Peace River Pulp Division

Sustainable. By Design.

Revised July 31, 2018

Bleached Aspen Hardwood Kraft

Grade Profile

Peace River Aspen Hardwood Kraft pulp has short, flat, uniform, ribbon-like fibres and thin walled, low proportion by weight non-interfering vessel elements. These attributes uniquely qualify the pulp to be used in a wide variety of applications including coated base stock and uncoated papers, tissue, photographic papers, technical and other specialty papers.

Typical Response to PFI Refining and Handsheet Data*

		PFI Revolutions: 0	500	1500	5000
Canadian Standard Freeness	mL	483	406	347	237
Opacity	%	78.9	77.5	76.4	74.0
Bulk	cm^3/g	1.60	1.55	1.50	1.41
Burst Index	kN/g	1.1	1.7	2.2	3.1
Multi ply Tear Index	$mN.m^2/g$	4.0	5.4	6.0	7
Single ply tear index	$mN.m^2/g$	3.0	4.1	4.8	6
Tensile	kNm/kg	26.7	35.6	42.0	53.6
Breaking Length	km	2.73	3.6	4.3	5.5
Stiffness	mNm/kg	5.1	5.8	6.2	7
Internal Bond	J/m^2	111	144	170	233

* Typical Characteristics

Typical Pulp Characteristics

Pulp Sheet Properties

Brightness (ISO)	≥ 88.0%
Dirt Specks (ISO)	≤ 5.0 mm^2/m^2 (ppm)
Moisture (Air Dry)	96 ± 2%

Chemical Properties

Ash Content	0.46%	Alpha Cellulose	93.0%
Extractives (Acetone)	0.33%	Beta Cellulose	5.65%
Viscosity (CED)	≥18.0 CPS	Gamma Cellulose	1.34%

Fibre Properties

Species	Aspen
Fibre length	0.75 mm
Fibre width	27.4 µm
Fibres per gram	14 million
Fibre coarseness	0.10 mg/m

Bale Dimensions

Size

Length	80.2 cm (31.5 in.)
Width	81.3 cm (32 in.)
Height	40.6 cm (16 in.)

Mass

Weight	263 kg (578 lb.) ± 2.5%

ISO 9001

ISO 14001

DAISHOWA-MARUBENI INTERNATIONAL LTD.

Peace River Pulp Division

Postal Bag 4400, #1 Pulp Mill Site Road

Peace River, AB T8S 1V7

www.dmi.ca

PEFC

SUSTAINABLE FORESTRY INITIATIVE

Figure A.3 Continued.

A.2 Chalk

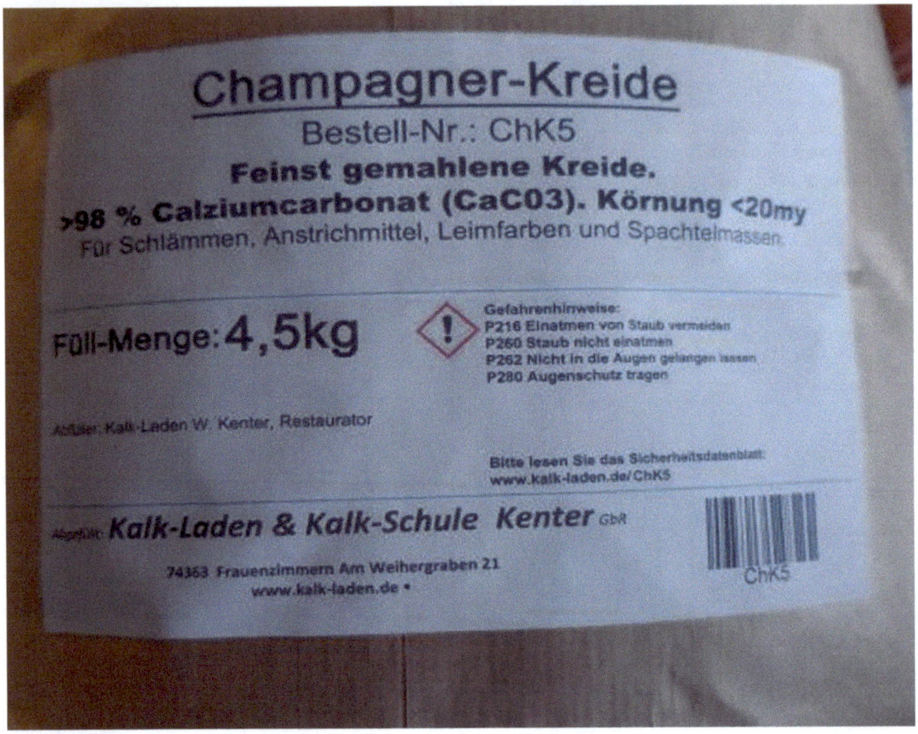

Figure A.4 Label cover of the used chalks.

A.3 Peel ply fabric

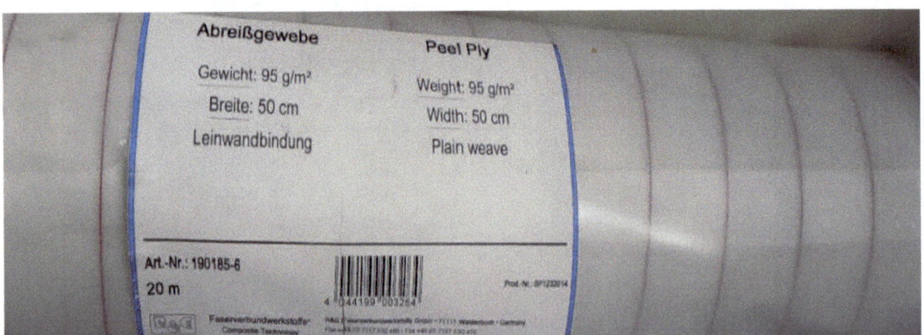

Figure A.5 Photo of used peel ply fabric.

A.4 Wood fibers

Gütekriteren für Holzfasern als Substratausgangsstoff (RAL-GZ 250/5-1)

RAL GÜTEZEICHEN

			Wertebereiche
1	**Physikalische Eigenschaften**		
1.1	Rohdichte trocken	[g/l]	wird analysiert
1.2	Trockenmasse	[Gew-%]	wird analysiert
2	**Chemische Eigenschaften**		
2.1	pH-Wert		≤ 6,5
2.2	Salzgehalt (KCl)	[g/l]	≤ 0,5
2.3	Lösliche Hauptnährstoffe		
2.3.1	Stickstoff (NH_4-N + NO_3-N)	[mg/l]	≤ 50
2.3.2	Phosphor (P_2O_5)	[mg/l]	≤ 50 (CAT-Extrakt) ≤ 100 (CAL-Extrakt)
2.3.3	Kalium (K_2O)	[mg/l]	≤ 100 (CAT-Extrakt) ≤ 150 (CAL-Extrakt)
2.4	Schwermetalle (Gesamtgehalte)		
2.4.1	Arsen (As)	[mg/kg TM]	≤ 40
2.4.2	Blei (Pb)	[mg/kg TM]	≤ 150
2.4.3	Cadmium (Cd)	[mg/kg TM]	≤ 1,5
2.4.4	Chrom (Cr)	[mg/kg TM]	≤ 300
2.4.5	Nickel (Ni)	[mg/kg TM]	≤ 80
2.4.6	Quecksilber (Hg)	[mg/kg TM]	≤ 1
2.4.7	Thallium (Tl)	[mg/kg TM]	≤ 1
3	**Biologische Eigenschaften**		
3.1	Stickstoffdynamik [1)] (N-Immobilisierung/N-Mineralisierung)	[mg N/l]	Δ N ≤ 200 : Mischkomponente für Kultursubstrate **max. 20 Vol.-%**
			Δ N ≤ 100 : Mischkomponente für Kultursubstrate **max. 40 Vol-%**
3.2	Pflanzenschädigende Stoffe		keine pflanzenschädigende Wirkung
3.3	Unkrautbesatz		max. 1 keimender Same oder austreibendes Pflanzenteil / Liter Holzfaser
4	**Weitere Anforderungen**		
4.1	Fremdstoffe > 2 mm (z.B. Kunststoffe, Metalle, Glas)		keine
4.2	Steine > 10 mm		keine
5	**Auf Veranlassung des Güteausschusses**		
	Weitere, selten zu bestimmende Parameter werden im Einzelfall auf Veranlassung des Güteausschusses unter Einbeziehung der beteiligten Prüfinstitute festgelegt.		
	Deklaration [2)]		
	Produktionsstätte		
	max. zulässiger Anteil im Substrat		gemäß Tab. 5-1 Ziffer 3.1

1) Die Stickstoffdynamik bzw. der daraus resultierende max. Volumenanteil Holzfasern im Substrat ist dem Anwender mitzuteilen
2) Es sind die Vorgaben der Düngemittelverordnung einzuhalten

Gütegemeinschaft Substrate für Pflanzen e.V. Wunstorfer Landstraße 8 D-30453 Hannover

Figure A.6 The Data sheet of wood fiber used.

Appendix B: Used Devices

B.1 Microscope

a b

Figure B.1 Pictures for the used microscope show details(a) and (b).

B.2 Rheometer

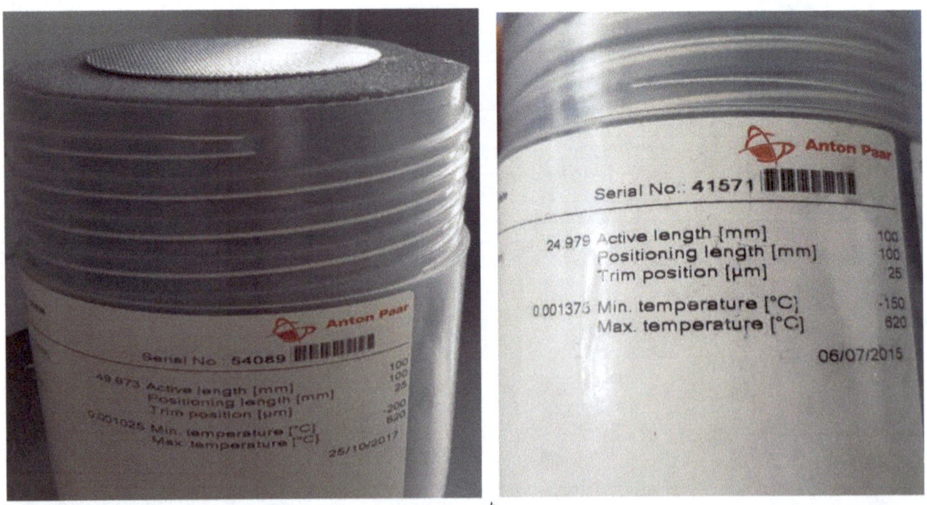

a b

Figure B.2 Pictures for the parallel plate of the used rheometer, show details(a) and (b).

Appendix C: WRV-Test Procedure

(10) International publication number WO 2015/054712 A2 By M.Sc. Frederic Kreplin / Department of Paper Manufacturing and Mechanical Process Engineering / TU Darmstadt.

WRV Test – PROTOKOLL
Allgemeine Informationen:

Anteil Klebefläche von Verpackung in %		Dicke µm	
g/m²			

Masse in g	23,28 Trocken				
Stoffdichte %					

WRV	Sieb leer [g]	Probe feucht mit Sieb [g]	Probe feucht ohne Sieb [g]	Probe trocken mit Sieb [g]	Probe trocken ohne Sieb [g]	WRV [%]
	0,565	1,147	0,582	0,888	0,323	55,498
	0,563	1,197	0,634	0,913	0,350	55,205
	0,522	1,124	0,602	0,855	0,333	55,316
	0,528	1,395	0,867	1,003	0,475	54,787
	0,562	1,187	0,625	0,911	0,349	55,840
	0,56	1,211	0,651	0,918	0,358	54,992
Mittelwert	0,55	1,21	0,66	0,91	0,36	55,27
Standardabw.	0,02	0,09	0,10	0,04	0,05	0,34
VK [%]	0,98	10,63	6,28	4,11	1,85	1879,85

Further remarks:
Nickel sieves: store at 105°C in a drying cabinet
Cooling down: 15 min in the desiccator then weigh out.
Centrifuge: speed with nickel screens: 5230 rpm,
Time: unground pulp 10 min, with ground pulp 15 min
Drying: 6 hours at 105°C in a drying cabinet, cooling down: 15 min in the desiccator then weigh out

$$\text{WRV } \% = \frac{\text{wet weight - dry weight}}{\text{dry weight}} \times 100$$

Appendix D: Design Details

a

b

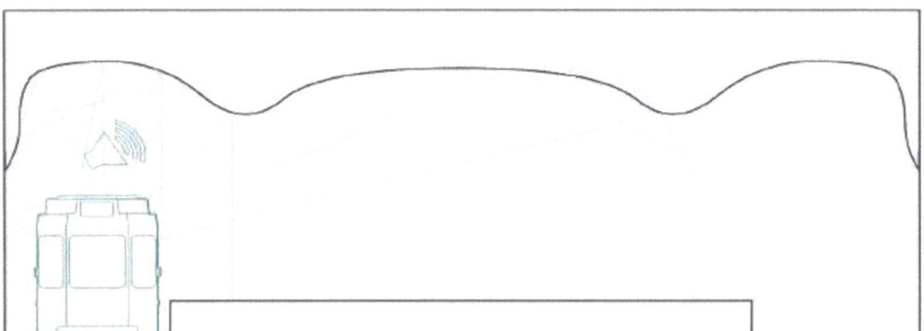

c

Figure D.1 Details for the printed specimen, a visualization image (a), front view (b), and sound orientation(c) these pictures are designed within the experimental façade course by Jonas Hartwig and Yannick Berold-Seguret.

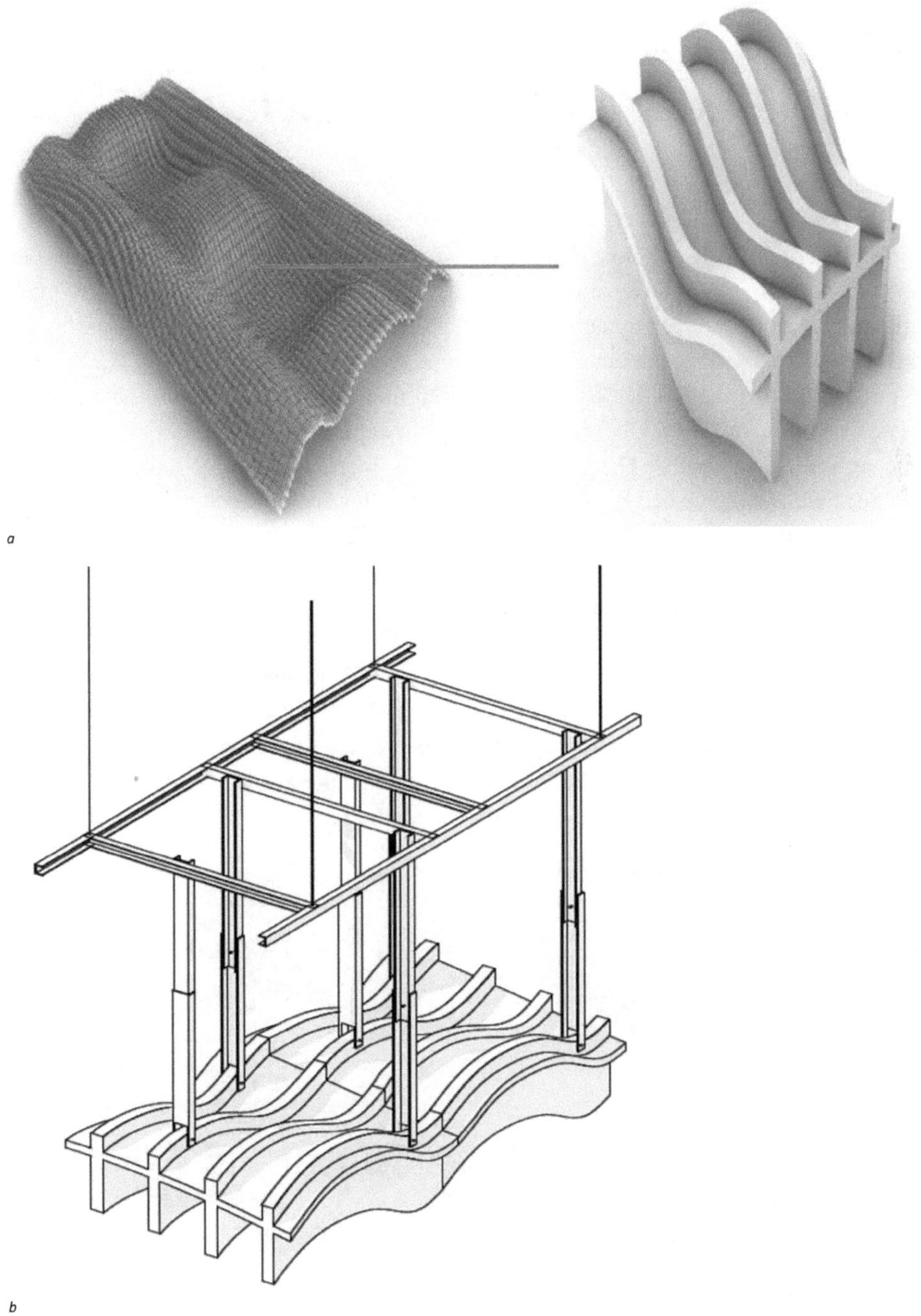

Figure D.2 Divided panel into printable elements(a) and the techniques of hanging(b), these pictures are designed within the experimental façade course by Jonas Hartwig and Yannick Berold-Seguret.

Appendix E: Adhesion Property of the Mixture to Other Surfaces

The adhesion properties of the mix describe its ability to adhere not only to itself but also to other surfaces, which is a critical factor in determining its potential uses. An innovative application of a highly adhesive mixture is as a welding material for paper products such as cardboard. In order to evaluate this property, a standard I-bone specimen made primarily of cardboard paper is used with a replaced portion of paper paste (R-paste) and then compared to a specimen made entirely of paper paste, as shown in Fig. E.2 (b). The R-paste specimen is modified by replacing a 20 mm intermediate section with a paper paste of type 12 in Table 3.9. The thickness of the replaced part is 2 mm more than the thickness of the cardboard. The paste overlaps both sides by 10 mm, as shown in Fig. E.2 (a). Specimens are prepared by hand and checked with a spatula and vernier. This means that the cellulose fibers are randomly distributed. The (Paste) sample has the same thickness as the central part of the paper paste of the (R-Paste) sample.
Figure 5.11(b) shows the stress-strain behavior under tensile loading for the paste section. As shown in figure 5.11 (c), the R-paste specimens broke at an ultimate stress of 1.25±0.1 MPa, 12% less than the paper-paste specimens. While the Young's modulus was approximately the same for both specimens. This indicates that the bond strength of the mixture within the replaced part in the (R-paste) specimen functions almost similarly to the specimen made whole from the paper paste, and that the bond strength of the mixture functions properly within the reference formulated mixture.

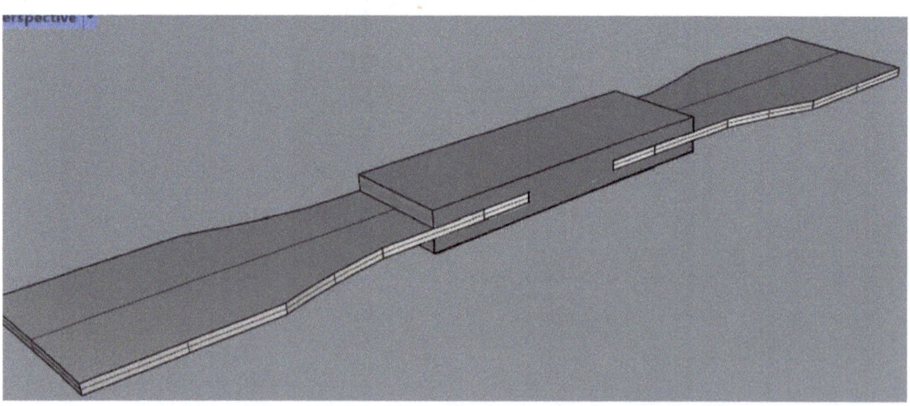

Figure E.1 specimen preparation to test the adhesion properties to other surfaces using tensile test.

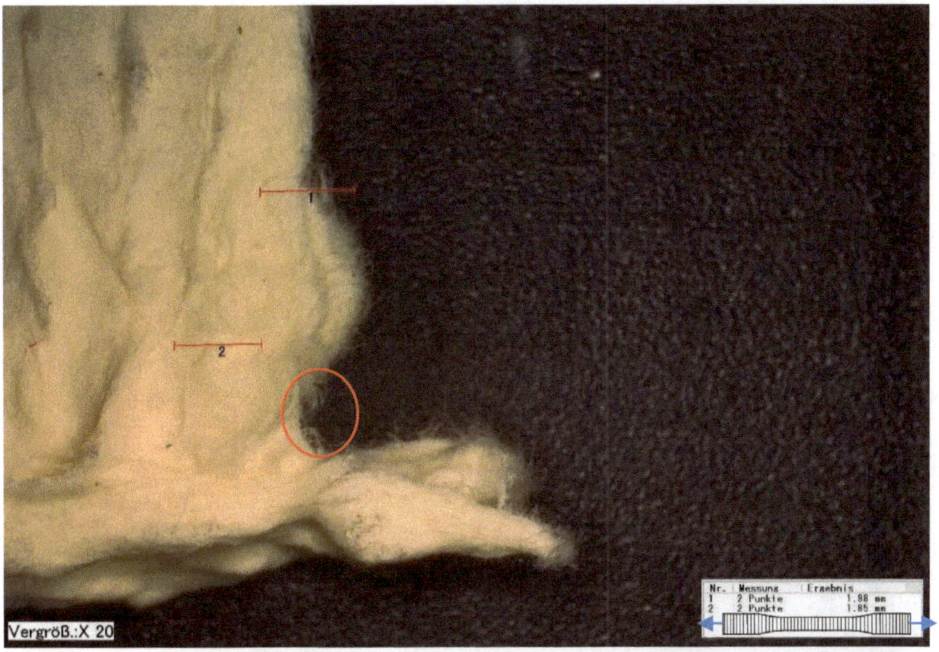

a

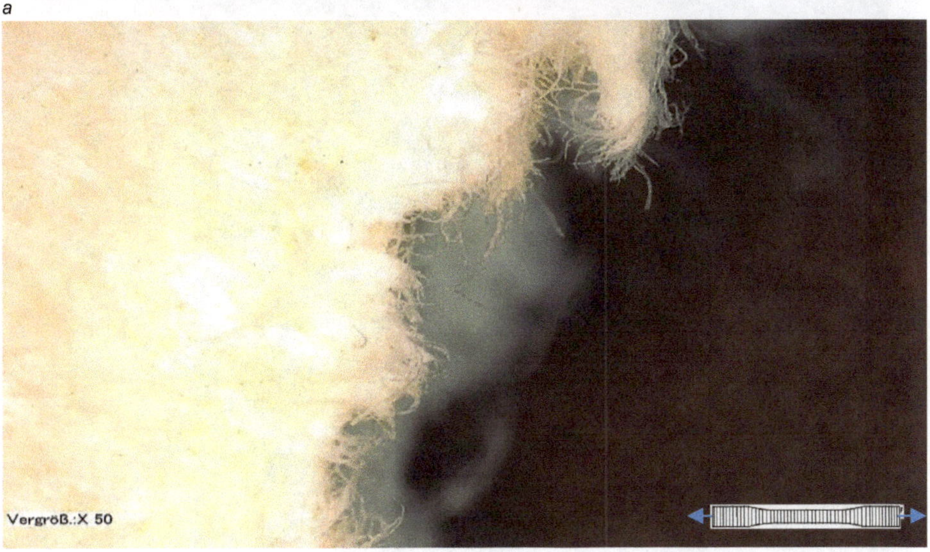

b

Figure E.2 Adhesion failure of the 90-degree filament orientation (a) and magnified area of adhesion failure (b).

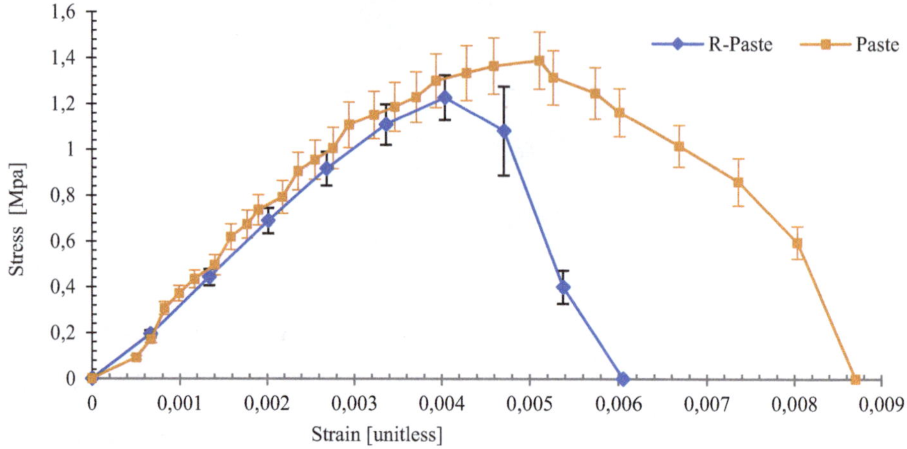

a

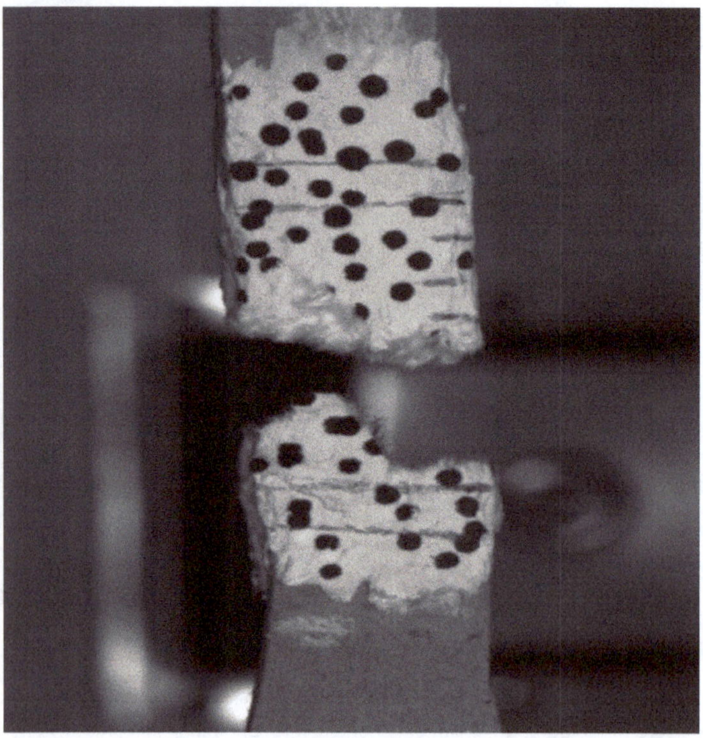

b

Figure E.3 The results of the tensile test for the adhesion properties to other surfaces (a) Stress-strain results and a picture for failure mode (b).

Appendix F: Acoustic Performance for AM Paper Material by Experimental Device

The acoustic properties for the 3D paper printing material are tested at the Department of System Reliability, Adaptive Structures, and Machine AcousticsSAM/TU Darmstadt by Dr.-Ing. Christian Adams. The tested samples are 4 solid samples with a diameter of ϕ9cm and a thickness of 1cm. Each sample has a different filling pattern.

a

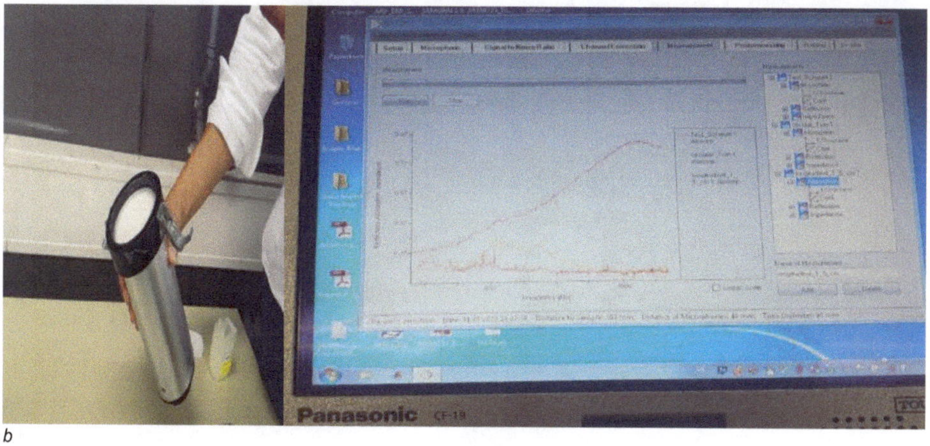

b

Figure F.1 The device and specimens for testing sound absorbing coefficient for 3D printed paper models.

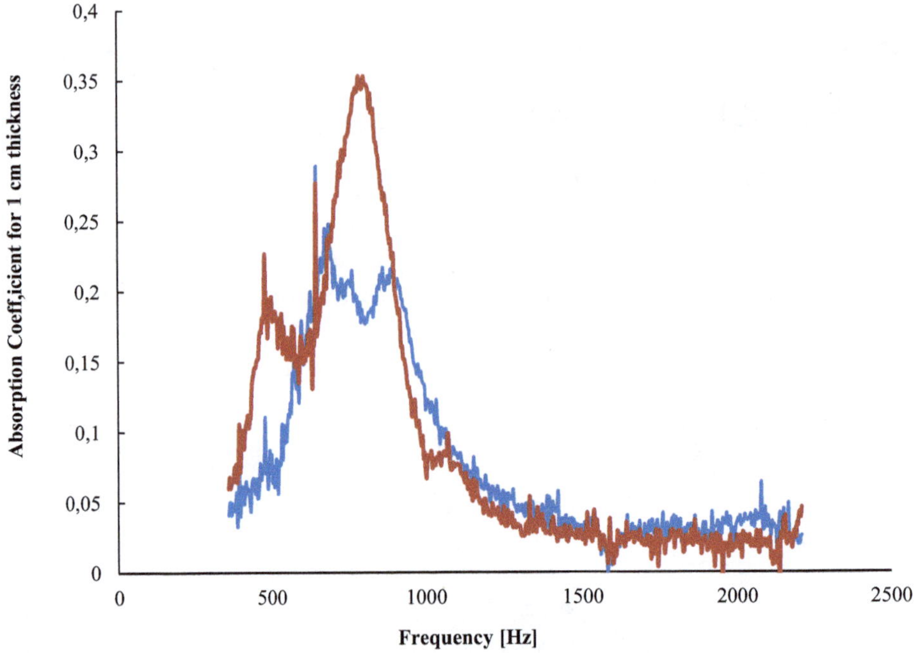

Figure F.2 The test results of the sound absorbing coefficient for 3D printed paper models.

Appendix G: 3D Models Details

This Appendix displays the 3D models' details. The geometrical designs were done by students of façade courses /TU Darmstadt during tutoring assignments. The design technique and principles of the façade construction of all the 3D model were based on the lectures by Professor Dr. Ing. Ulrich Knaack, which could be referred to as:[Meijs, M. and Knaack, U. 2012], [Knaack, U. et al 2014], [Knaack, U. and Koenders, E. 2018].

G.1 Acoustic Simulation

The simulation for the 3D models was done using Grasshopper & Pachyderm; as shown in Fig. G.2, to find or adjust the geometry in a specific way that give a better performance in diffusing the acoustic. The animation of the sound progress path is captured in Fig. G.1 and G.3.

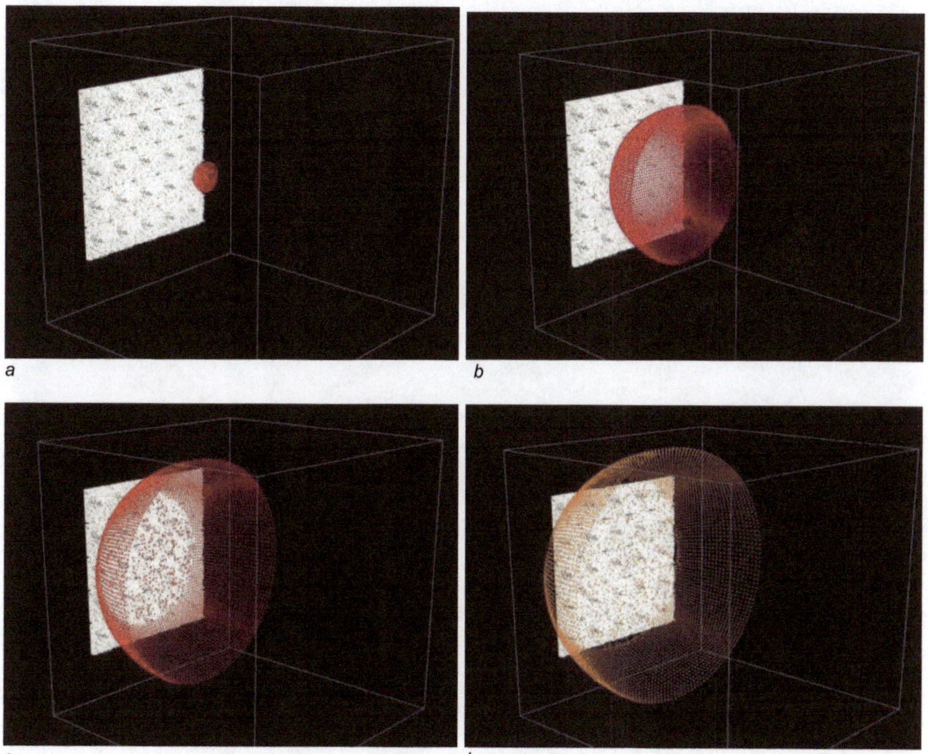

a b c d

Figure G.1 The animation of the acoustic progress on the perspective view (a), before hitting the panel (b), after hitting the surface (c), diffusing execution (d).

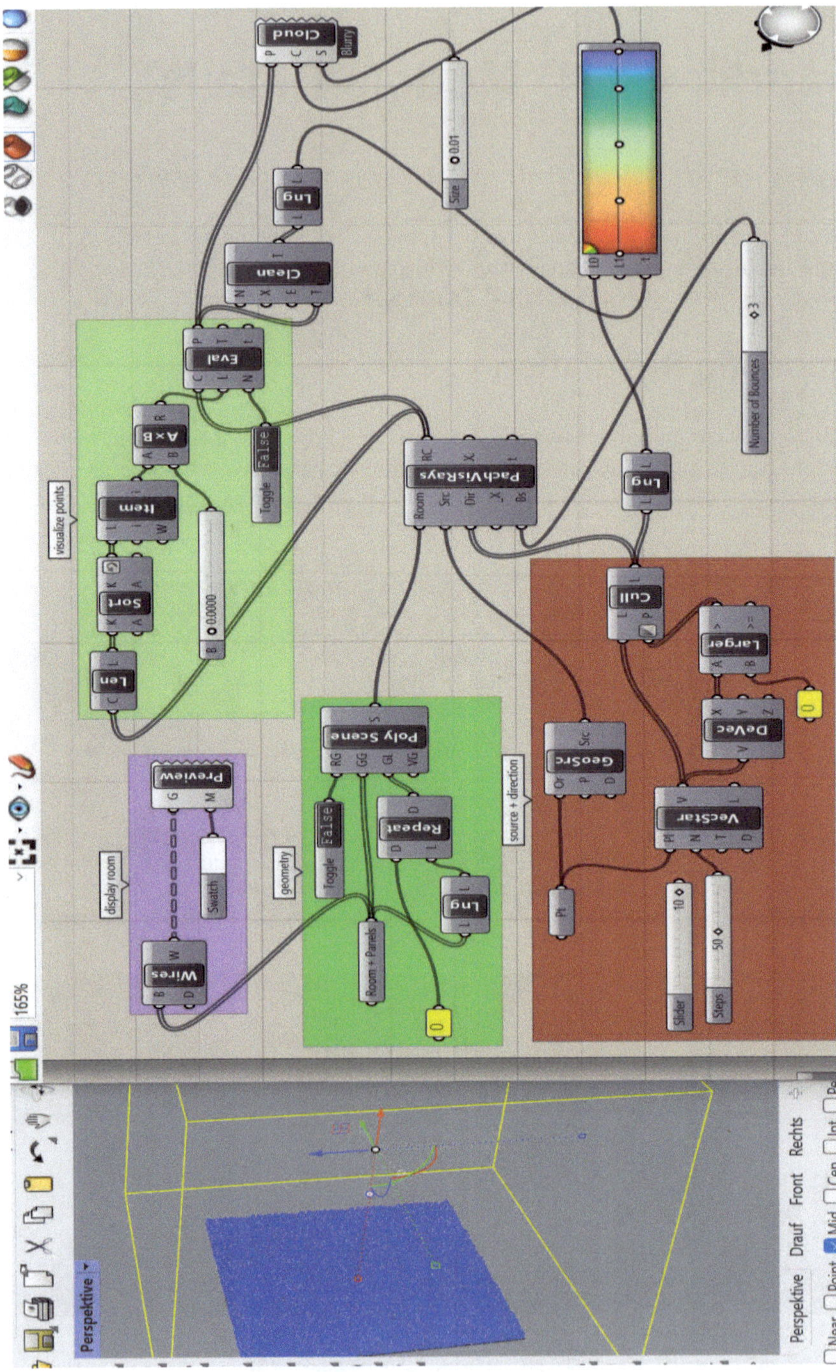

Figure G.1 Acoustic simulation using Grasshopper& Pachyderm.

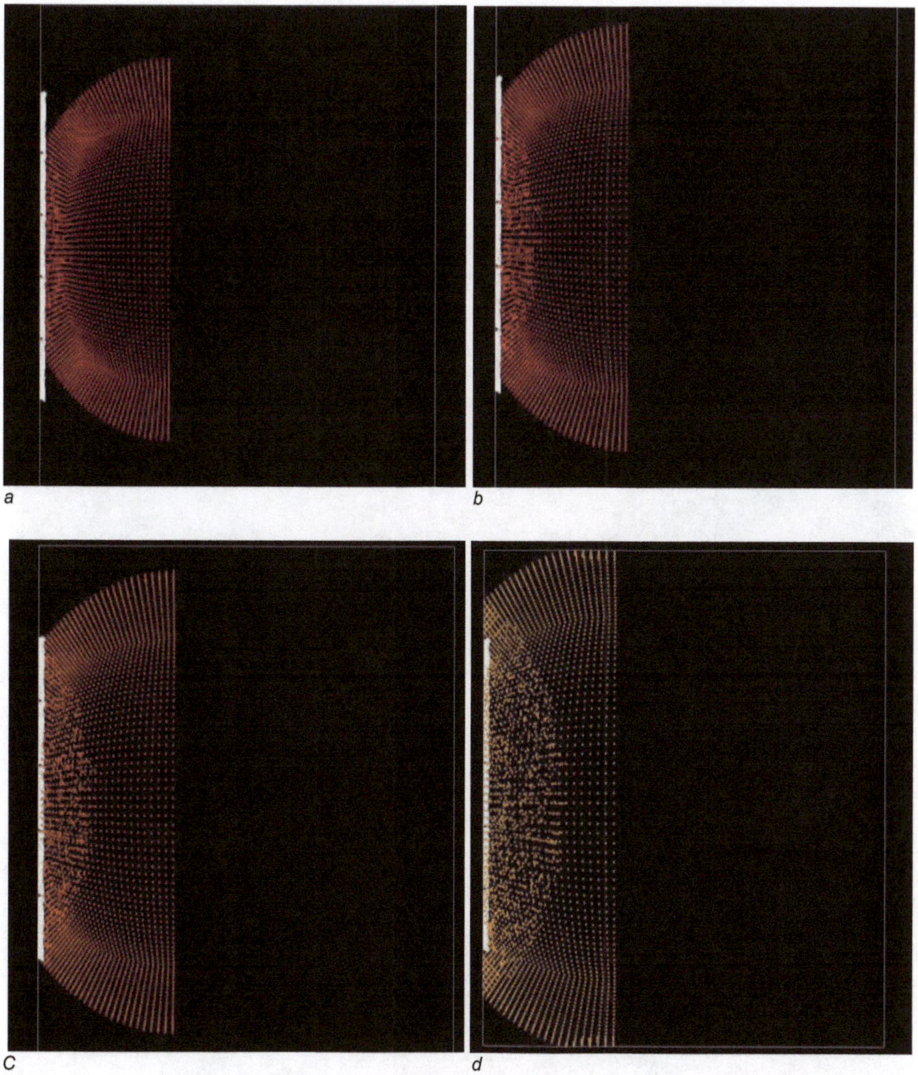

Figure G.3 The animation of the acoustic progress hit the surface on the side view (a) before hitting the panel (b), after hitting the surface (c), and diffusing execution (f).

G1.1 Gills panel

The details of the parametric design are listed in Fig. B.2, B.3, B.4, B.5, and B.6. The design of the geometry was made by Katja Heilingbrunner/ student within façade course at TU Darmstadt.

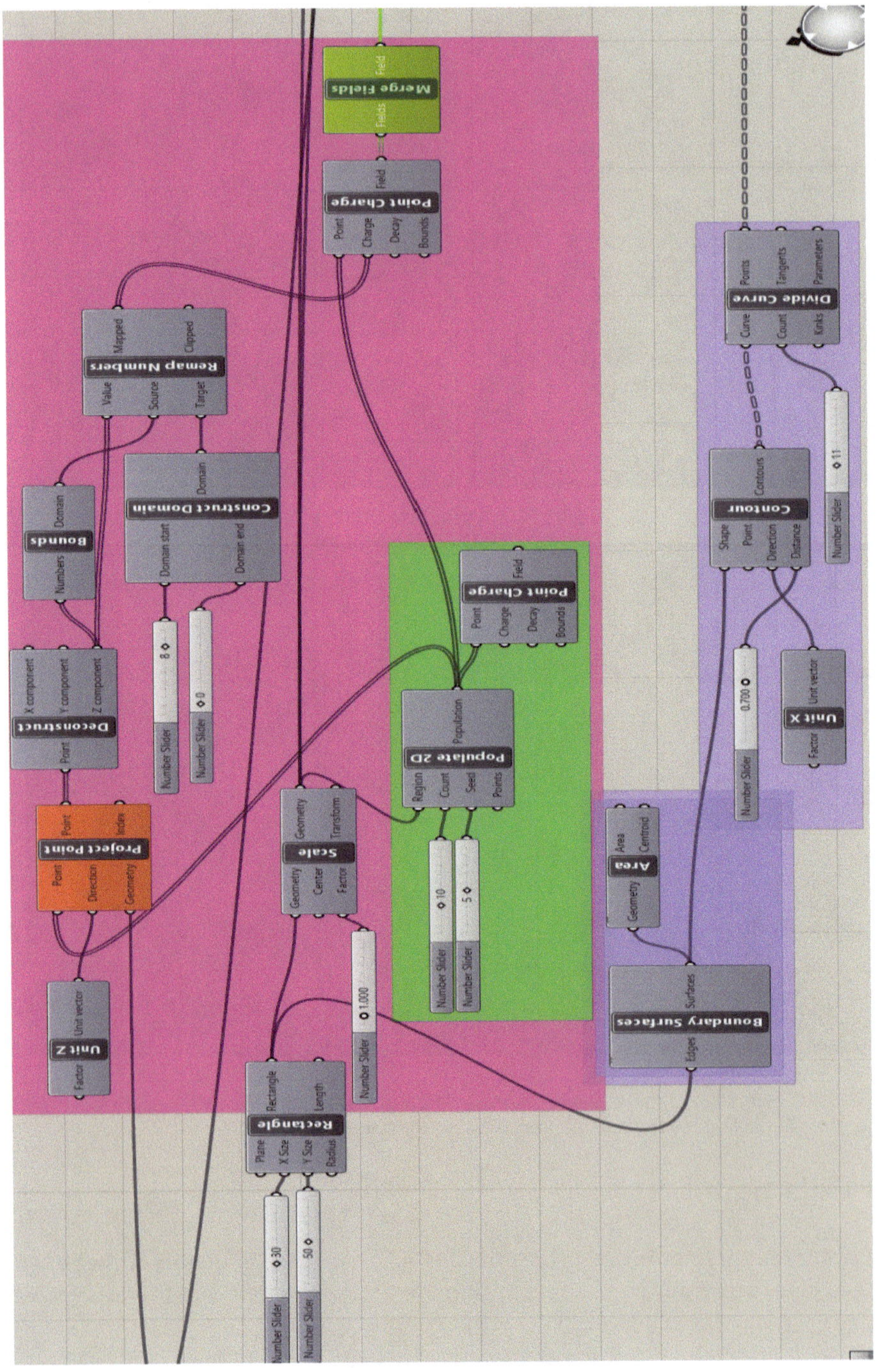

Figure G.4 The parametric design of the acoustic gills/3D model, generating the frame area (a); to be continued in the next page.

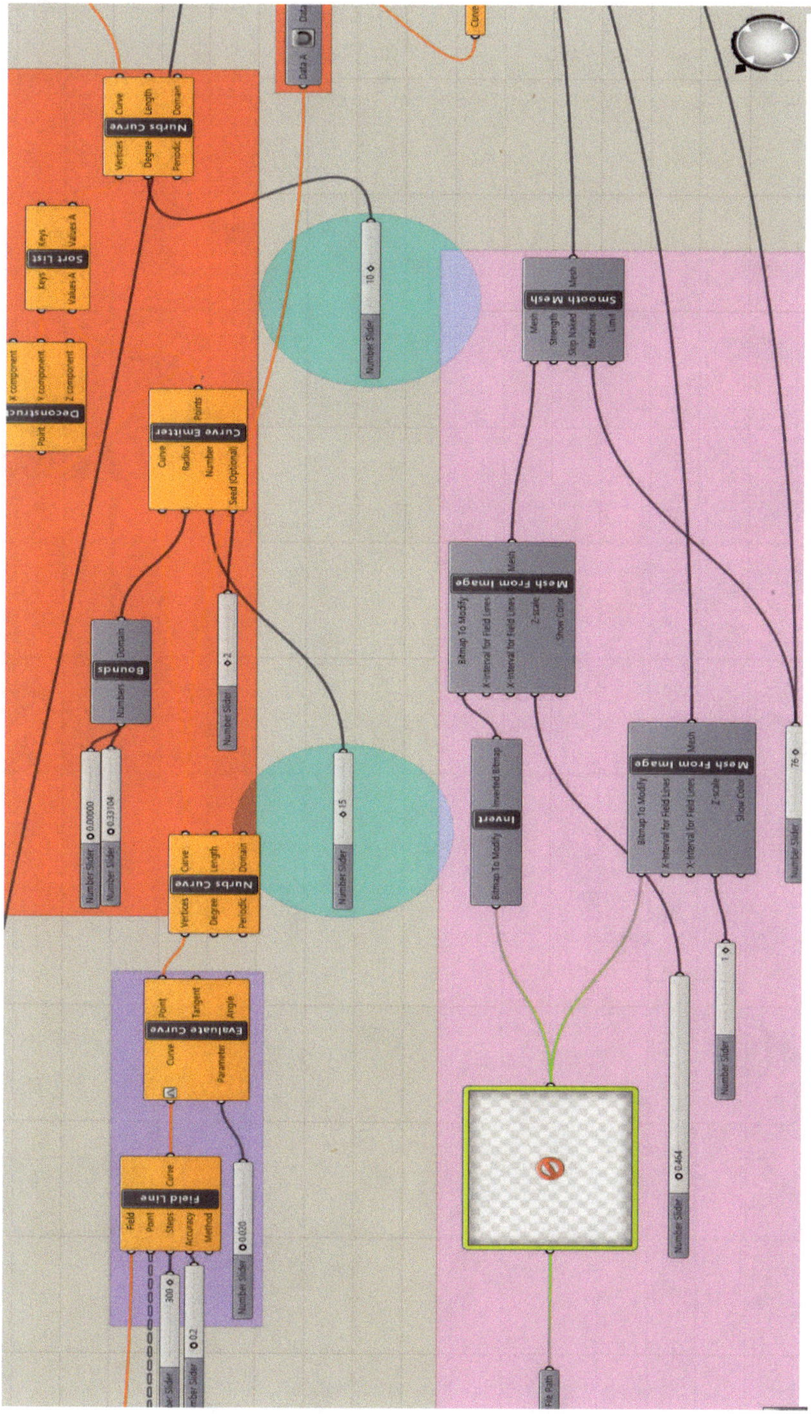

Figure G.4 continued.

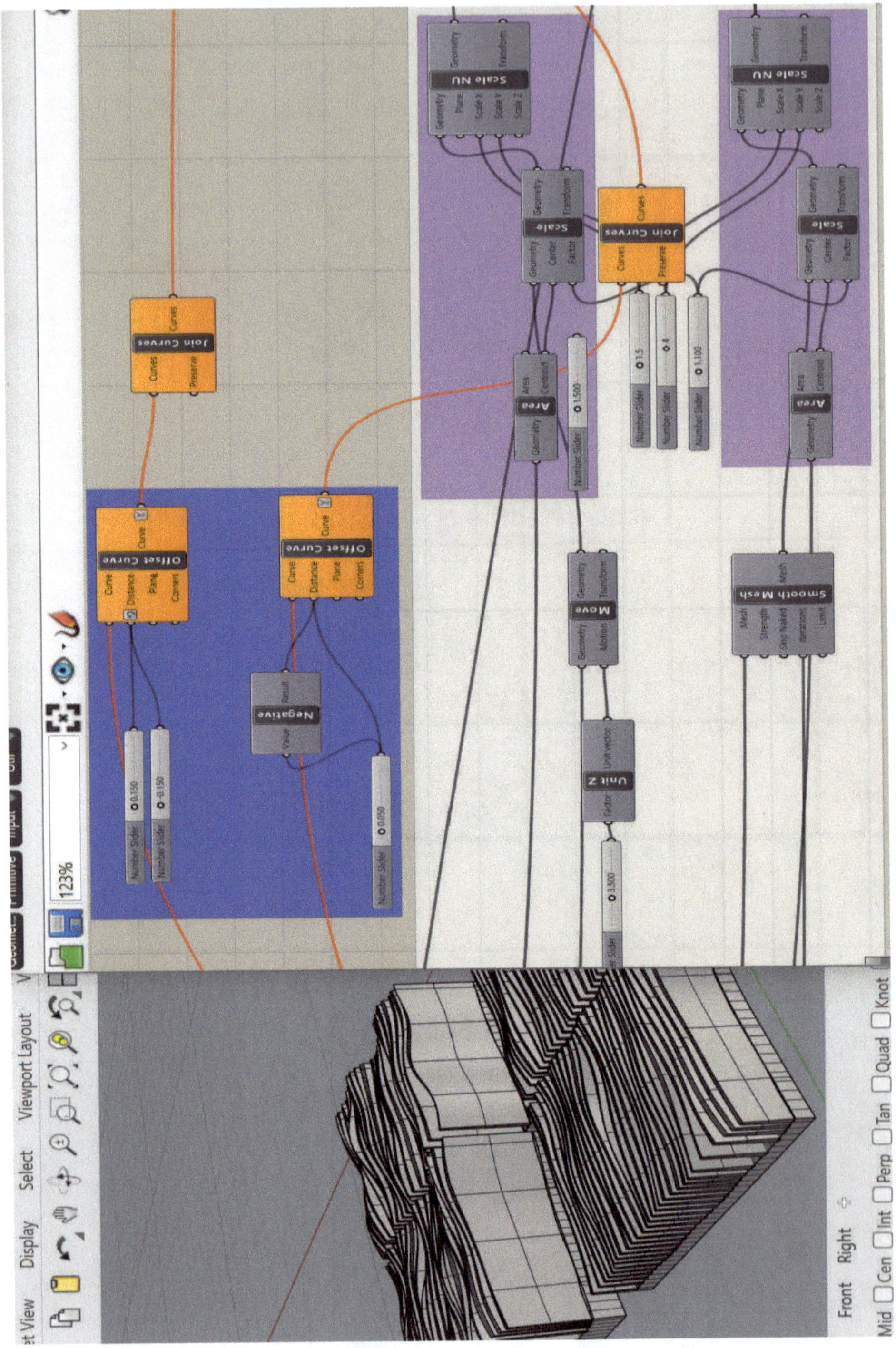

Figure G.4 continued.

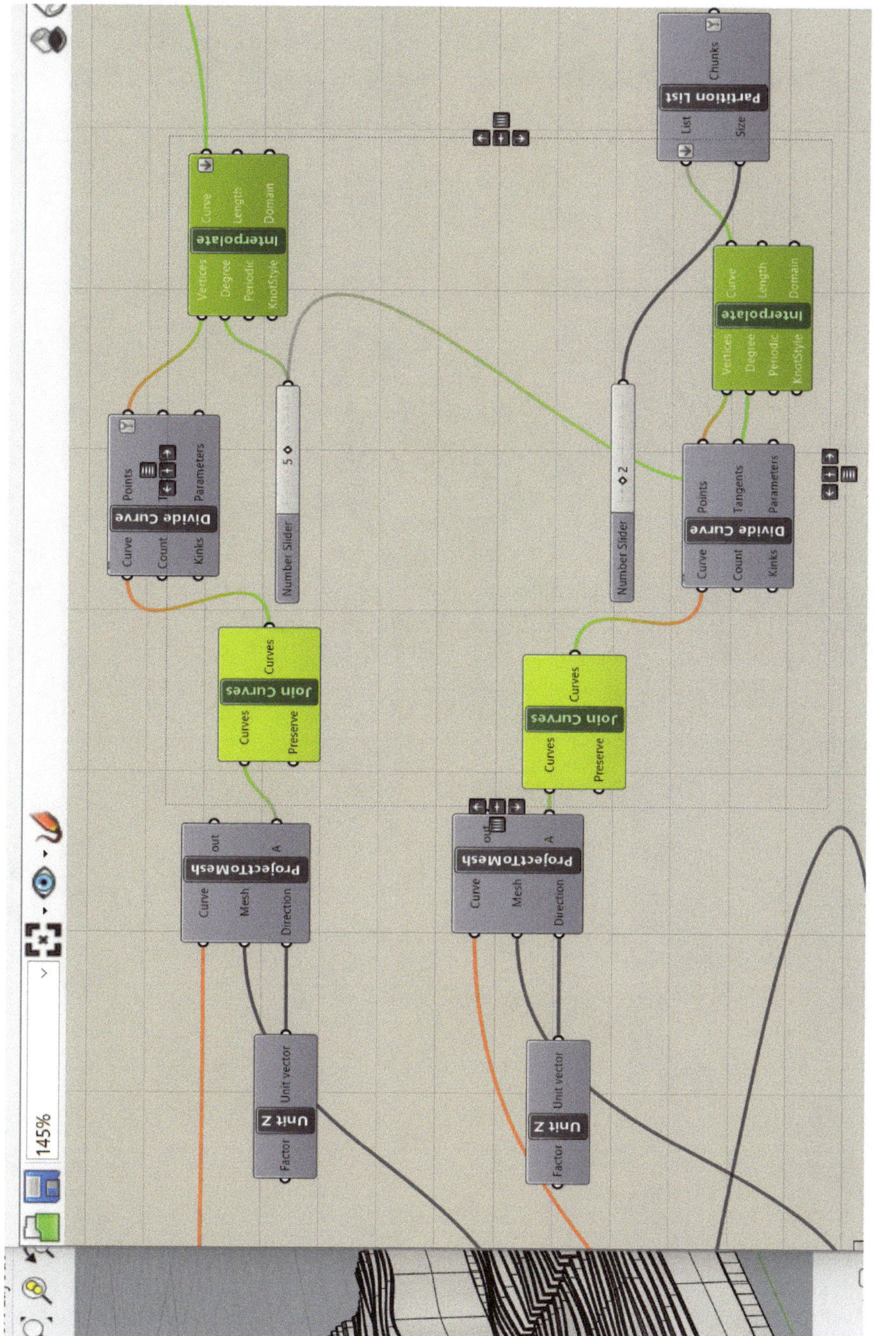

Figure G.4 continued.

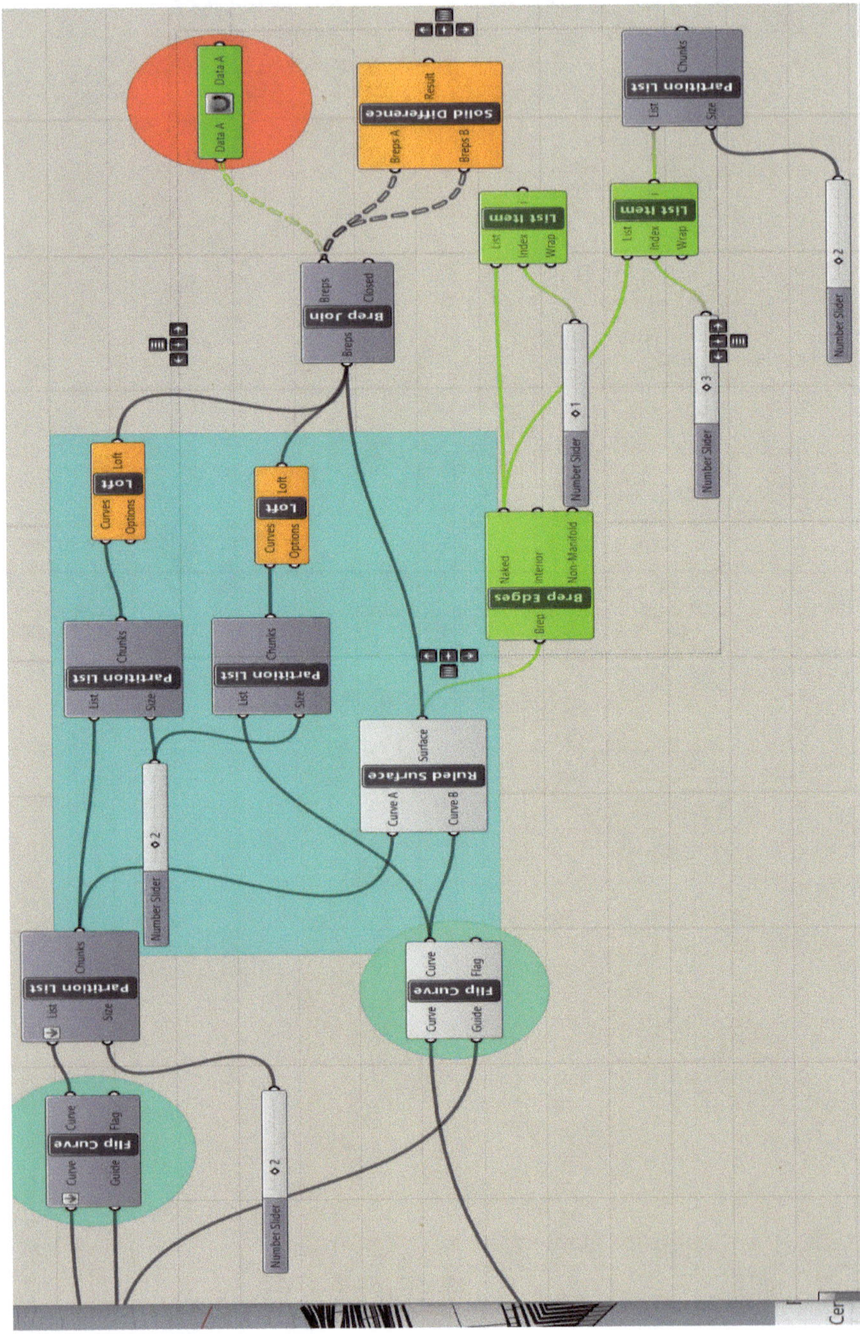

Figure G.4 continued.

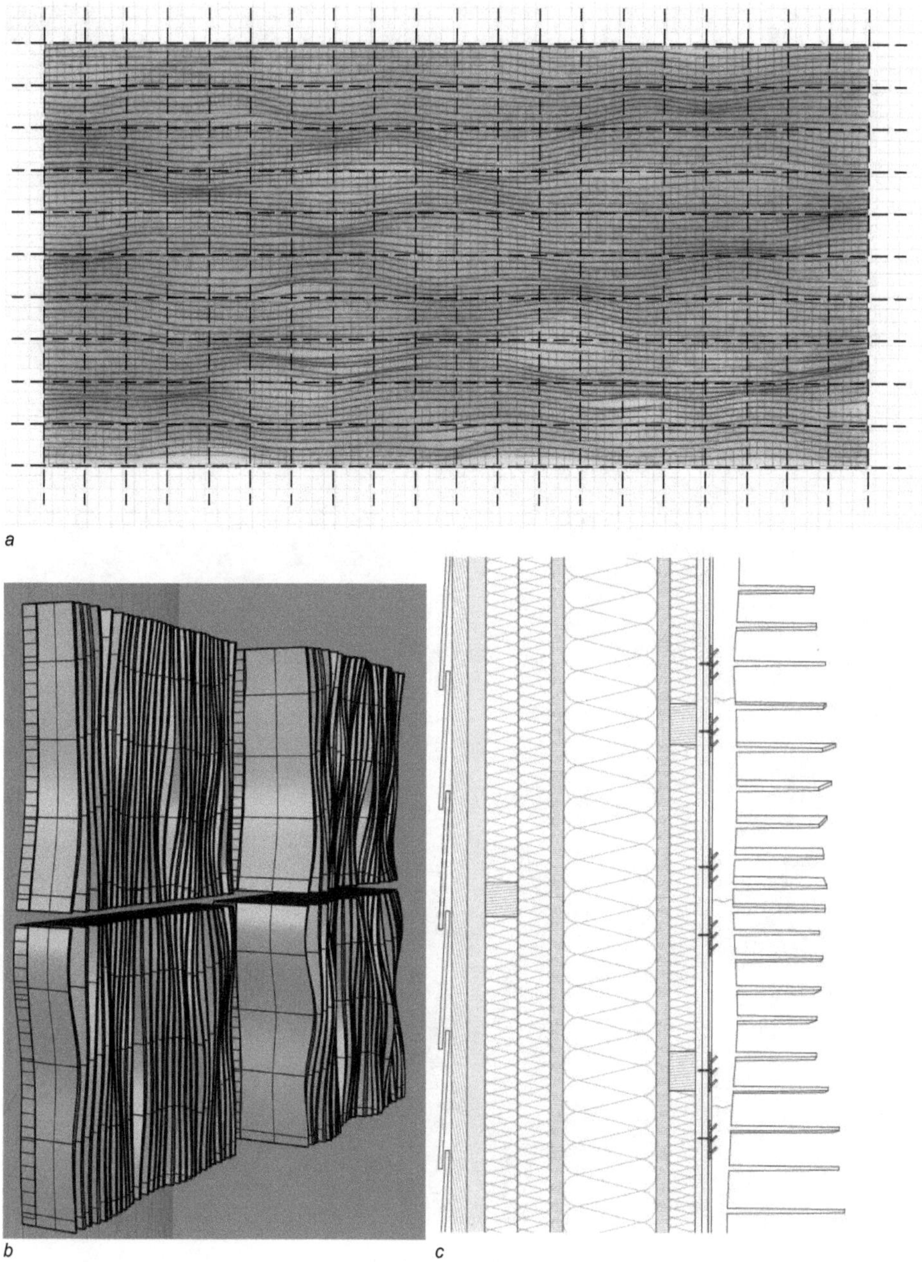

Figure G.5 Preparation of the panel for printing by division into possible printed parts (a), 4 divided parts (b) and installation on wood construction wall (c).

Figure G.6 The visualization of the acoustic gills model.

Figure G.7 The printing of second part from the geometry.

G1.2 Coral Panel

This panel was generated using the object "Growth", and Plug-in "Parakeet". The details of the design and preparation are shown in Fig. G.8. The design of the geometry was done by Lindmir Zubaki and Maximilian Schäfer.

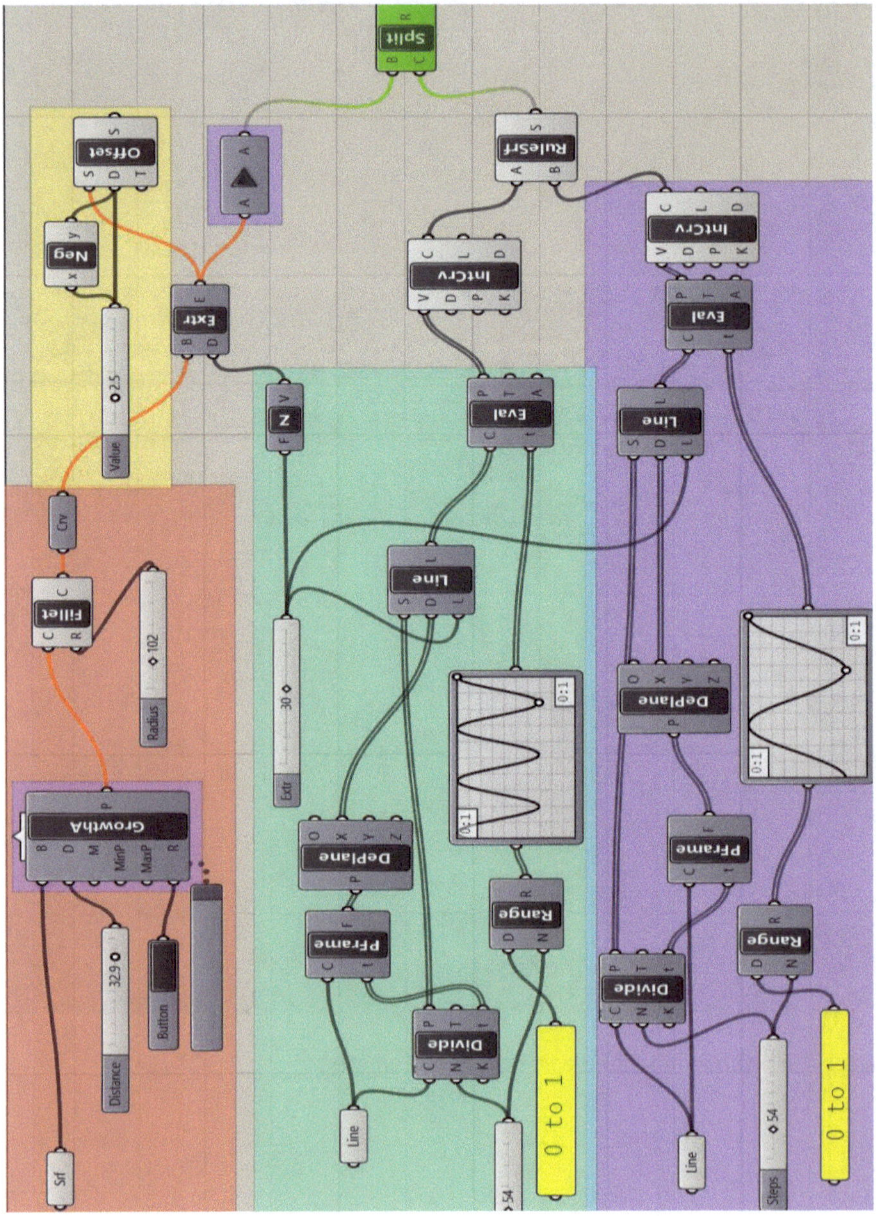

Figure G.8 The parametric design of the coral panel.

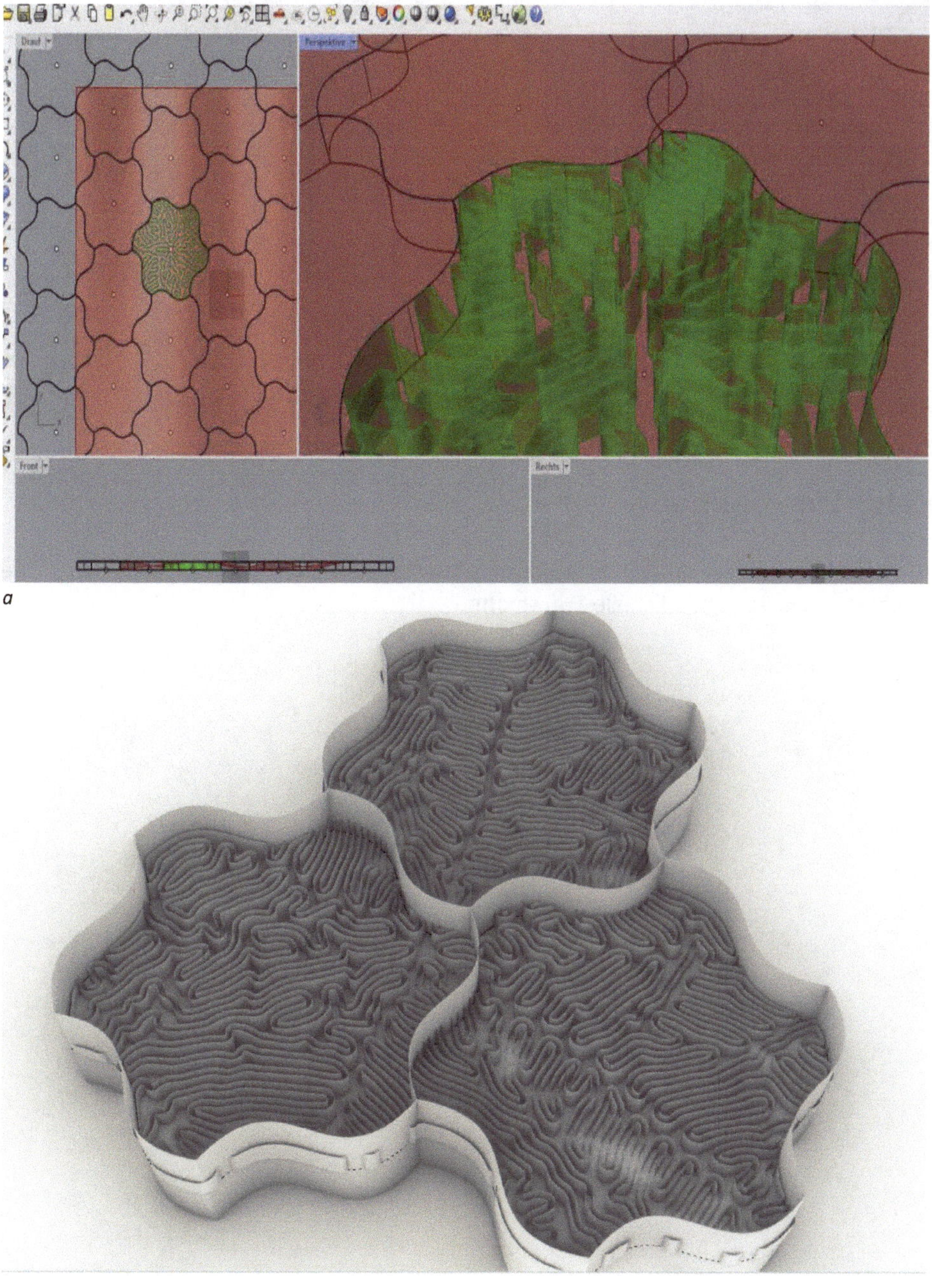

b

Figure G.10 The preview of the parametric design of the coral panel(a) and there adjacent elements (b).

Figure G.11 A side view for the printed element.

G1.3 Tides panel

The details of the design and its visualization are pictured in Fig. B.7. The design of the geometry was done by Michelle Trautmann.

Figure G.12 The visualization of the acoustic tides panel.

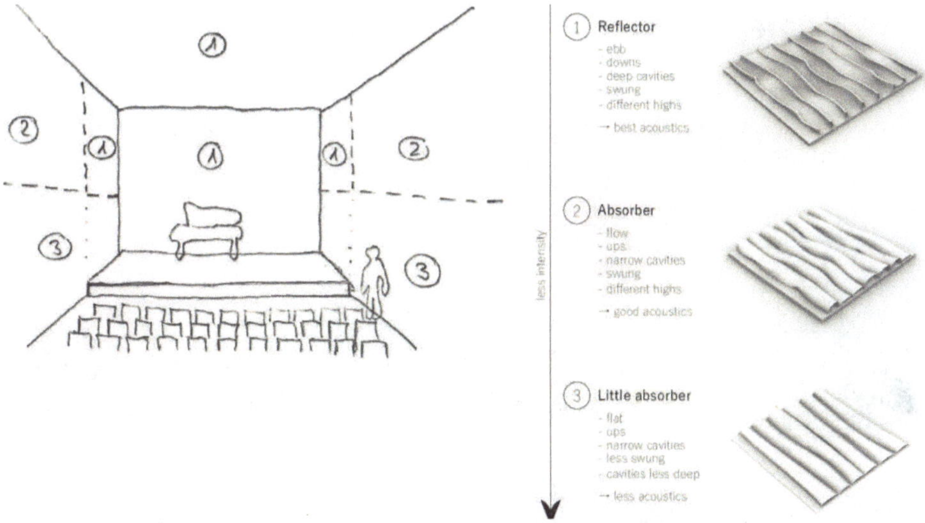

a

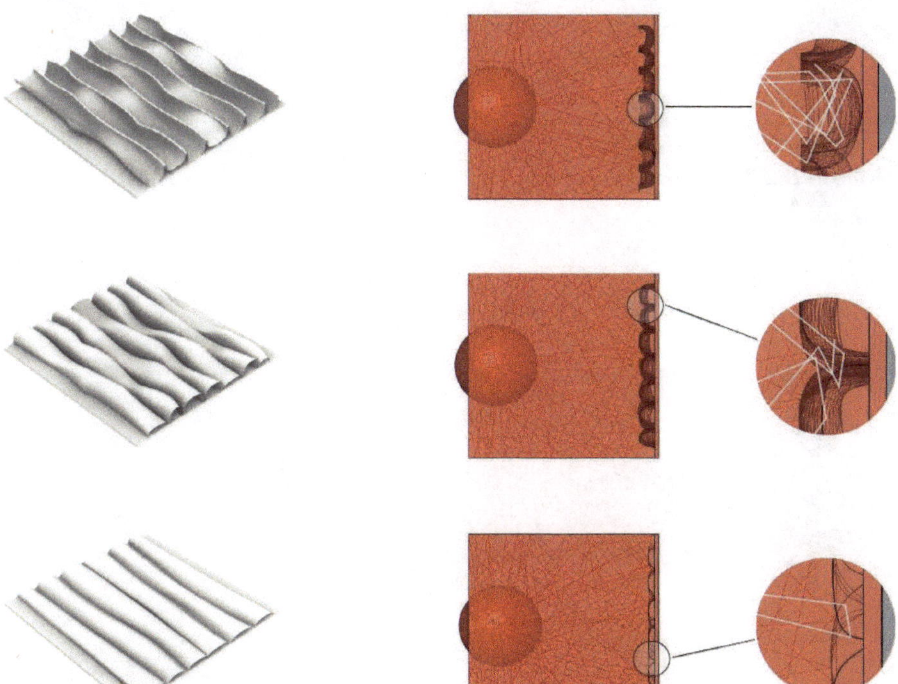

b

Figure G.13 The design process for tides panel (a) and the acoustic behavior for each pattern.

G1.4 3D map panel

The design geometry is done by Nita Morina.

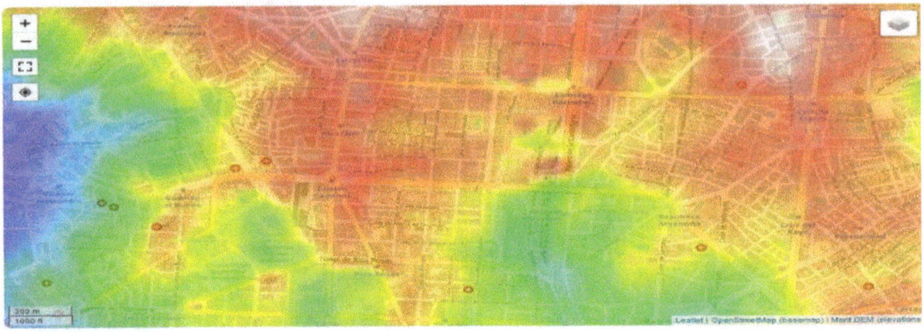

a

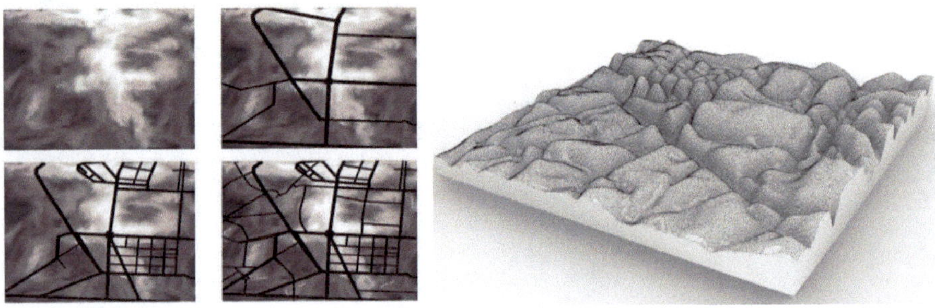

b

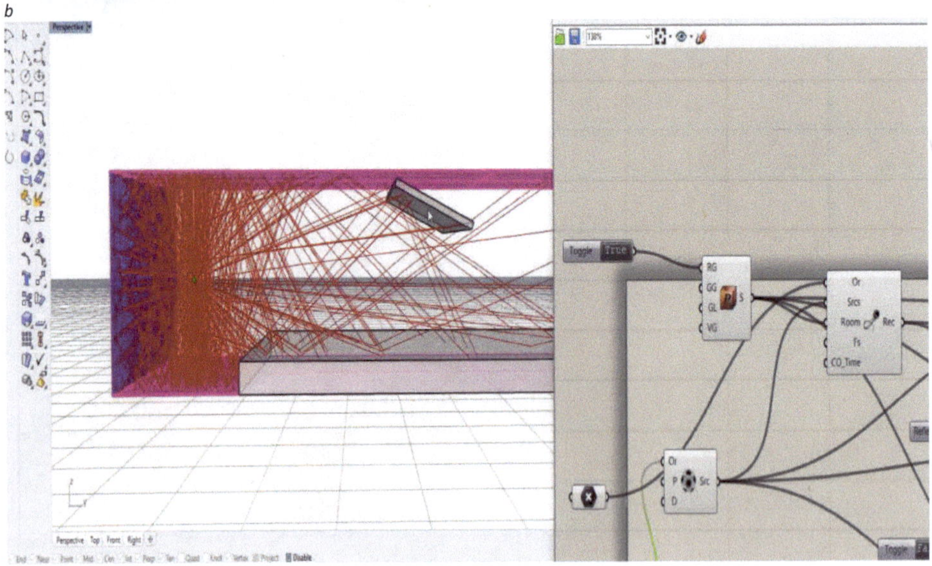

c

Figure G.14 The design inspiration (a), build-in design (b), and the outcome of the simulation (c).

G1.5 Shell/floral leaf shape of triangles

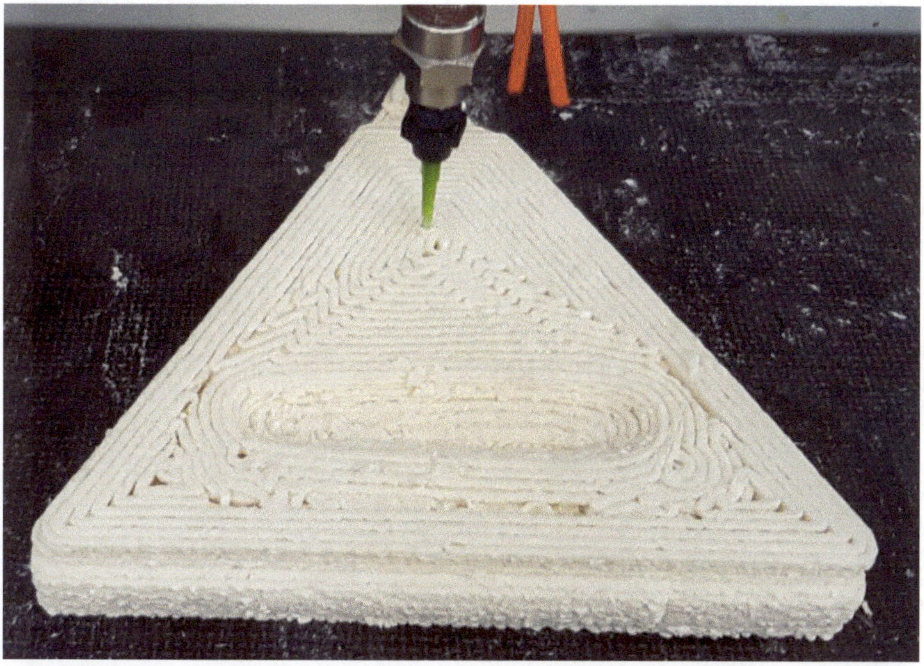

a

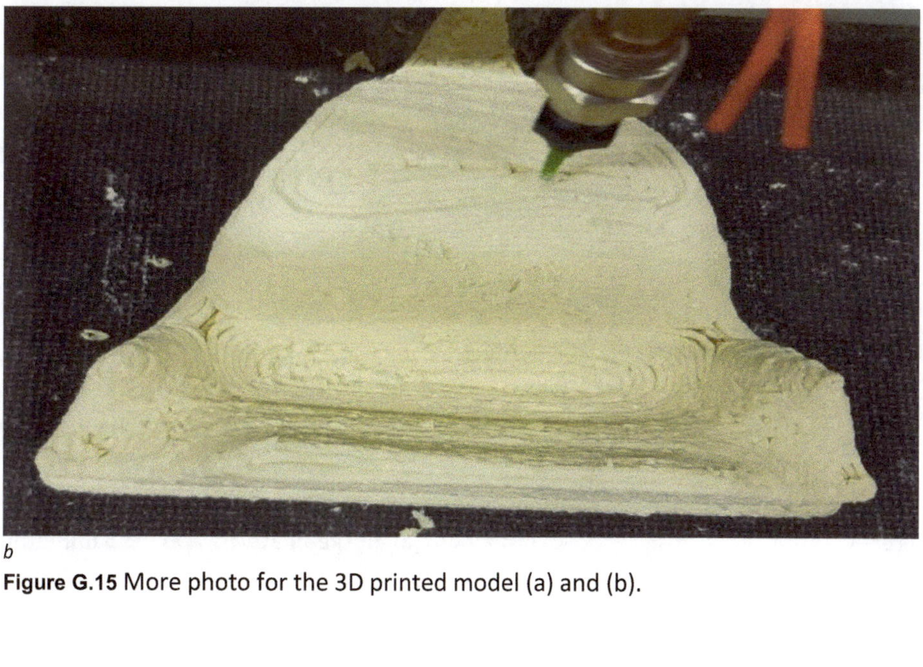

b

Figure G.15 More photo for the 3D printed model (a) and (b).

G1.6 Bubbles panel

The geometry was designed by Aytug Dogan.

a

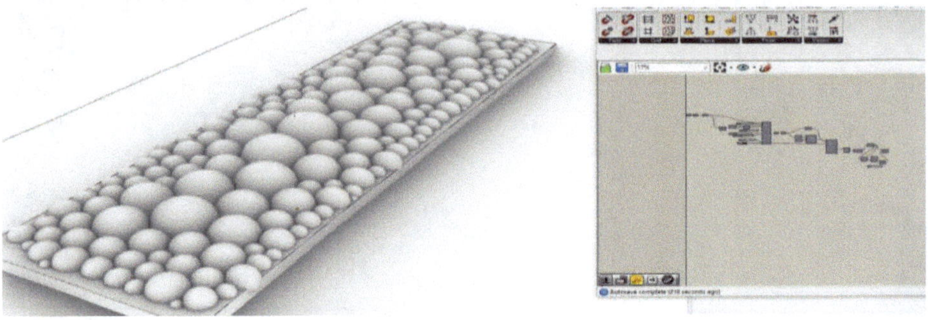

b

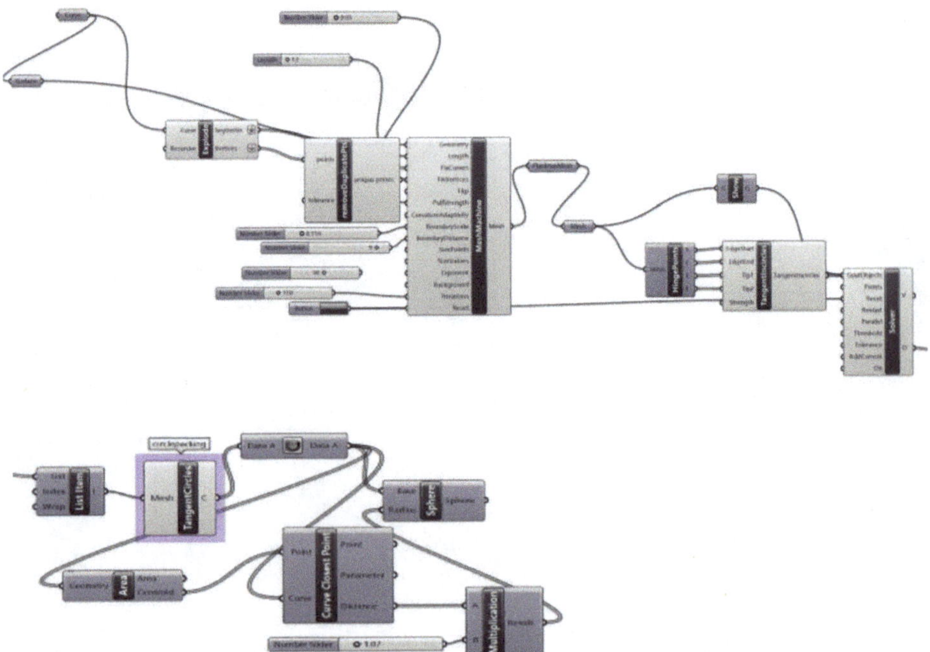

c

Figure G.16 The design inspiration ©Nestlé Aero (a), the geometry (b) and the parametric design by Grasshopper (c)

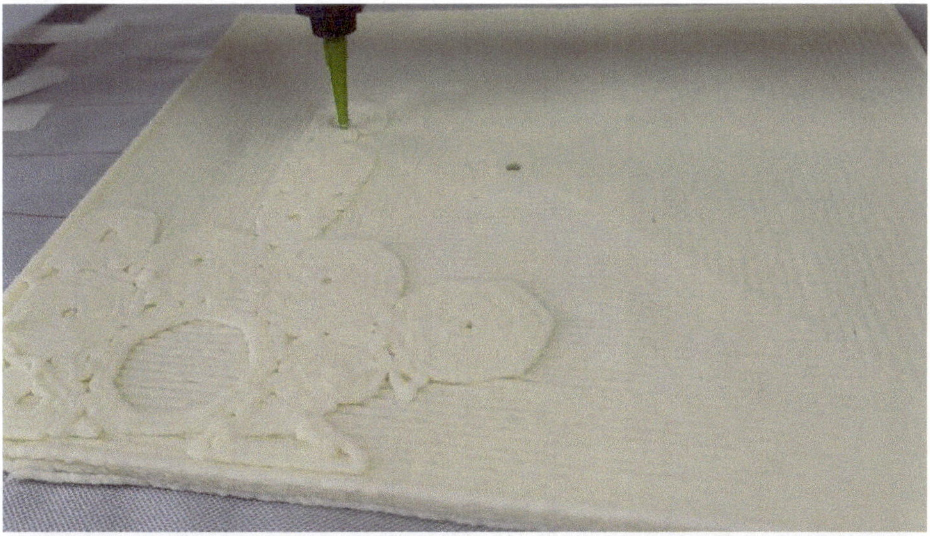

a

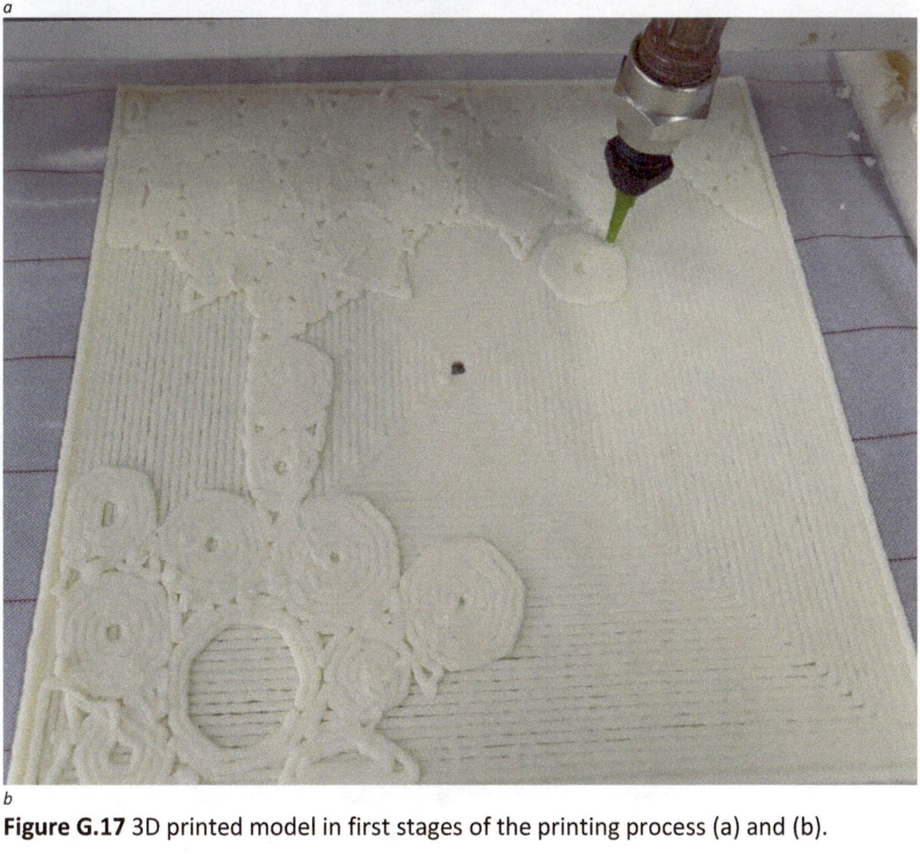

b

Figure G.17 3D printed model in first stages of the printing process (a) and (b).

G1.7 3D crowned Buffel model

The geometry was designed by Lisa Menzel.

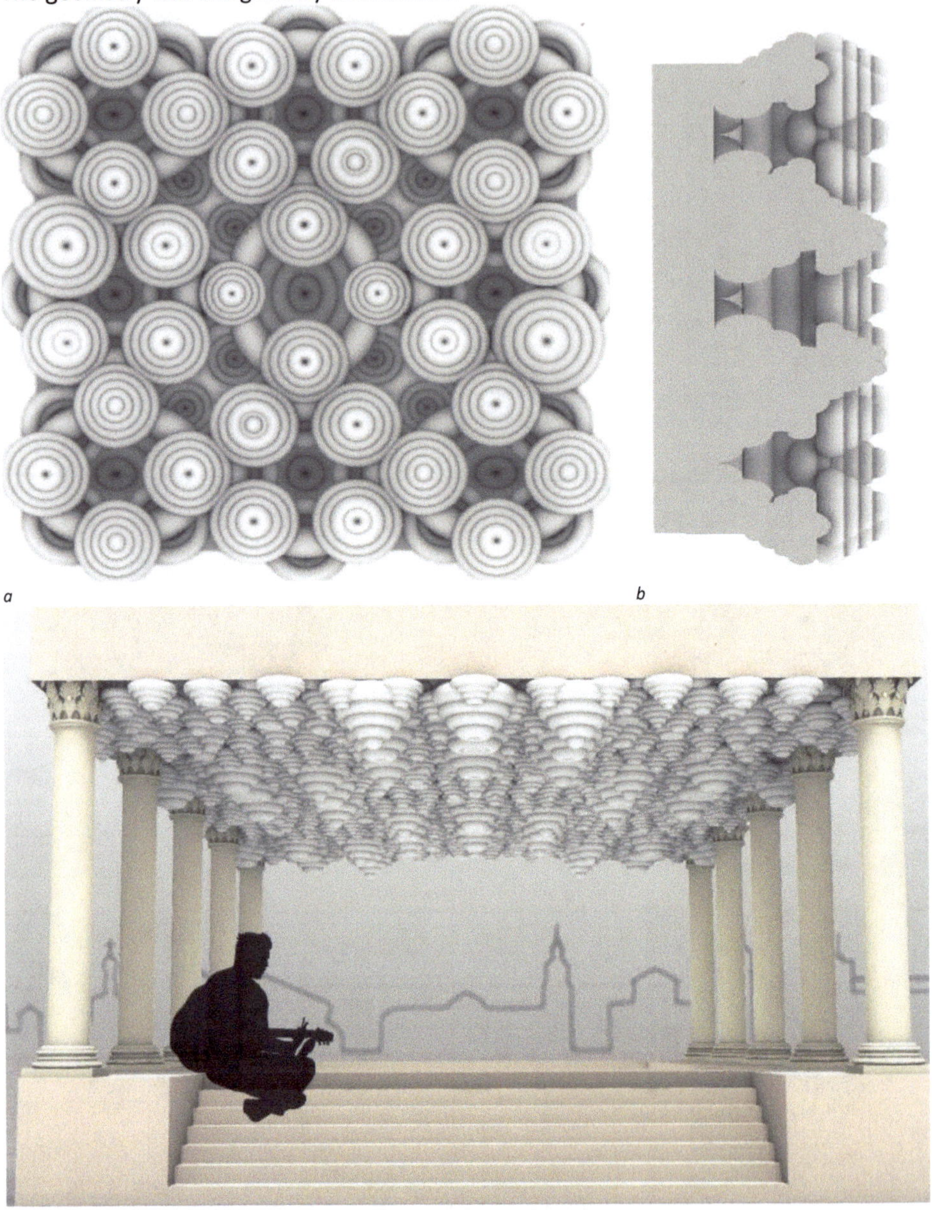

Figure G.18 The geometry of the 3D model (a), cross-section(b), and its visualization.

a

b

Figure G.19 More Photo for the printed sample.

G1.8 Acoustic reef

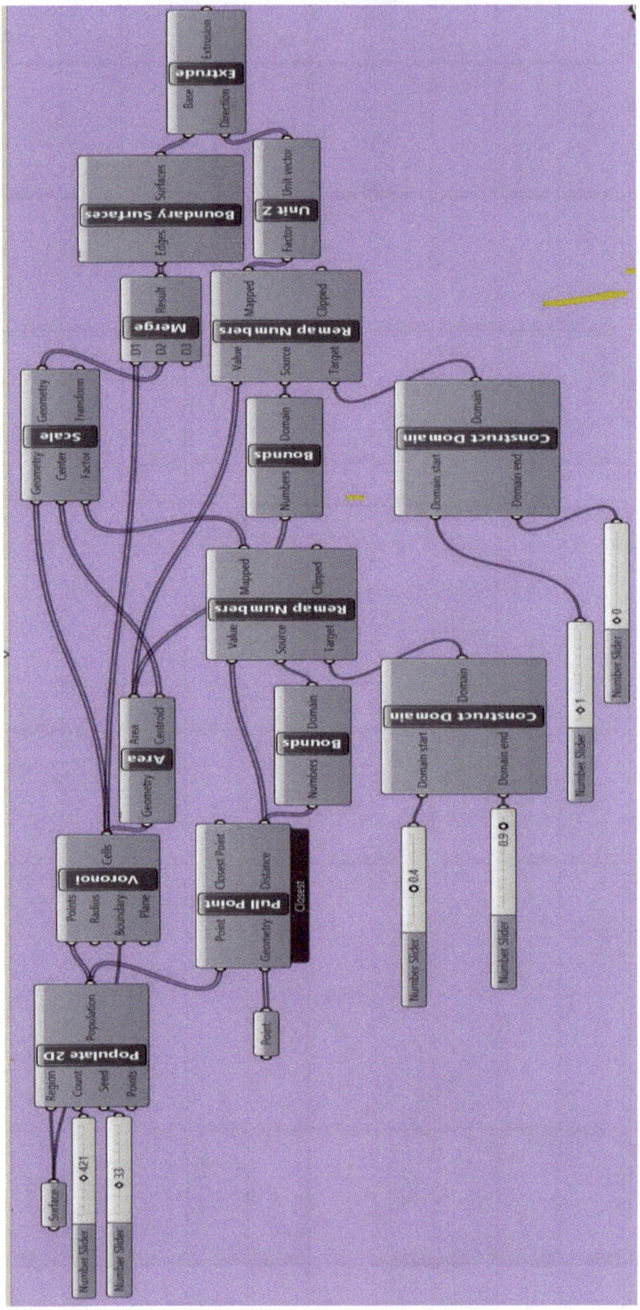

Figure G.20 The parametric design of the panel. This was designed by Samira Cramm.

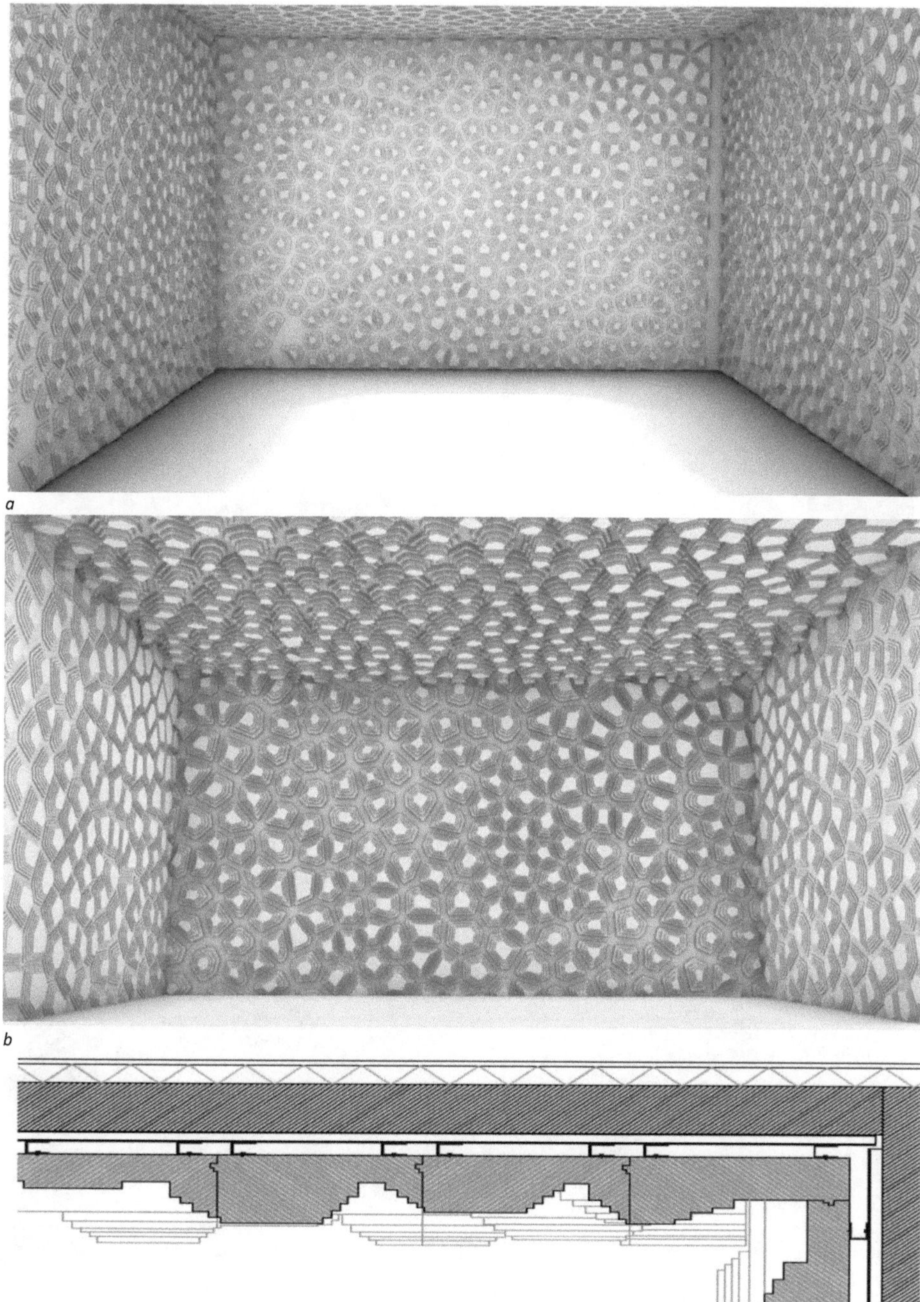

Figure G.21 The visualization of first 3D model (a) and the improved one (b), and the across section of the corner(c).

a

b

Figure G.22 The visualization of another two pattern for 3D acoustic reef model(a) and (b).

Figure G.23 Side view for the 3D printed paper model. γδσϒ

G.2 Sun Shading elements details (Solar radiation analysis)

Comparing the level of translucency and sunlight entrance by solar radiation analysis. Solar radiation analysis simulates sun-path by using a place-specific weather data (Frankfurt am Main) with variation of date (winter-summer) and daytime.

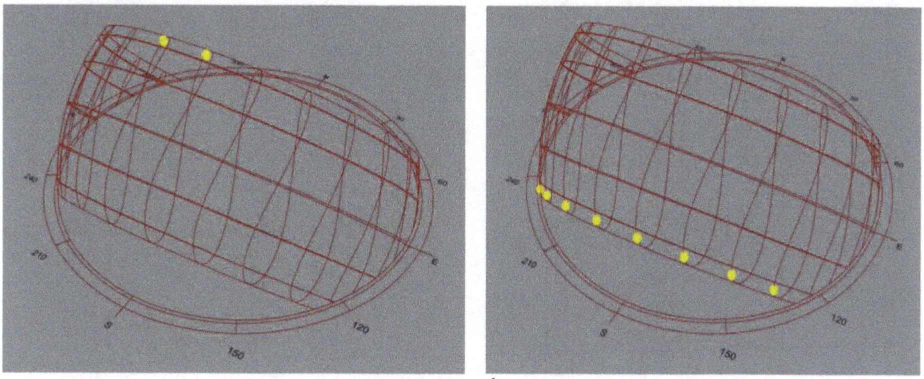

a *b*

Figure G.24 Sun location/ June 1st, 12pm –2pm (a), and December1st, 7am –9pm (b).

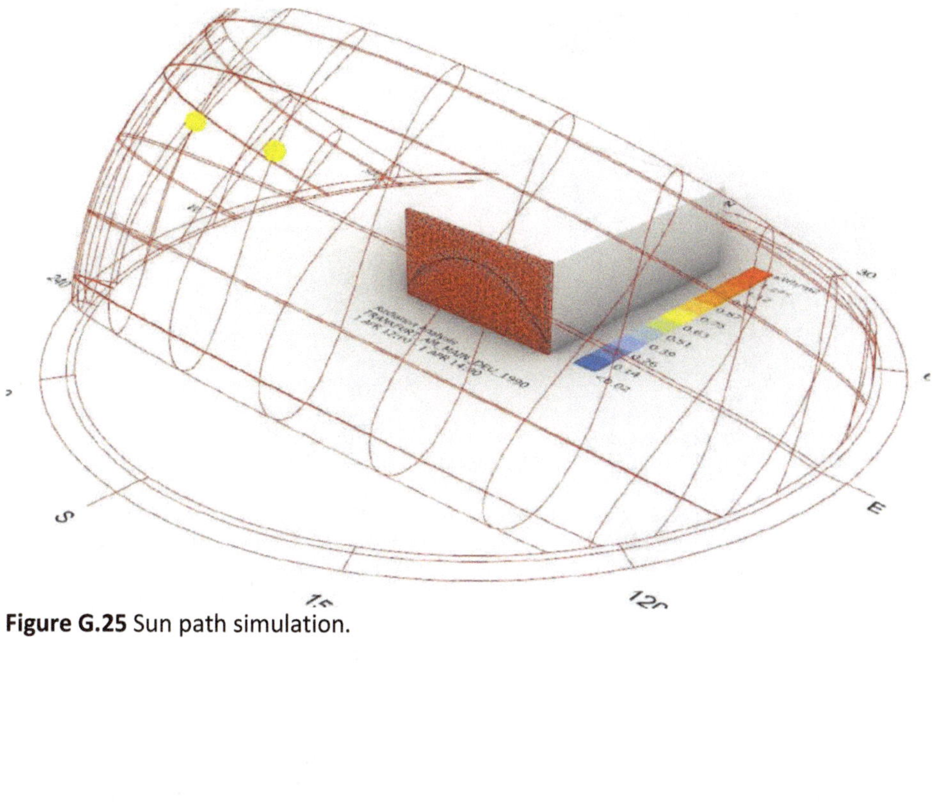

Figure G.25 Sun path simulation.

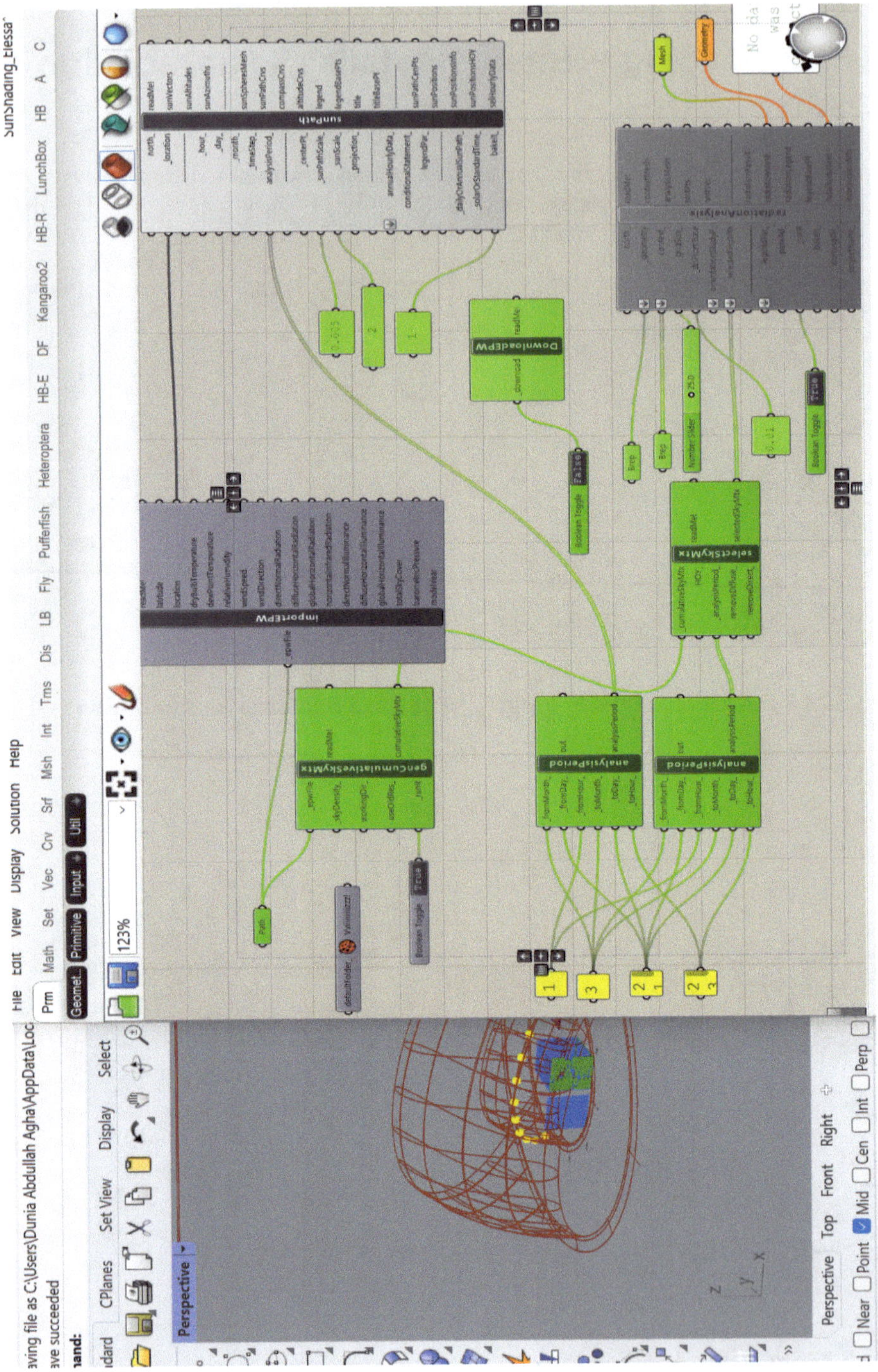

Figure G.26 The parametric design for analysis the radiation energy .

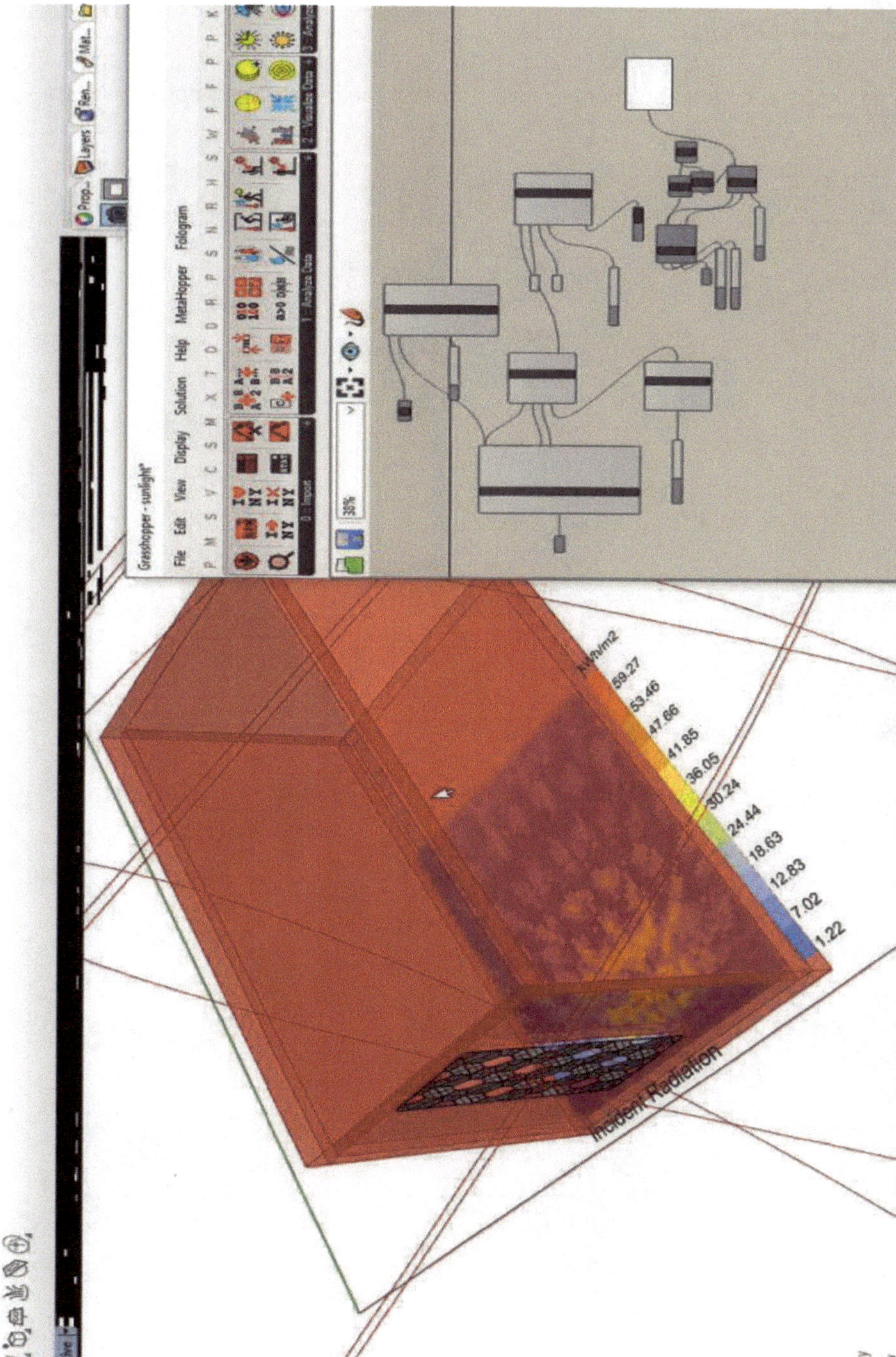

Figure G.27 The r analysis of the radiation energy for the Mais 3D model by Anna Nist and Felix Schmitt for calculation.

G2.1 Cell Design

The geometry and radiation energy simulation with the calculation is done by Alessa Heiden.

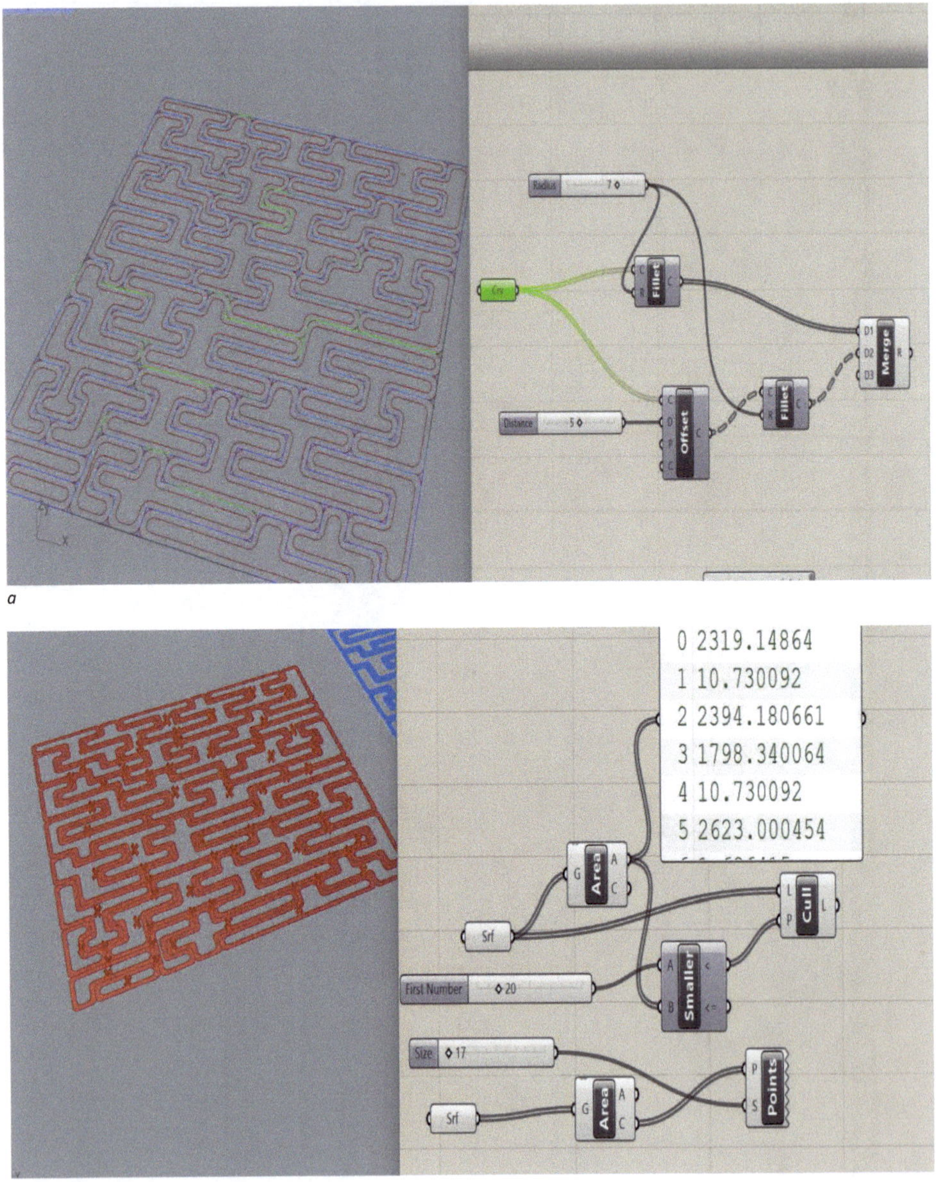

a

b

Figure G.28 The parametric design for the 3D model.

Figure G.29 Specimens to study better outcomes of radiation energy entrance by no angle of High density (a) medium density(b), and least density(c), and same specimens by rotation them of 90 degrees (d), (e), and (f).

Table G.1 Radiation analysis calculation, input Data: Frankfurt am Main as place, time:12am-1pm.

	1st January kWh/m²	**1st July** kWh/m²	**1st October** kWh/m²
No panel	0,008201	0,030919	0,168911
High density - No angle	0,001181	0,003609	0,032172
High density – 90-degree angle	0,001709	0,005822	0,05726
Medium density –No angle	0,002233	0,007254	0,062229
Medium density –90-degree angle	0,002981	0,01061	0,086434
Least density –No angle	0,002929	0,009868	0,07759
Least density –90-degree angle	0,004262	0,015266	0,111422

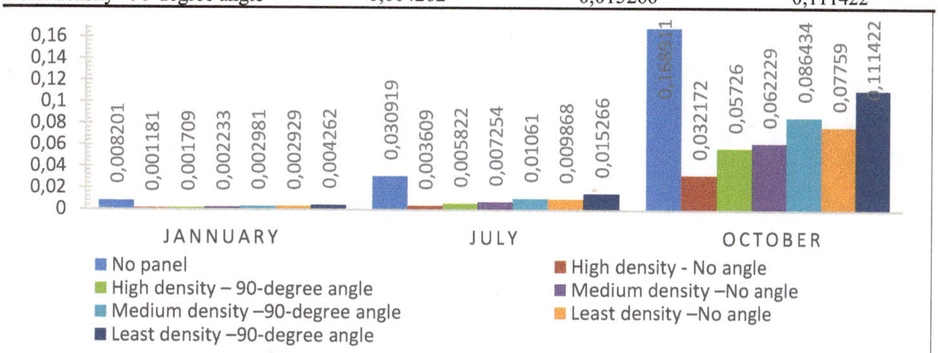

Figure G.30 Comparing the radiation energy for January, July, October.

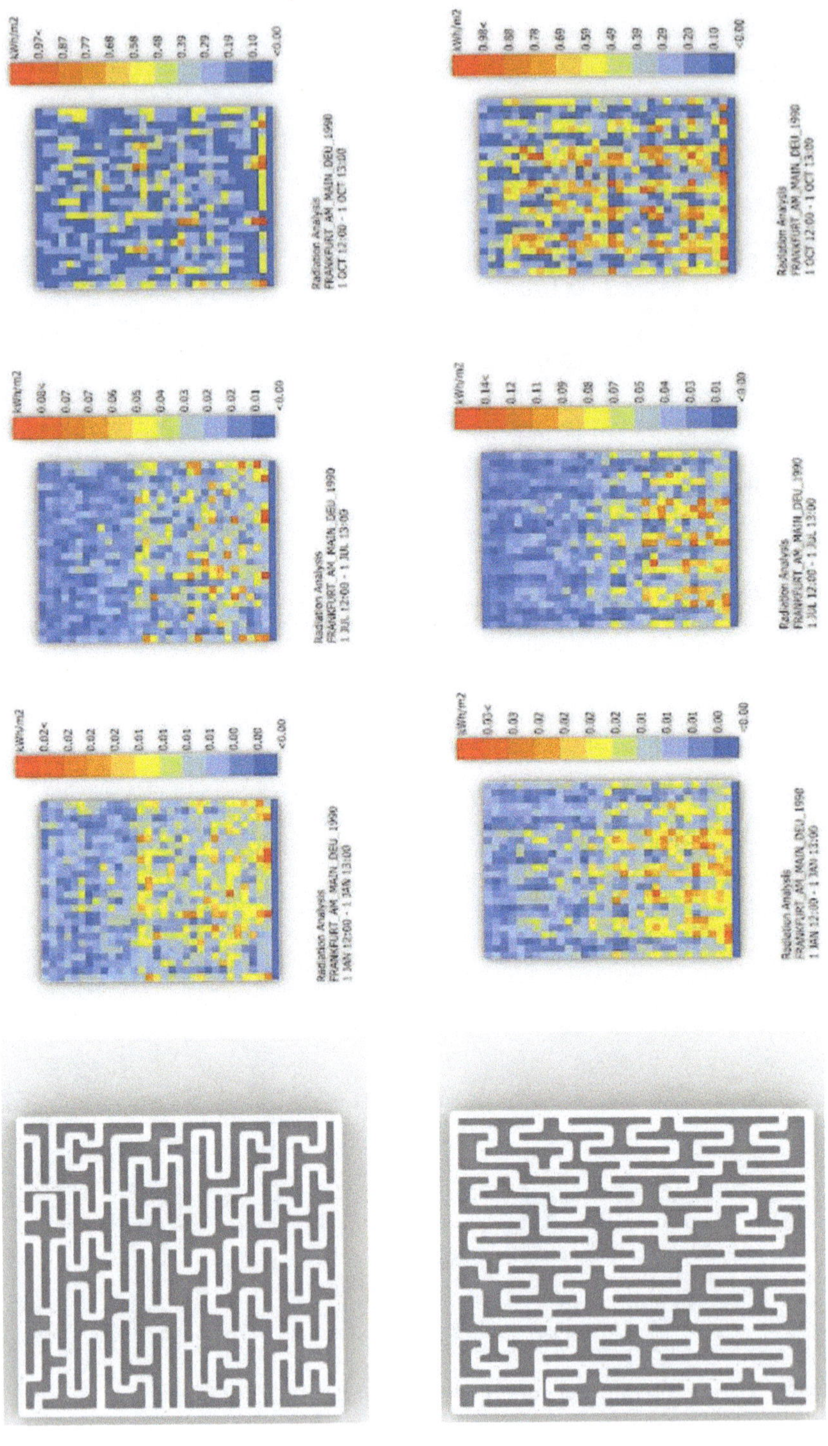

Figure G.31 The results of the radiation energy analysis for two specimens (normal installed panel and 90 degree rotated specimen for the month January , July and October.

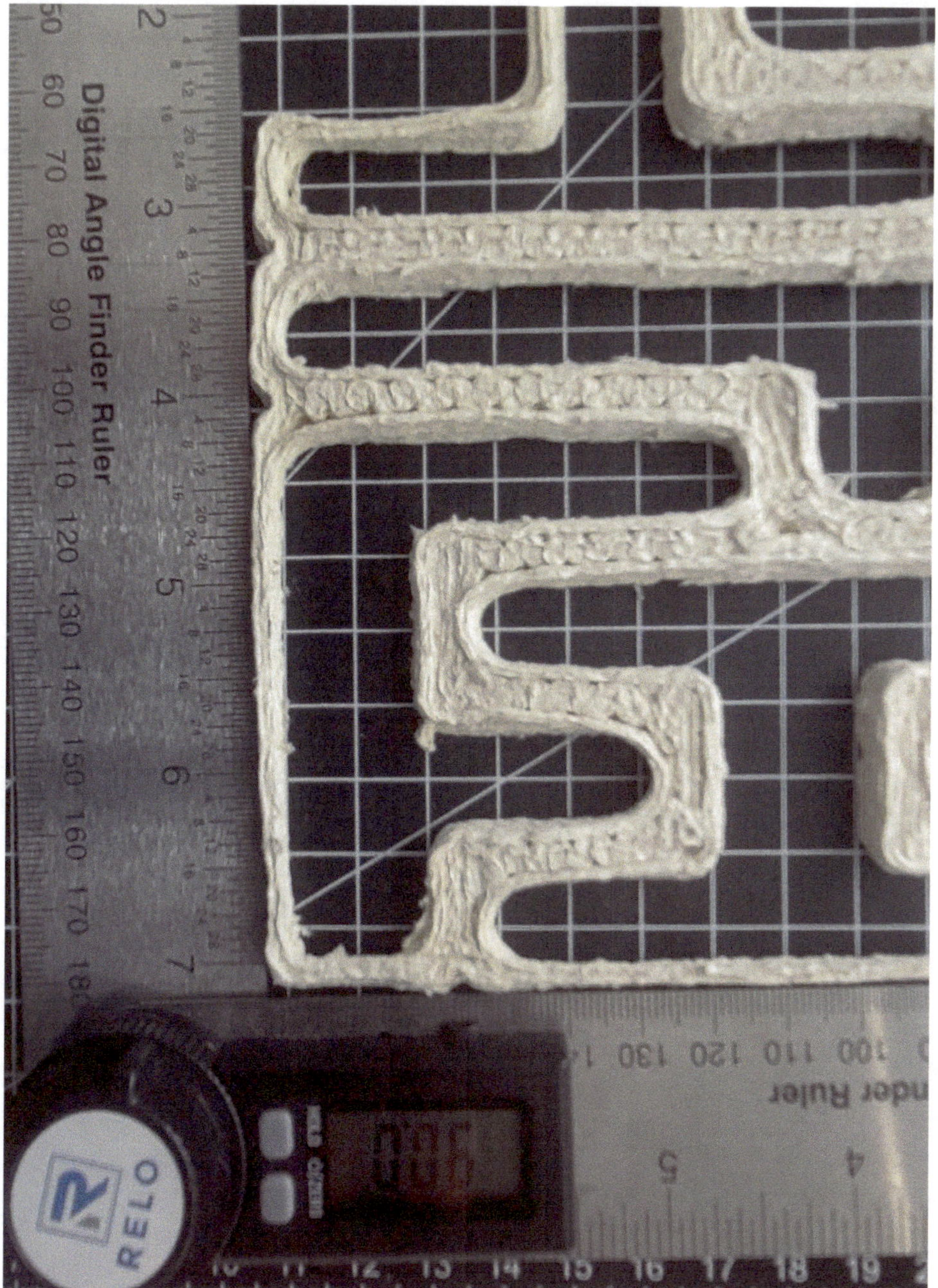

Figure G.32 A zoomed photo for the 3 D printed paper model.

G2.2 3D honeycomb model

The geometry was designed by Zoe Feiler and Jan Kalkkuhl.

Figure G.33 The designed 3D model (a) its functionality (b) and it visualization(c).

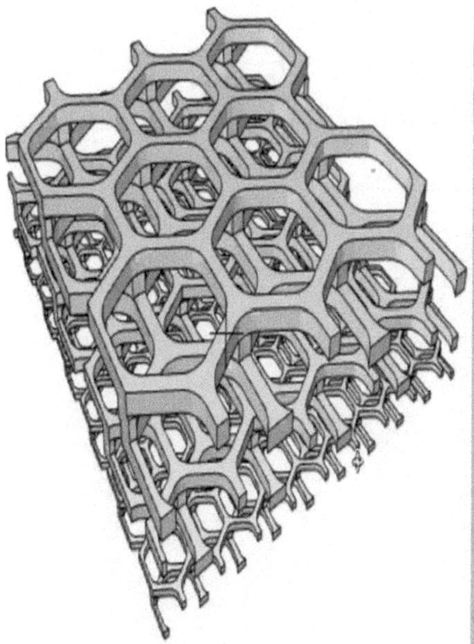
a

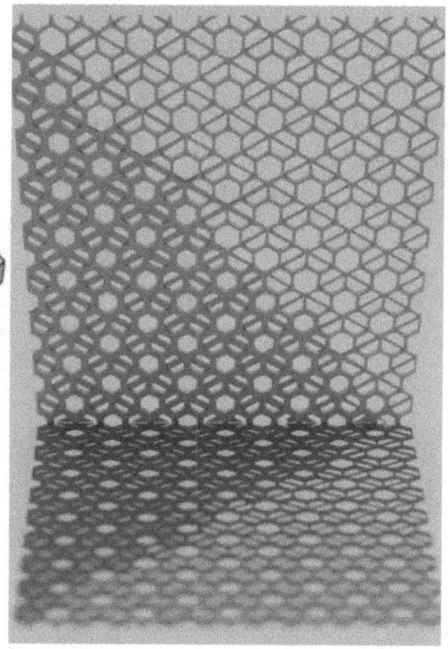
b

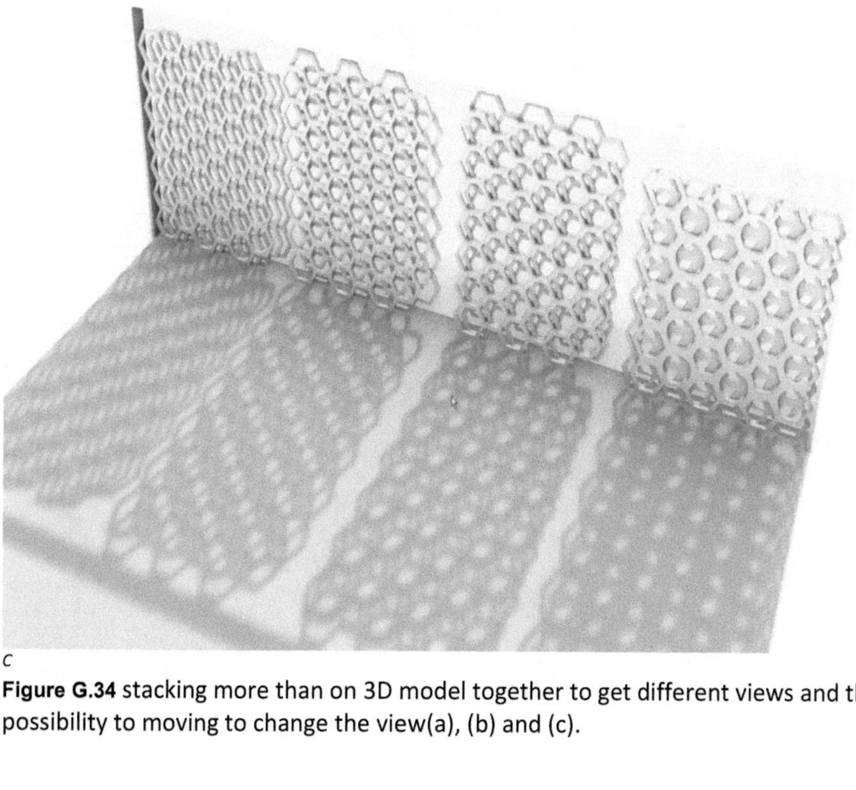
c

Figure G.34 stacking more than on 3D model together to get different views and the possibility to moving to change the view(a), (b) and (c).

G2.3 String wall partition

The geometry was designed by Zoe Feiler and Jan Kalkkuhl.

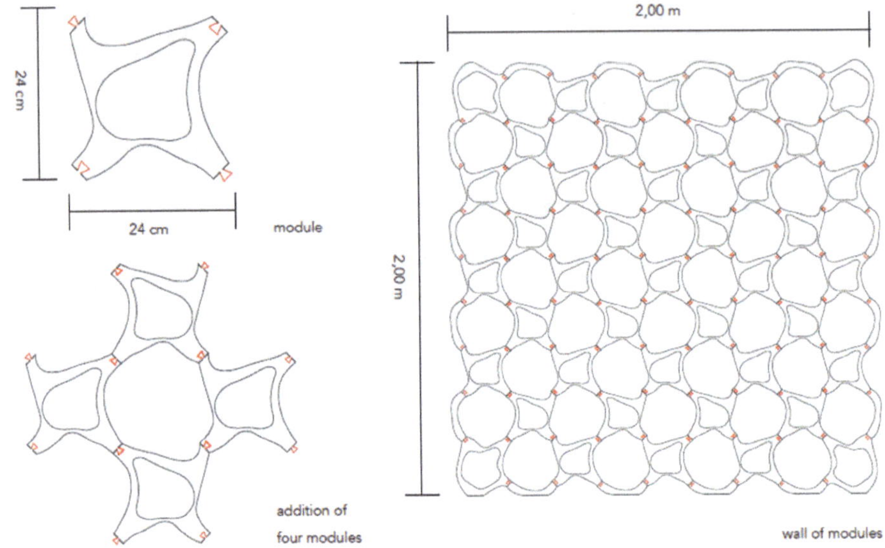

a *b*

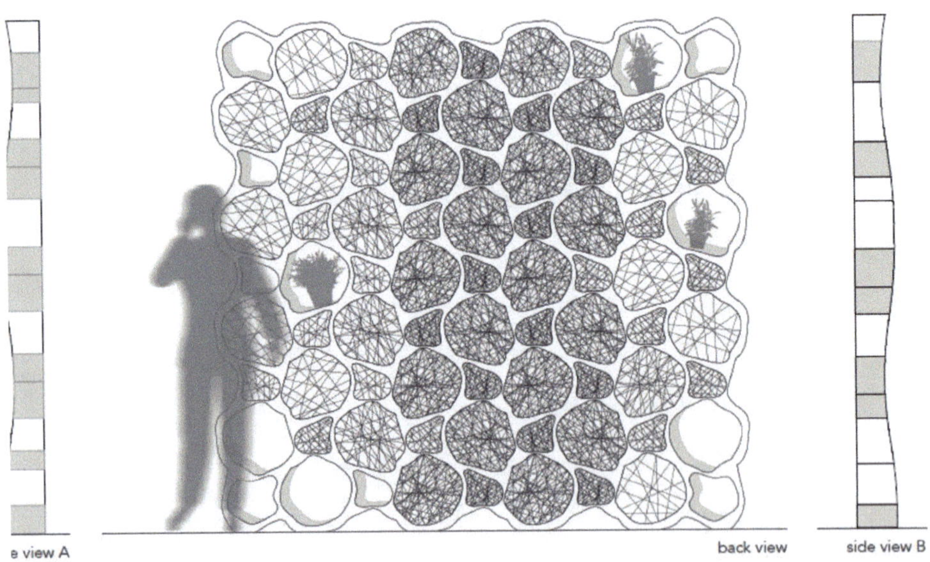

c

Figure G.35 The designed 3D model (a) its completed panel (b) and it visualization(c).

Figure G.35 continued.

G2.4 Tree Shade panel

The geometry of the 3D model is designed by Simon Wolber.

Figure G.36 First part of 3D printed model.

Figure G.37 The panel with different patterns(a) and (b), and it visualization(c).

G2.5 Mais panel

The geometry of the model is designed by Anna Nist and Felix Schmitt.

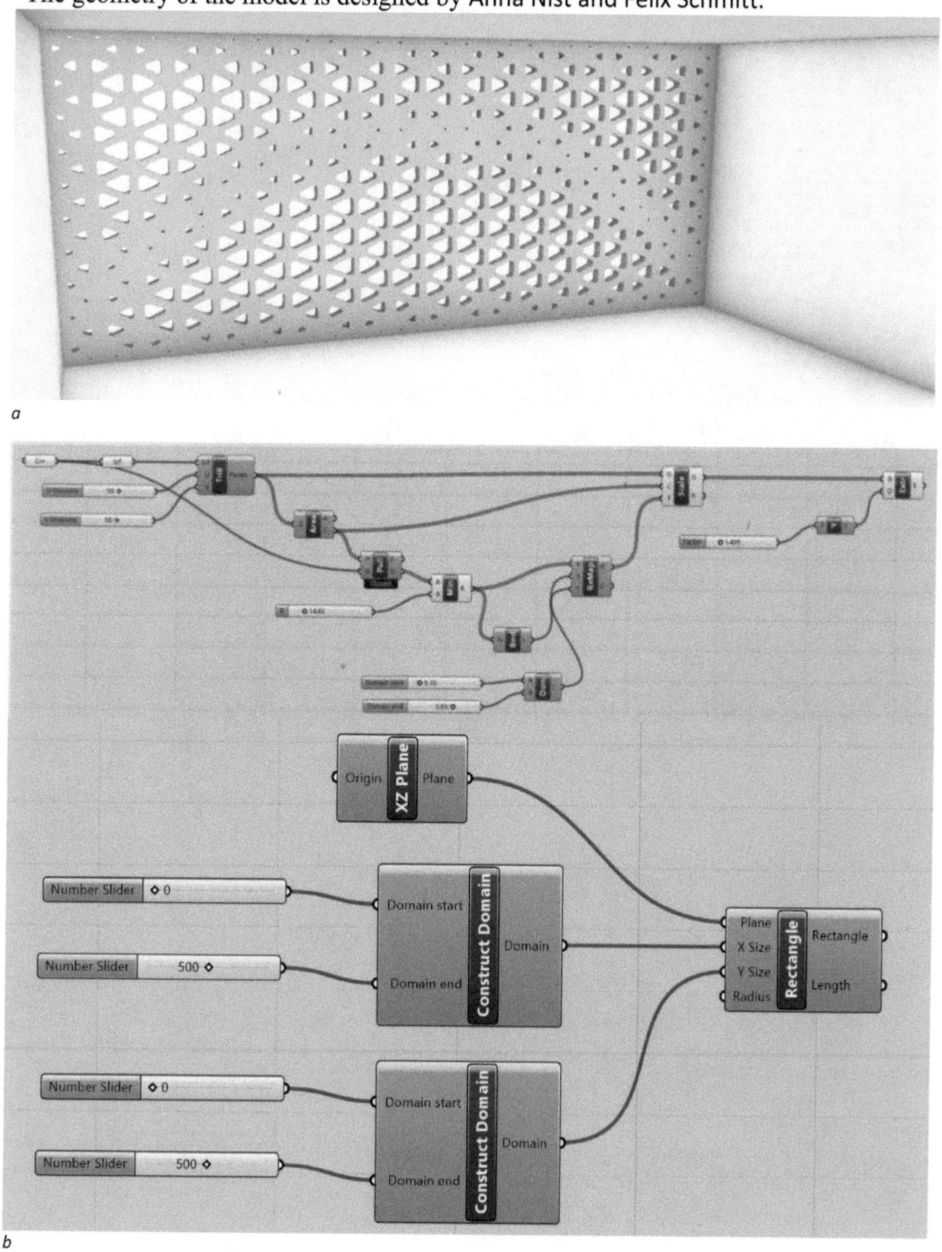

Figure G.38 The visualization of panel (a) and the parametric design (b).

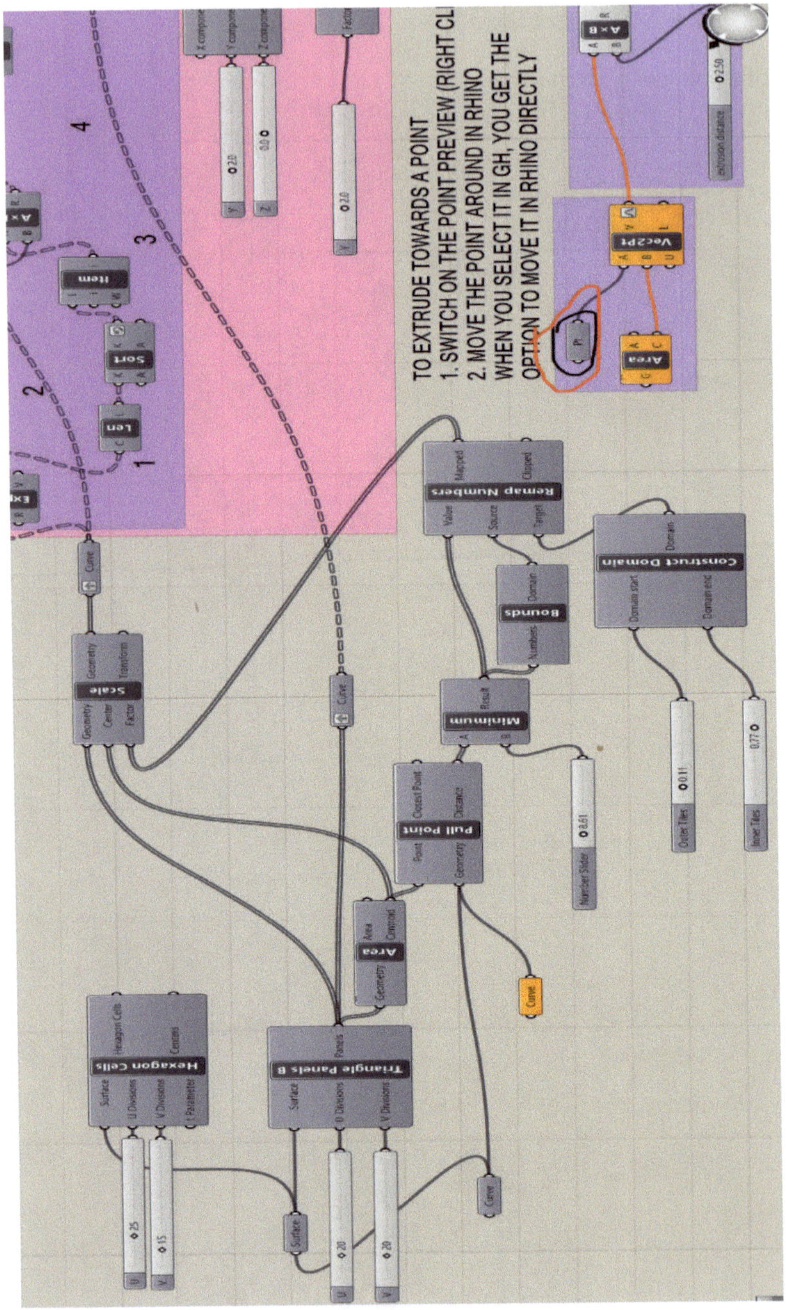

Figure G.38 continued the parametric design by Rhino and grasshopper.

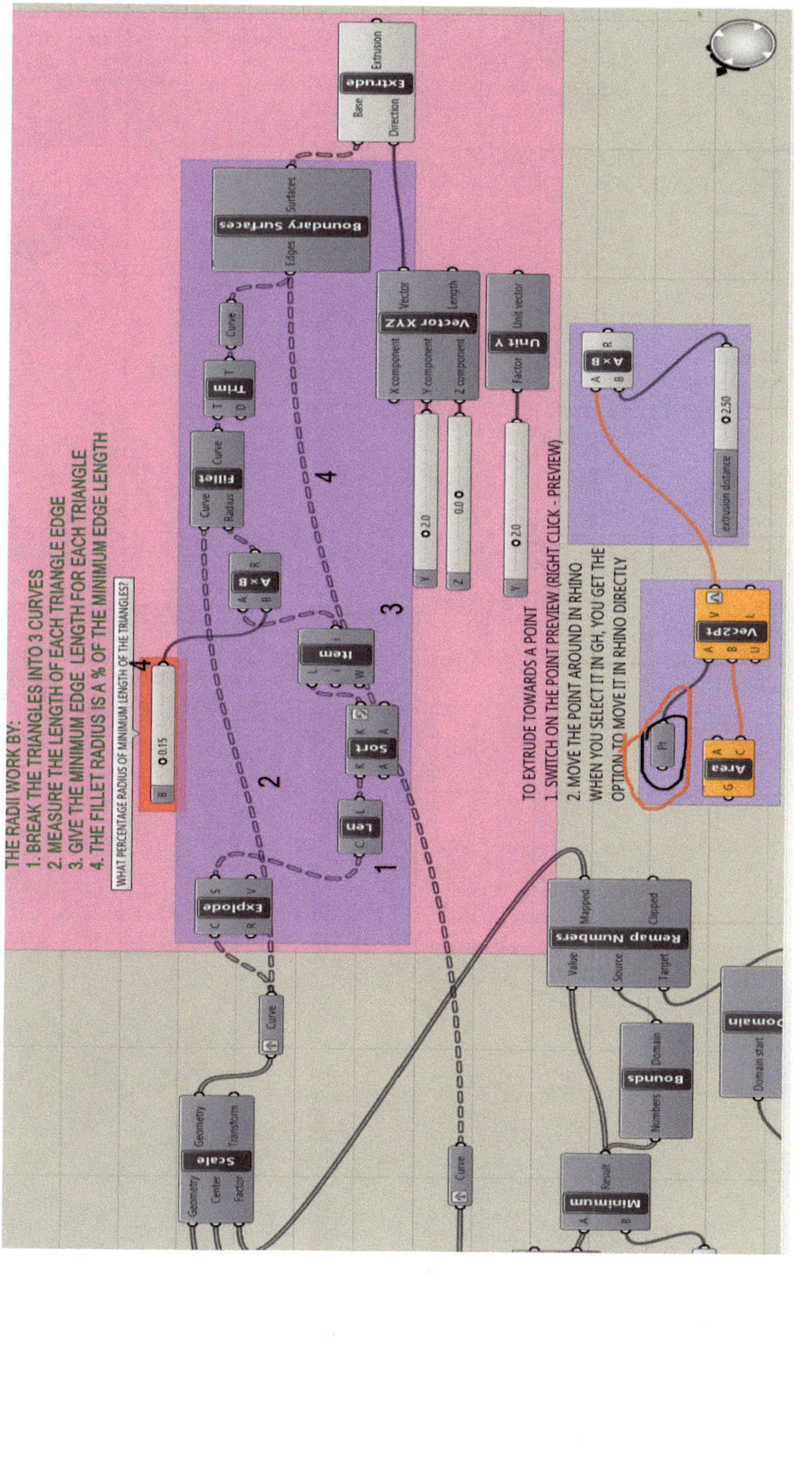

Figure G.38 continued .

a

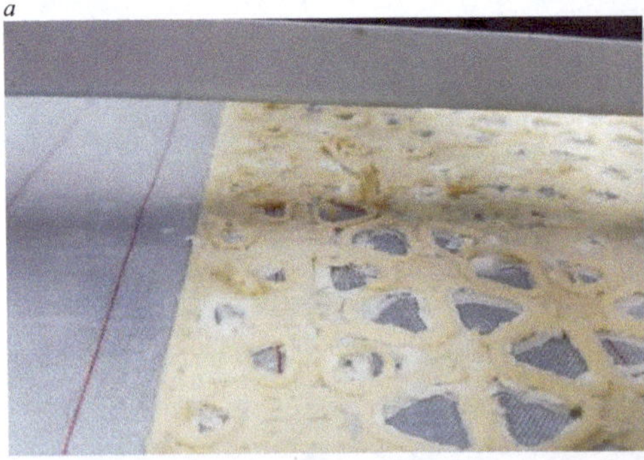

b

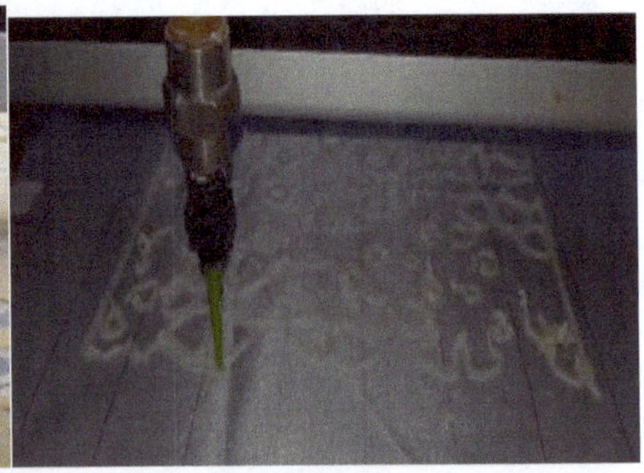

c

d

Figure G.39 The 3D printed paper model during printing (a), some defects by printing (b), (c), and final outcome due to warping (d).

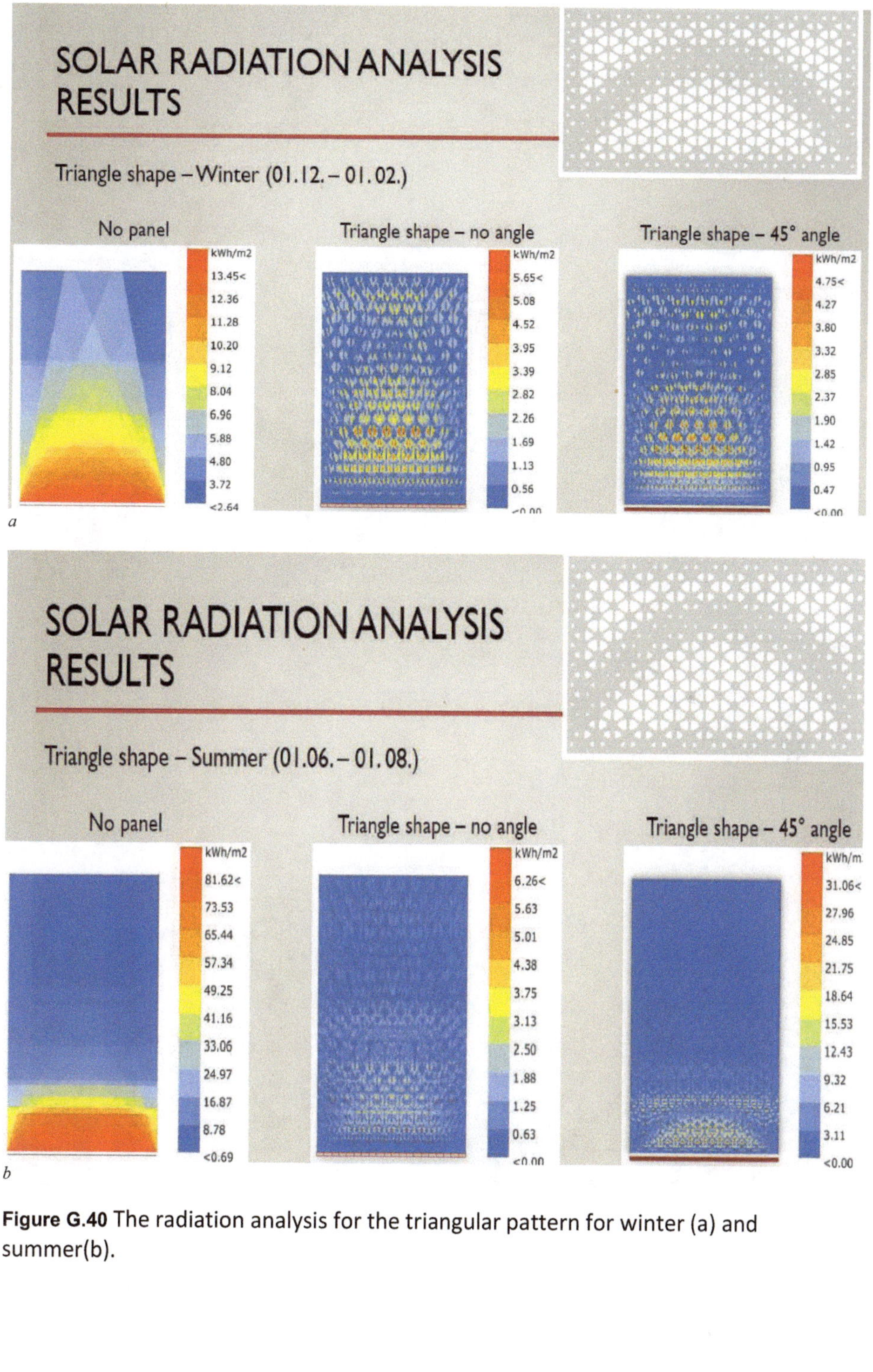

Figure G.40 The radiation analysis for the triangular pattern for winter (a) and summer(b).

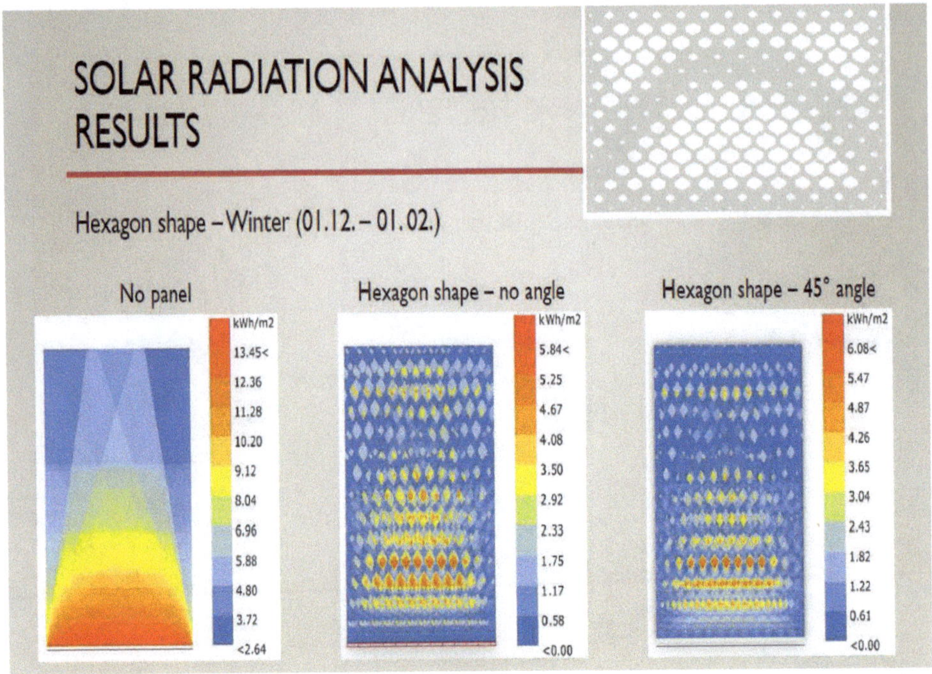

a

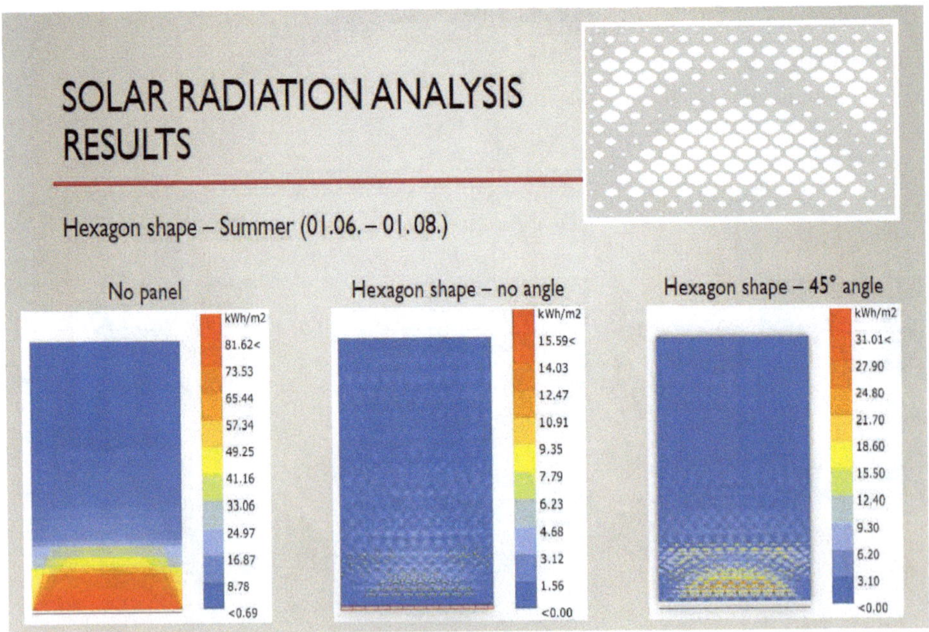

b

Figure G.41 The radiation analysis for hexagon pattern for winter (a) and summer(b).

G2.6 Floral leaf Buffel

This wonderful design and simulations were done by Joshua Schaefer and Akylzhan Kassymova.

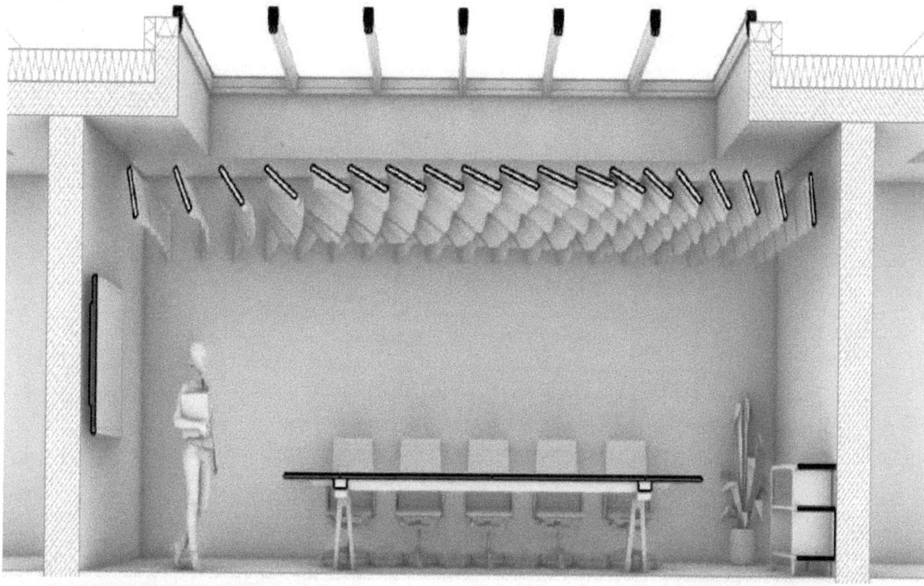

a

b

Figure G.42 The Design and its visualization side-view (a) and front-view (b).

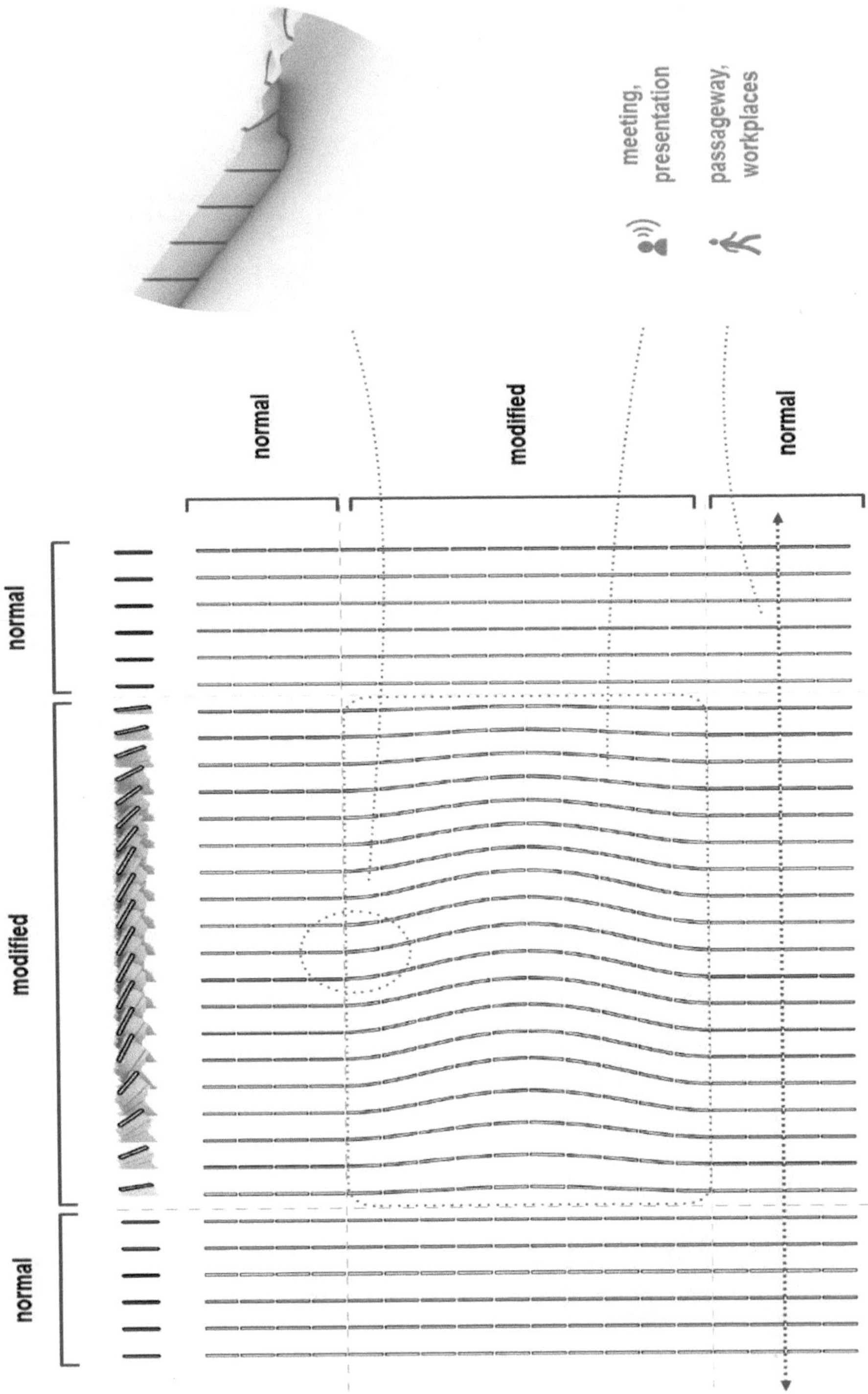

Figure G.43 continued.

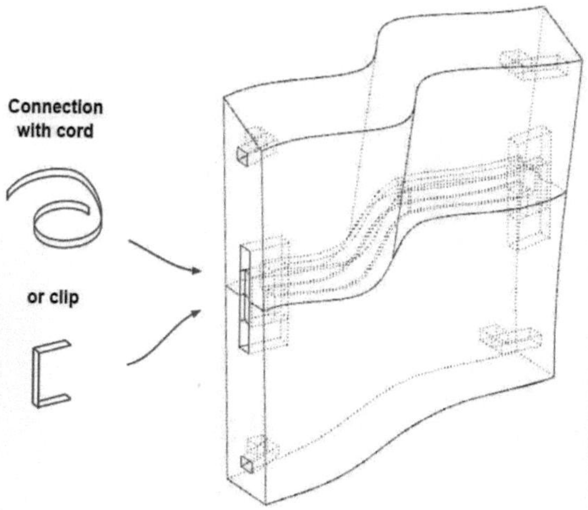

a

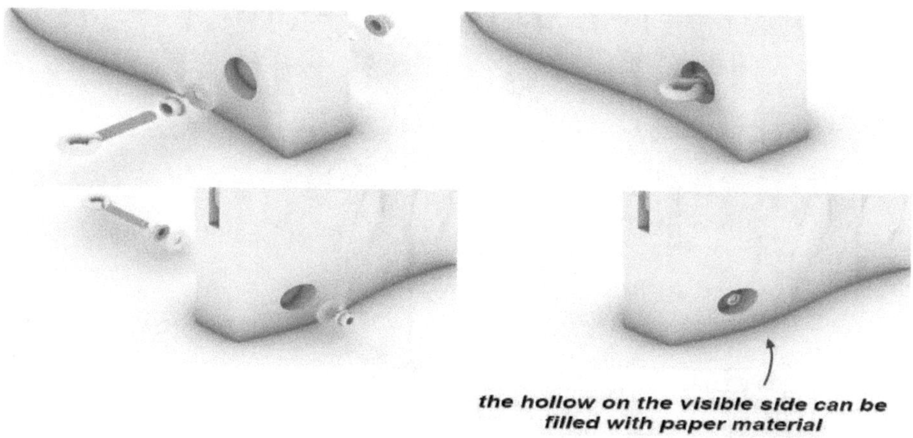

b

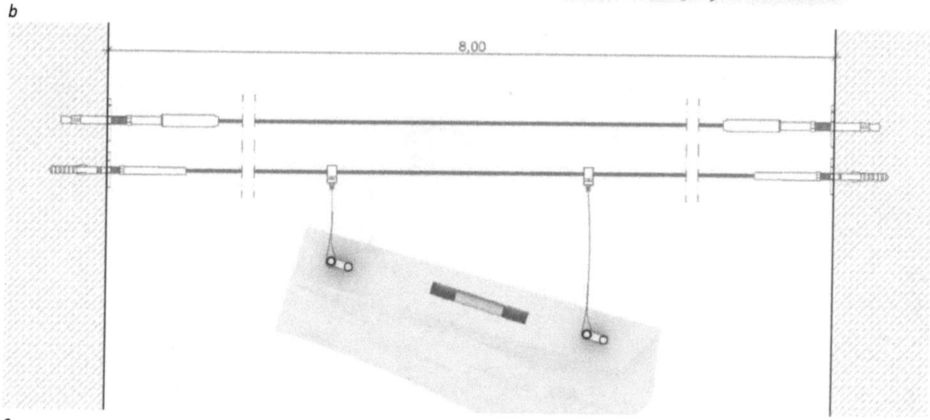

c

Figure G.44 Connection (a), installation (b) and suspension (c) details.

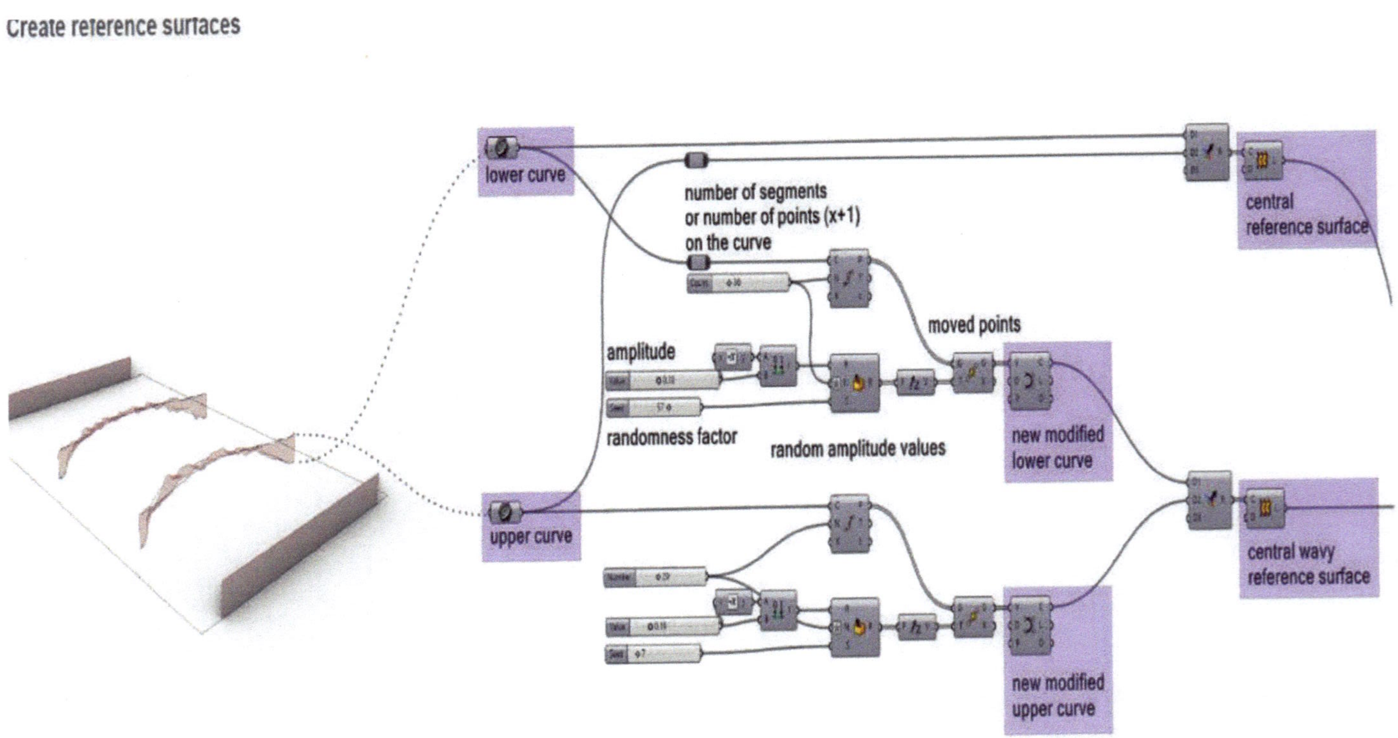

Figure G.45 Parametric design for the geometry.

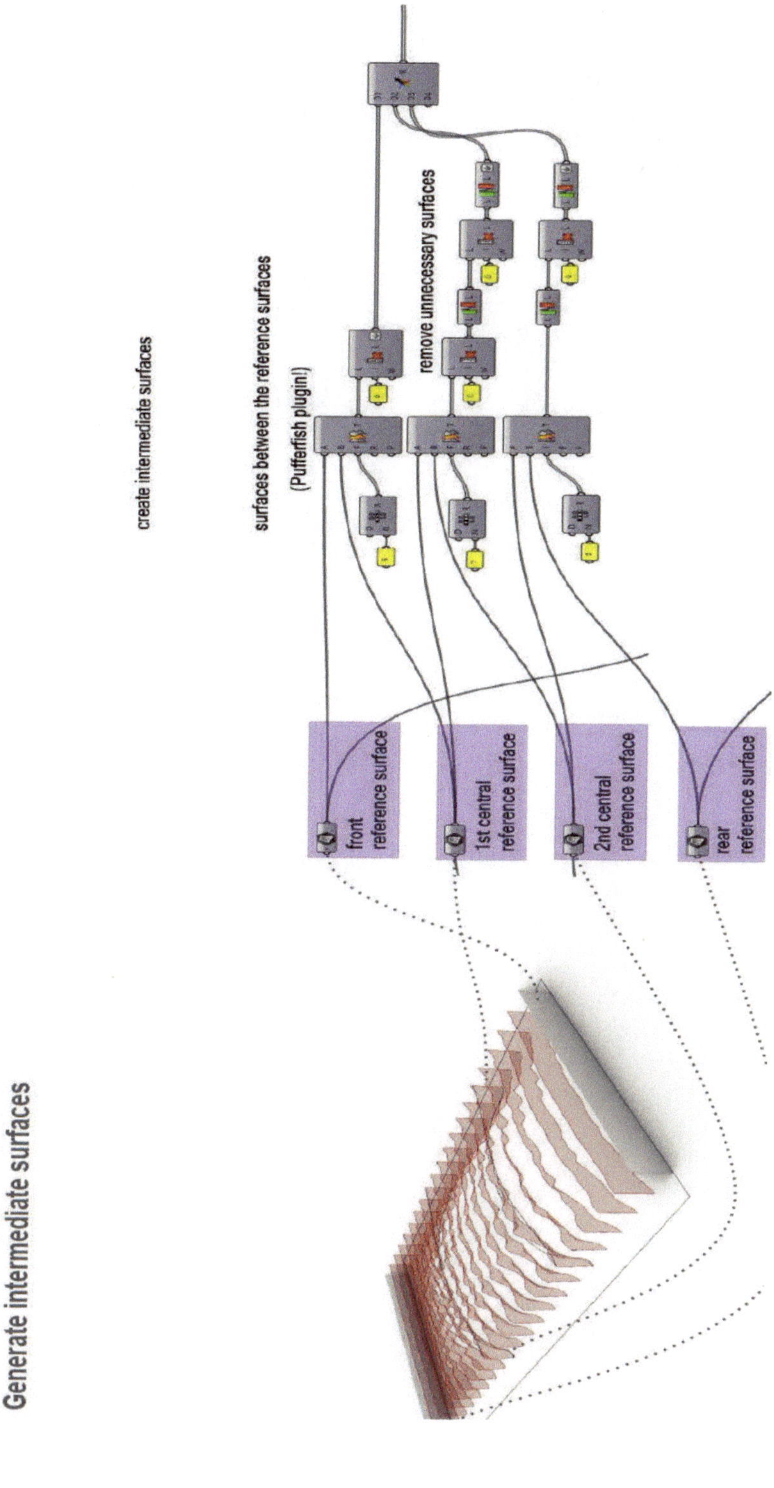

Figure G.45 continued.

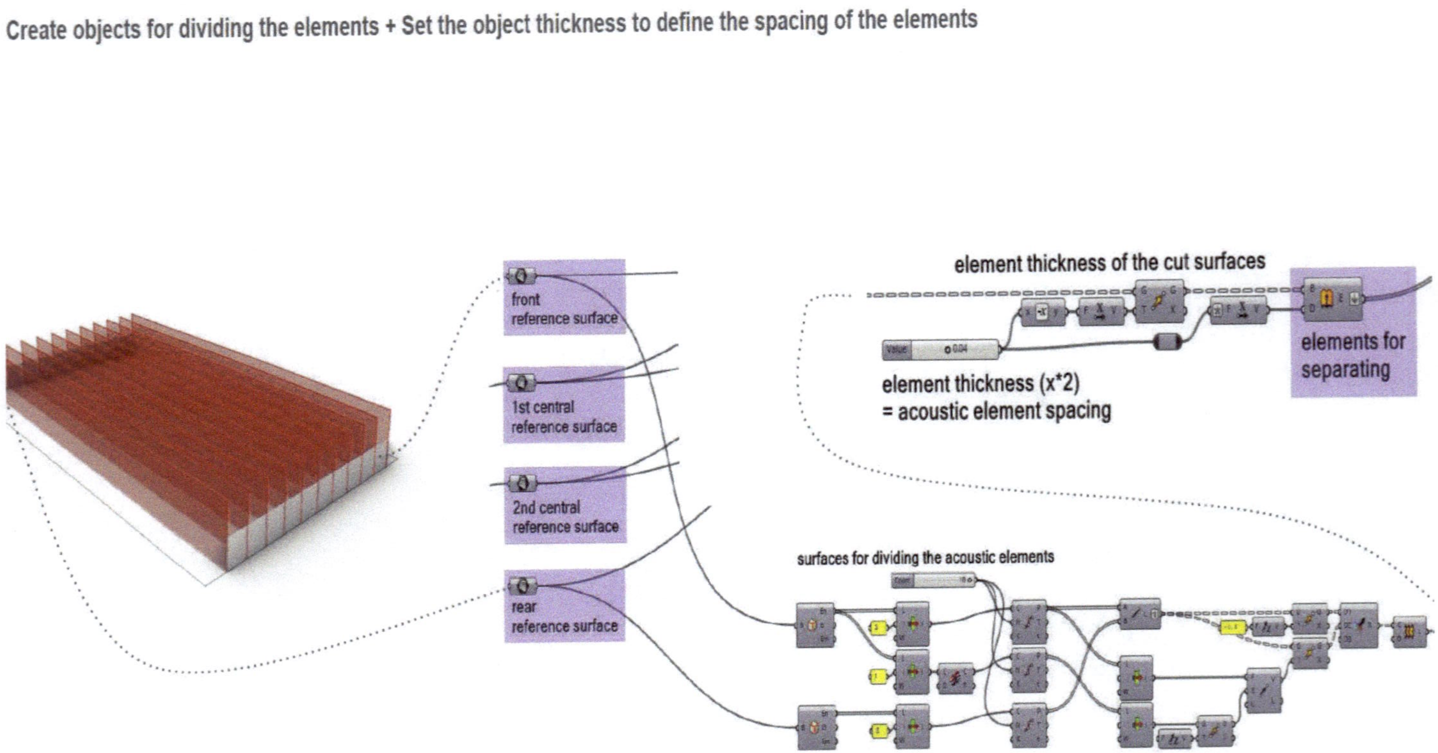

Figure G.45 continued.

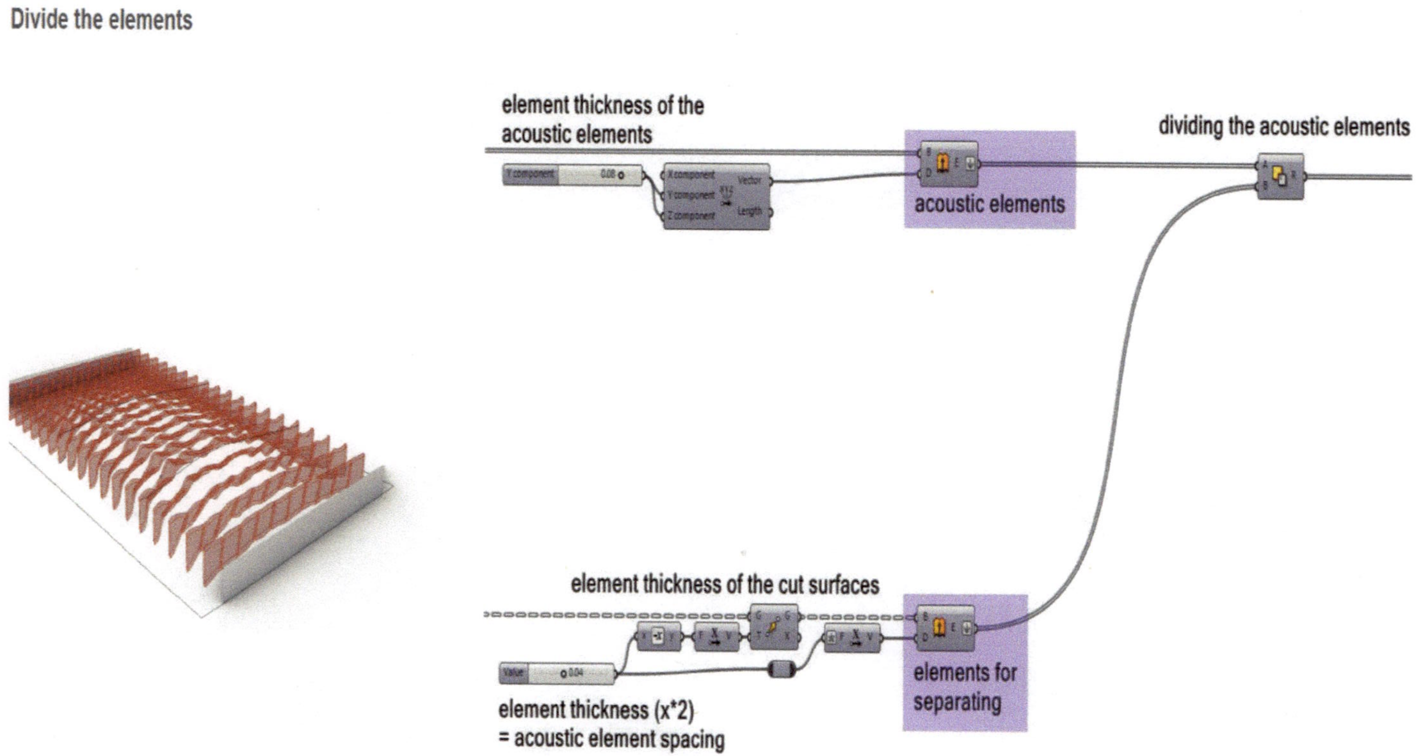

Figure G.45 continued.

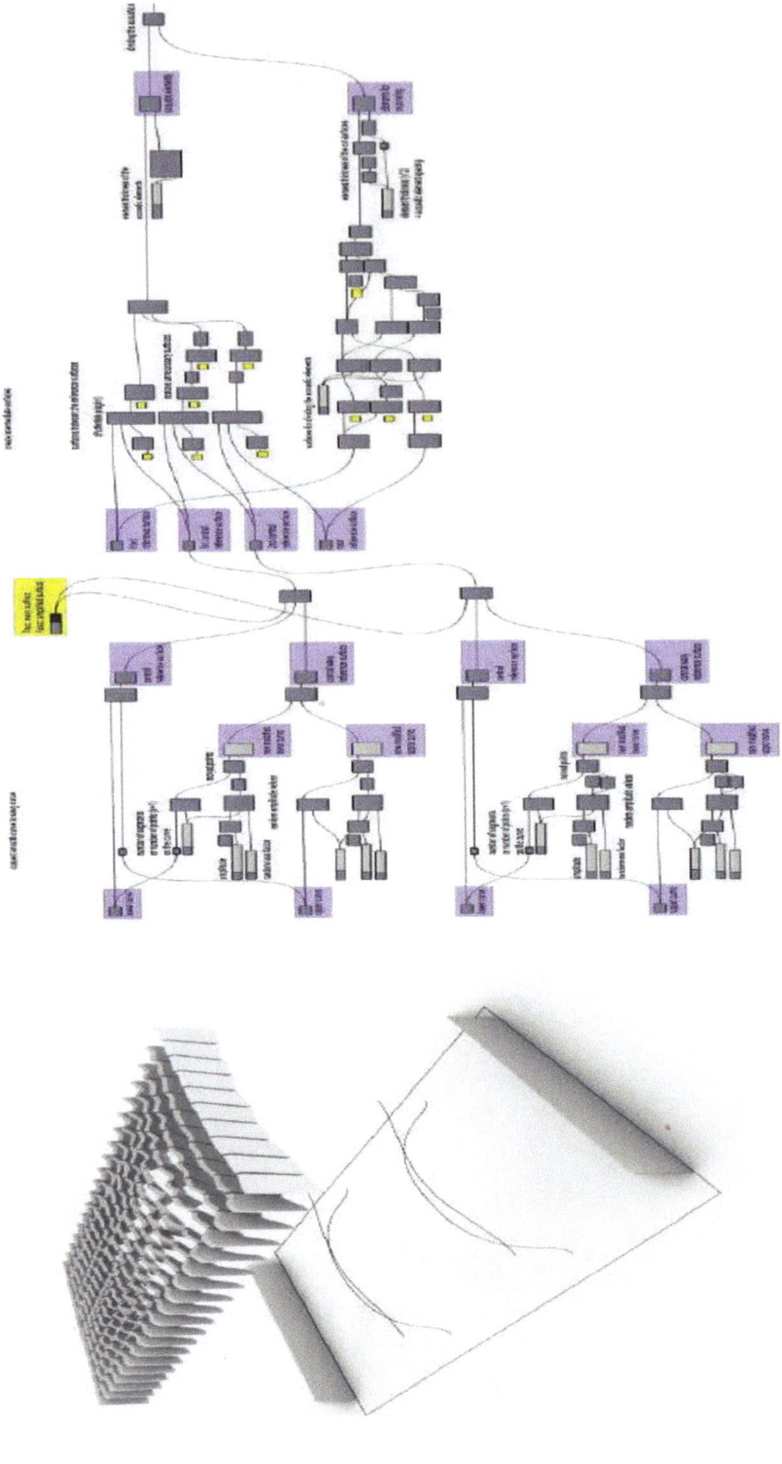

Figure G.45 continued.

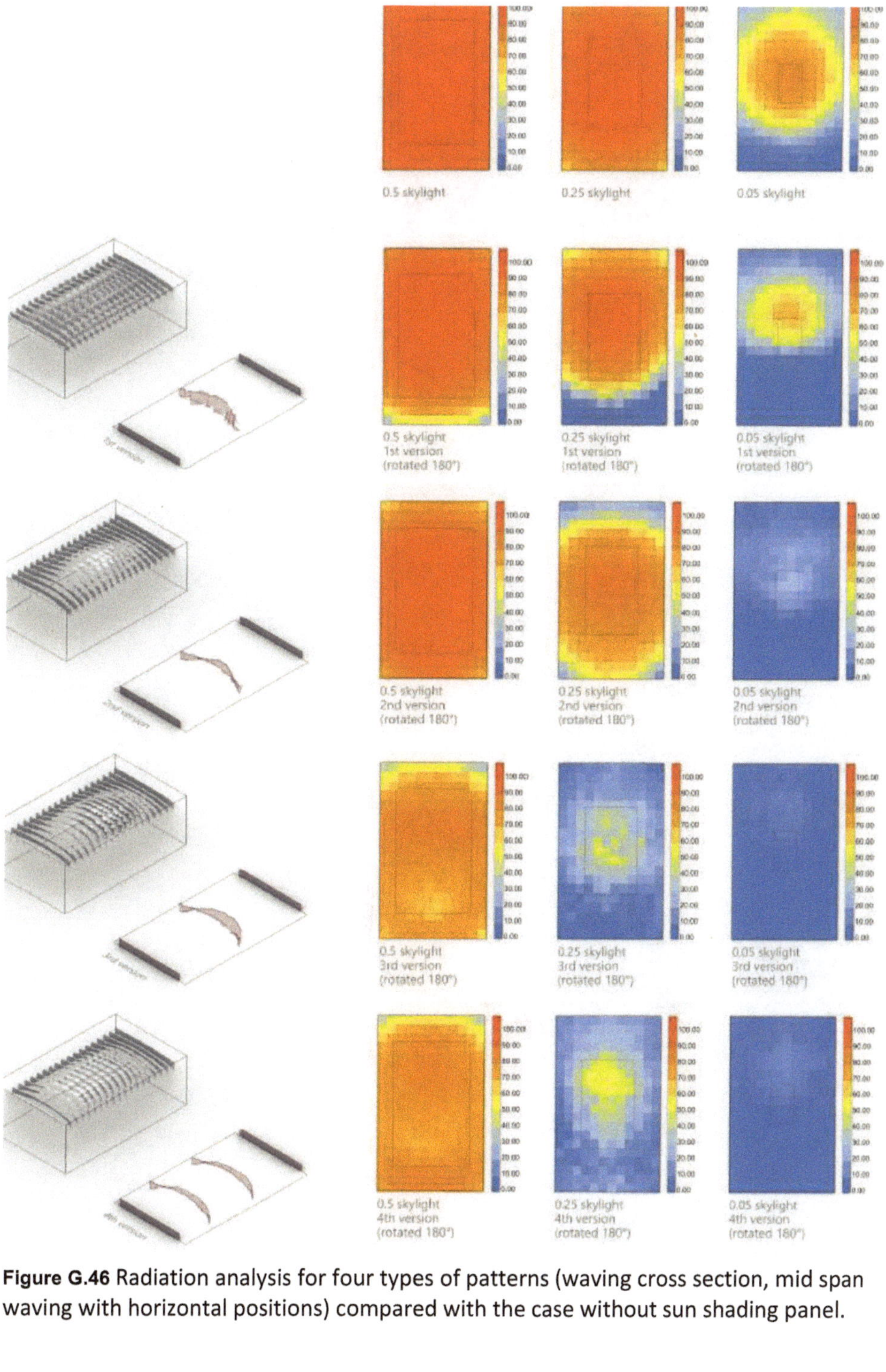

Figure G.46 Radiation analysis for four types of patterns (waving cross section, mid span waving with horizontal positions) compared with the case without sun shading panel.

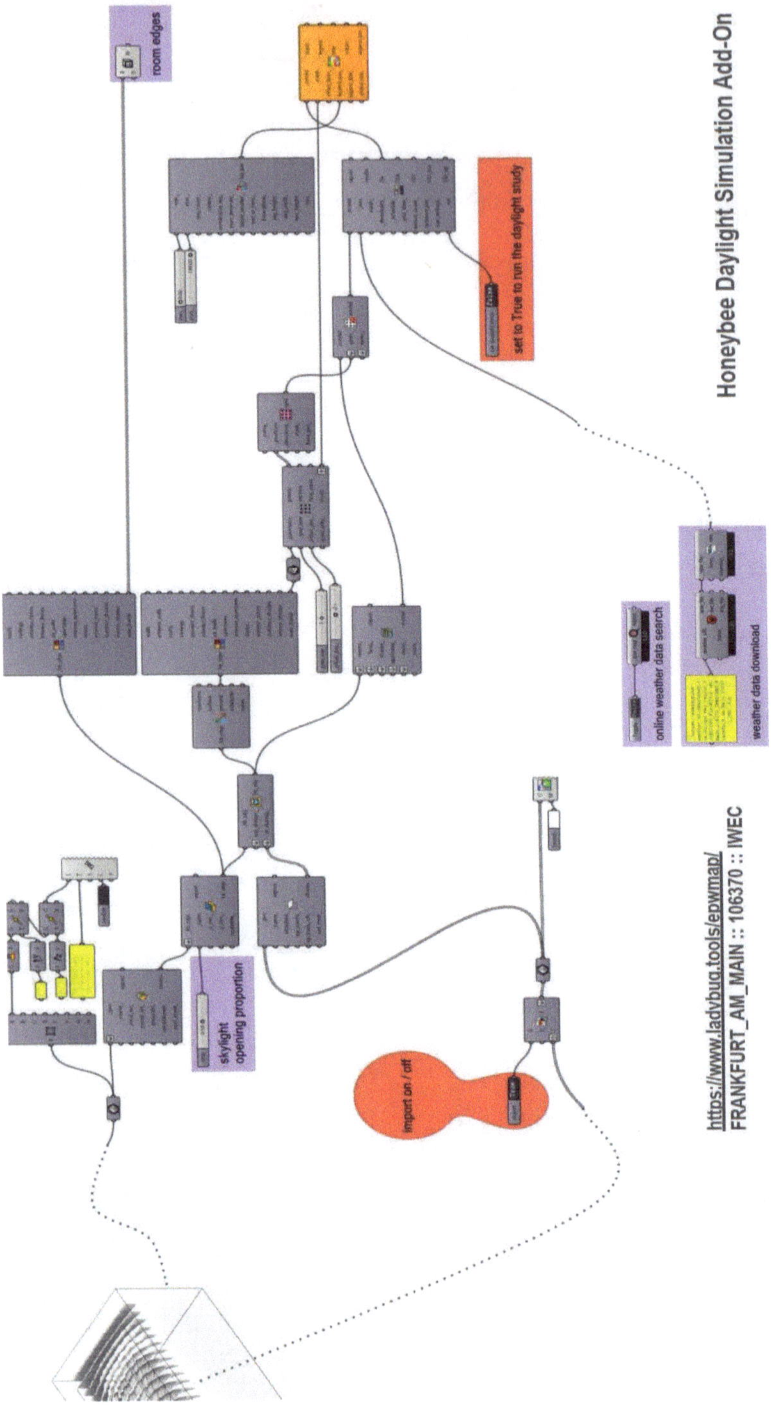

Figure G.47 The parametric design for the Radiation analysis.

Version 1
- hanging vertically

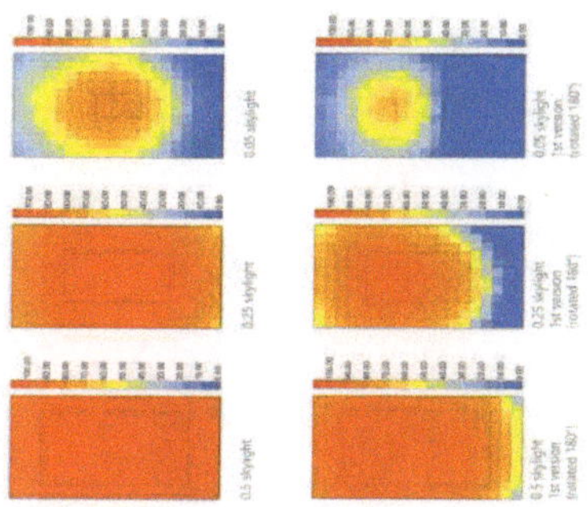

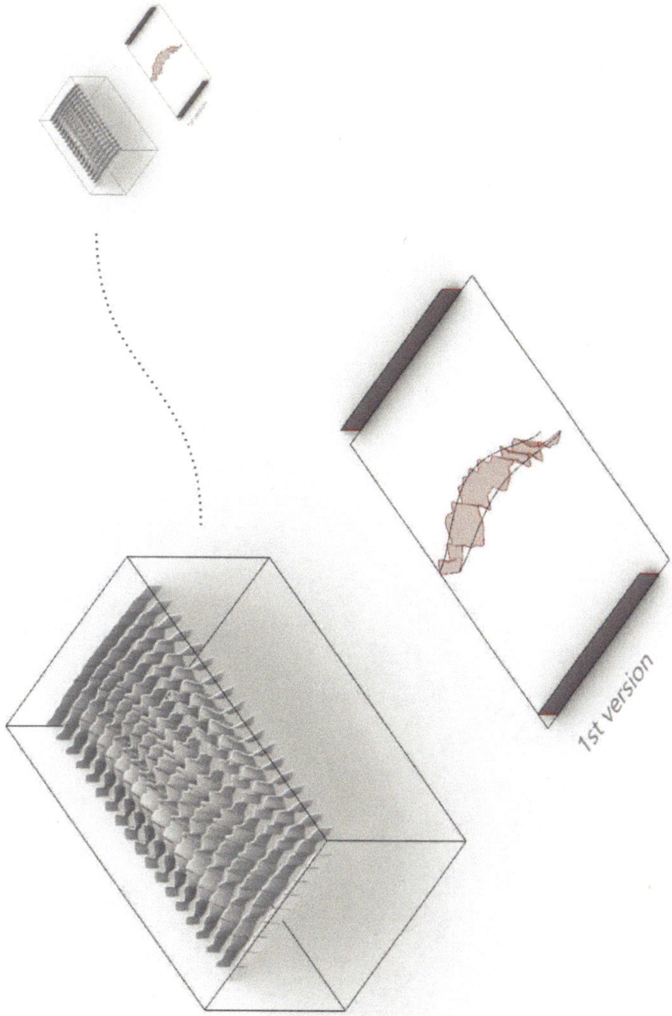

Figure G.47 continued.

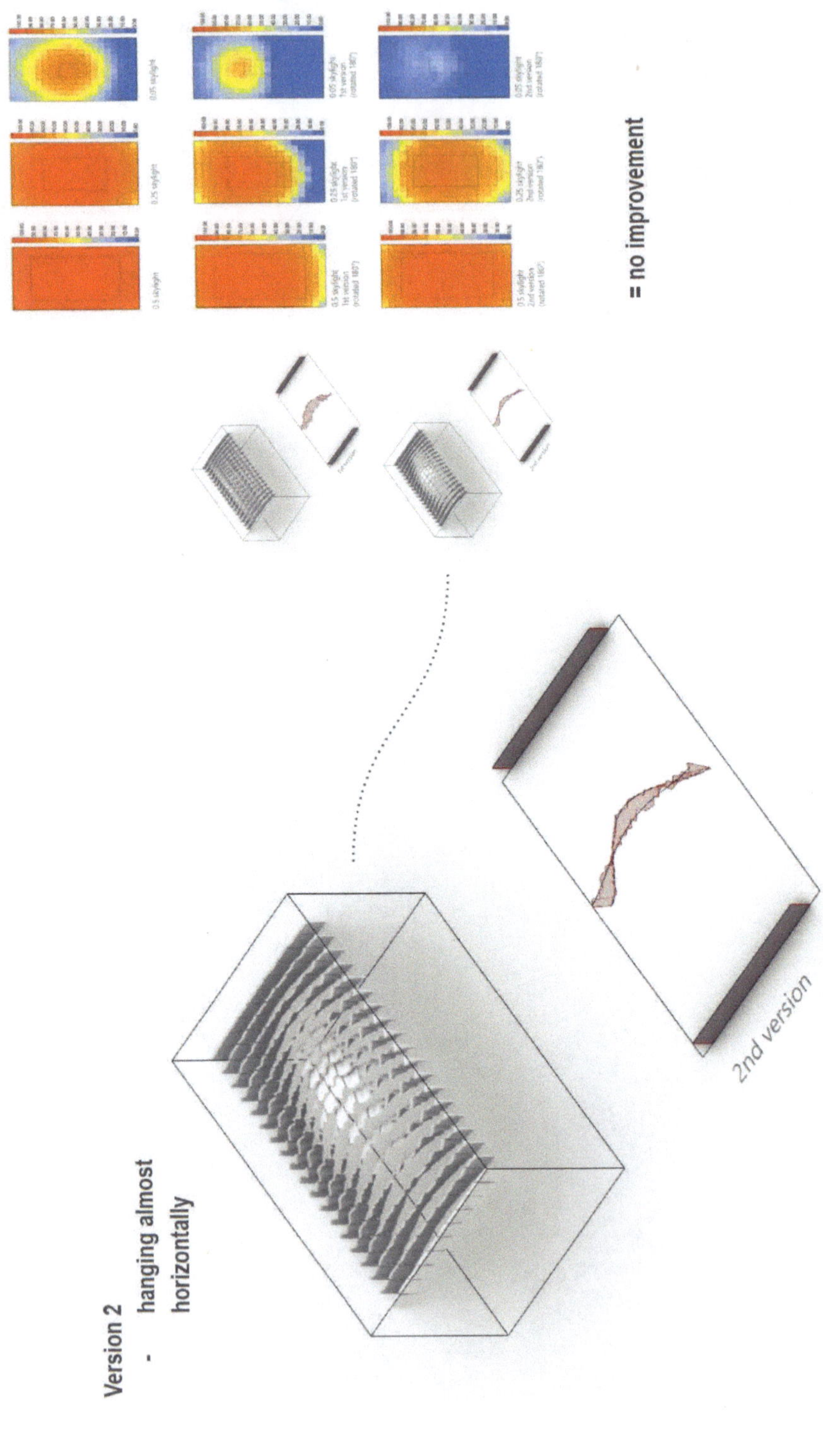

Figure G.47 continued.

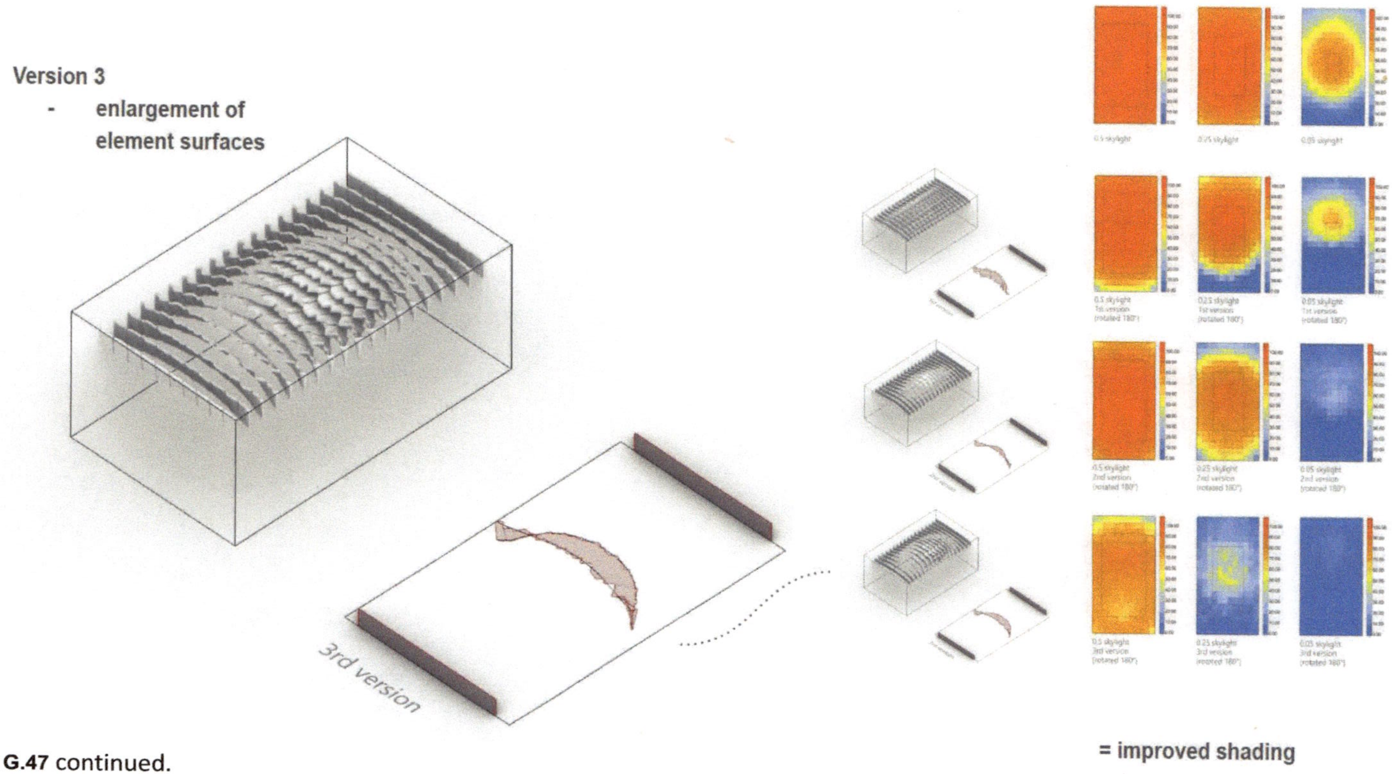

Figure G.47 continued.

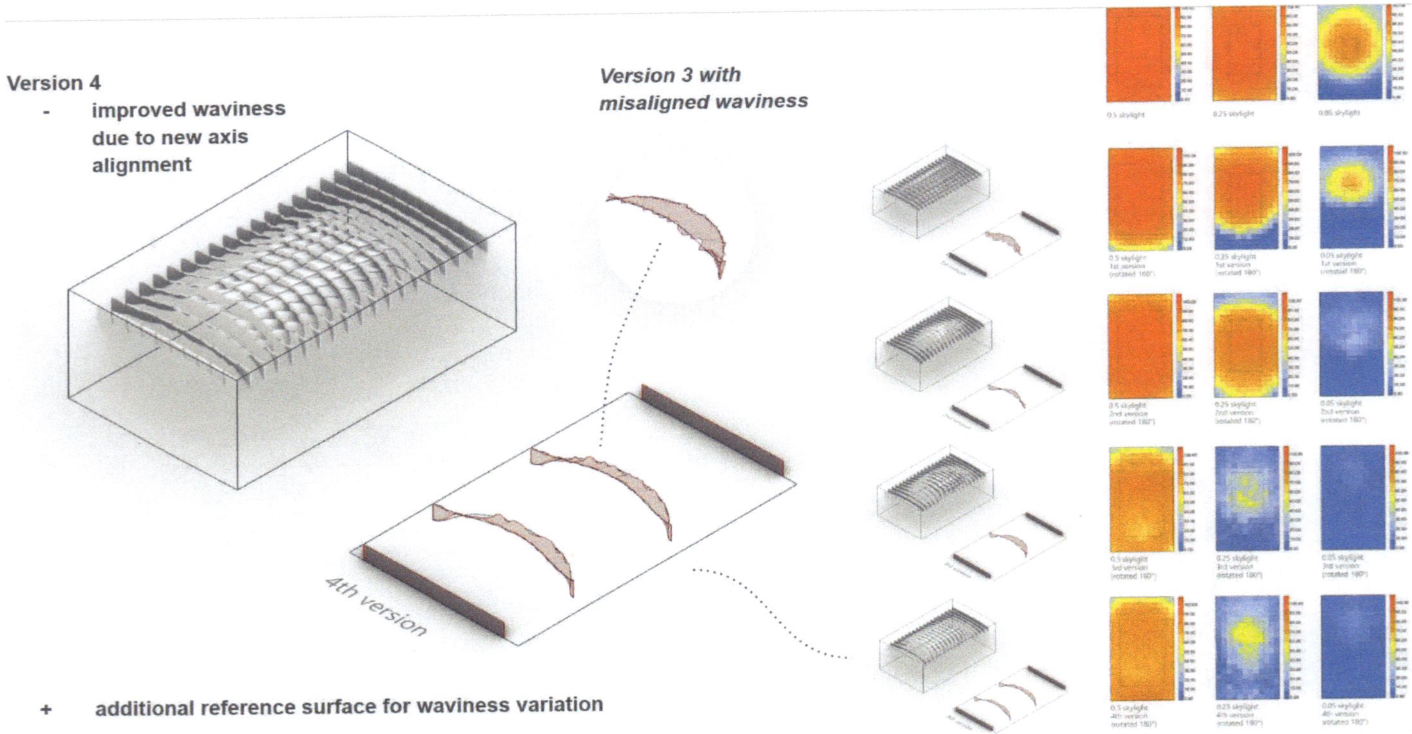

Figure G.47 continued.

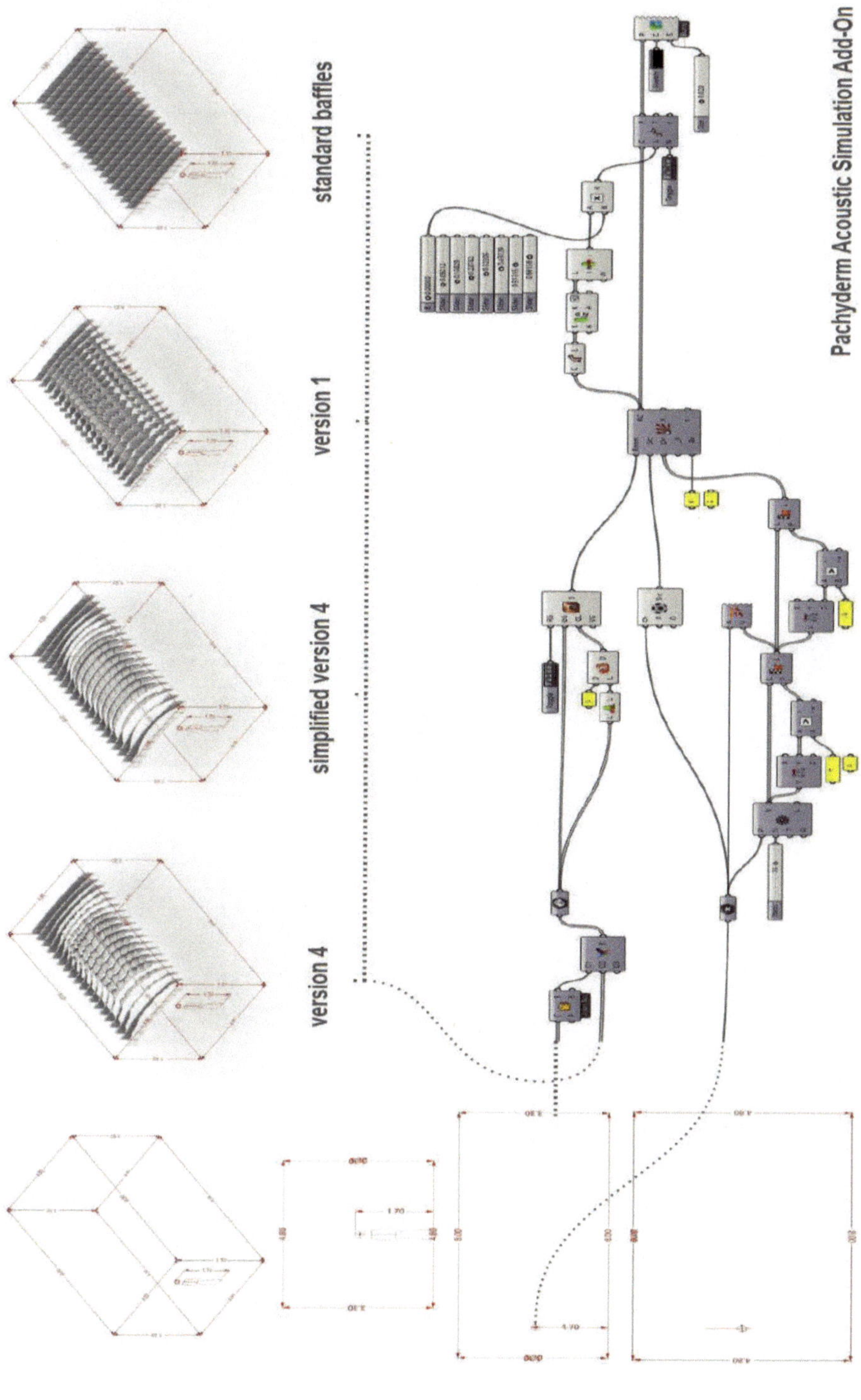

Figure G.48 The parametric design for the acoustic simulation.

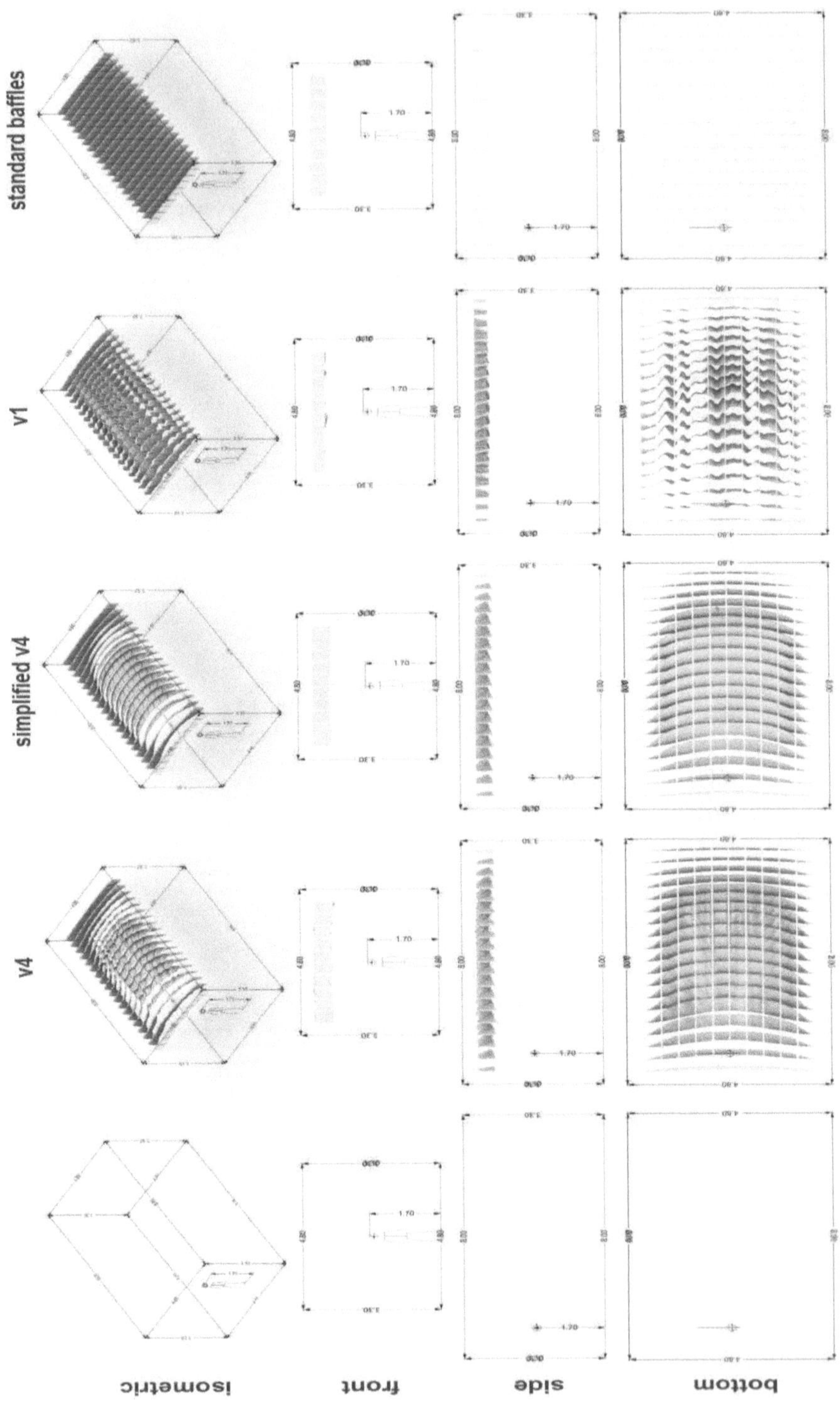

Figure G.48 Continued.

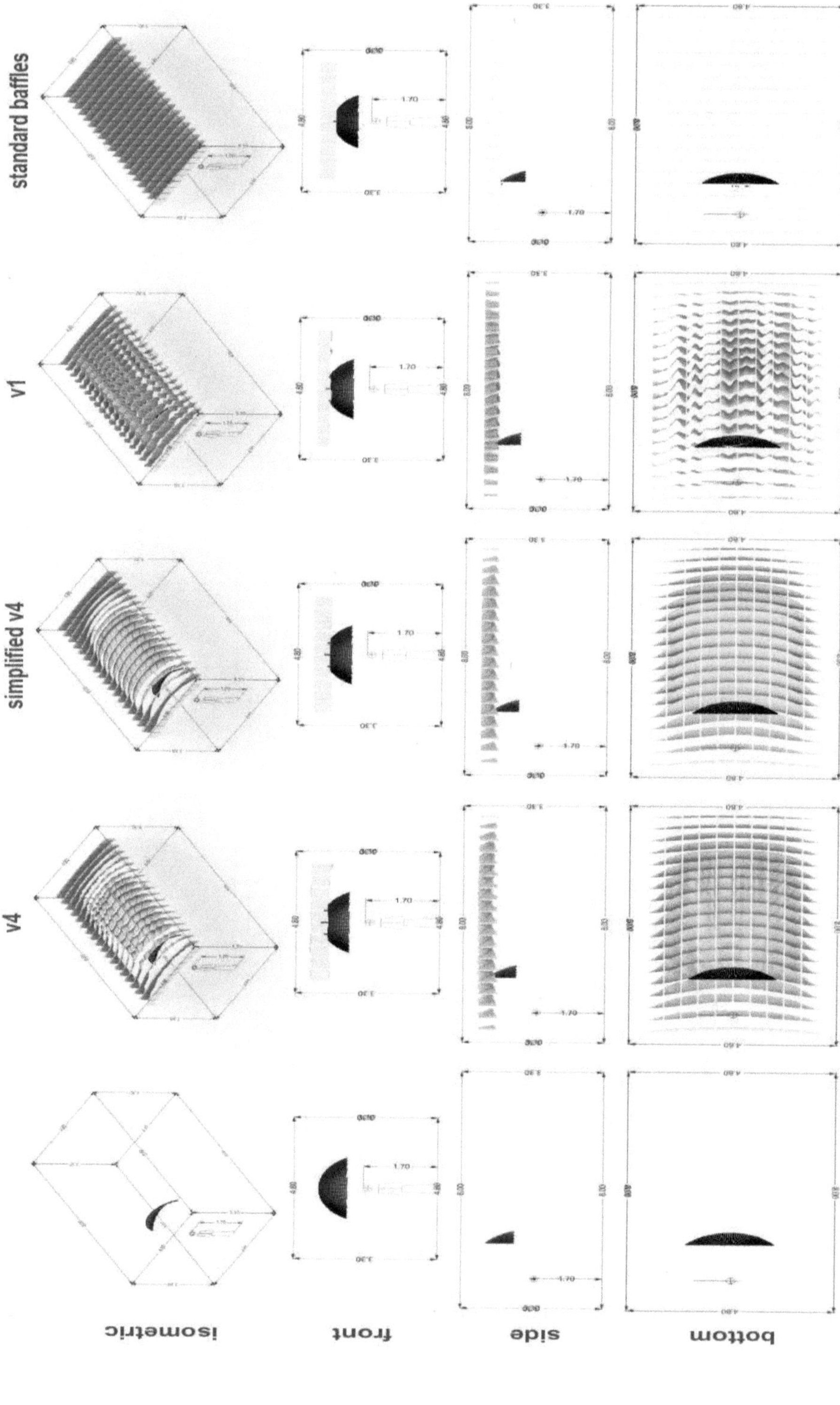

Figure G.48 Continued. The moment when the sound first hits the surface of the baffels.

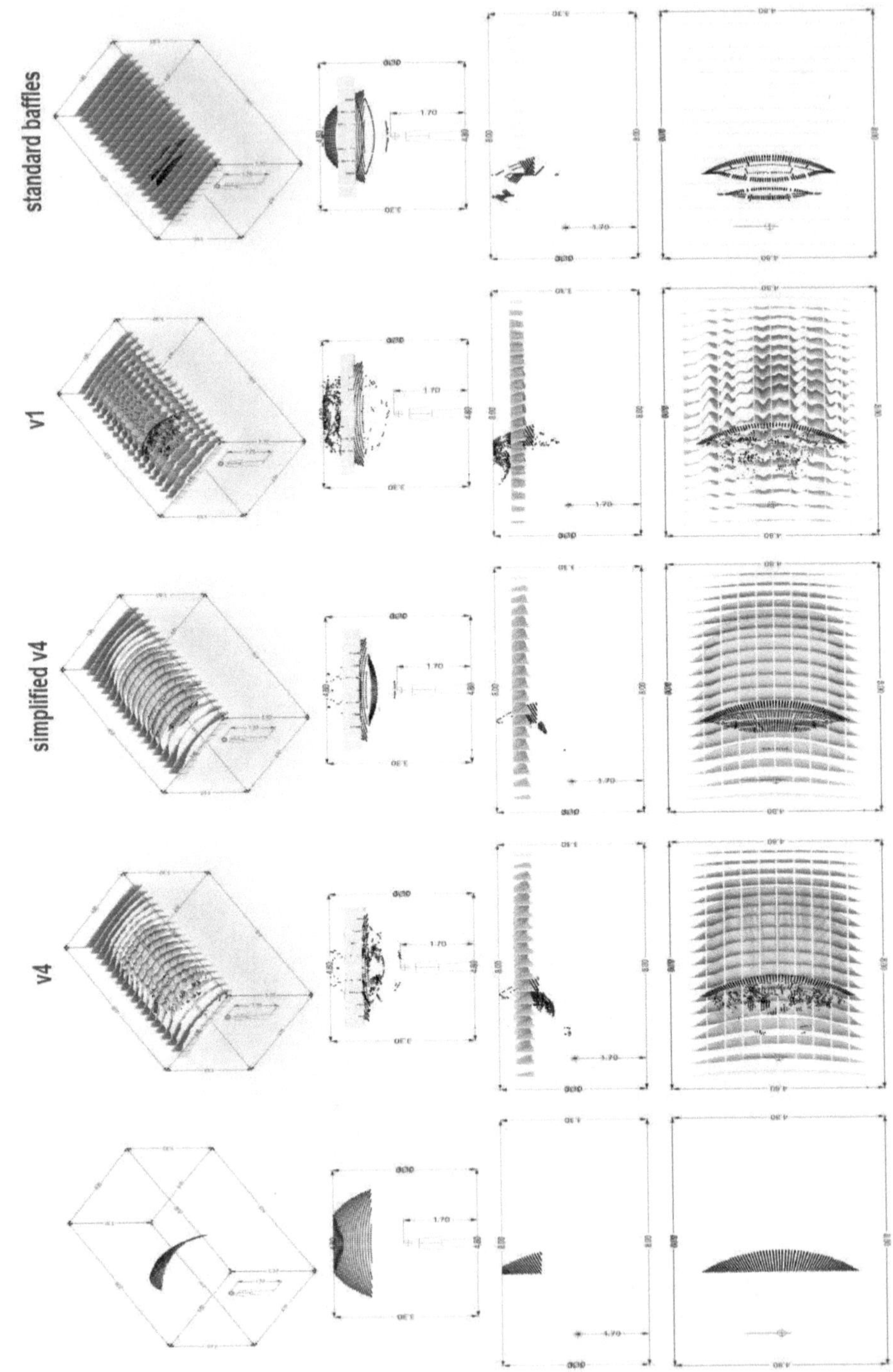

Figure G.48 Continued. The moment when the sound first hits the roof.

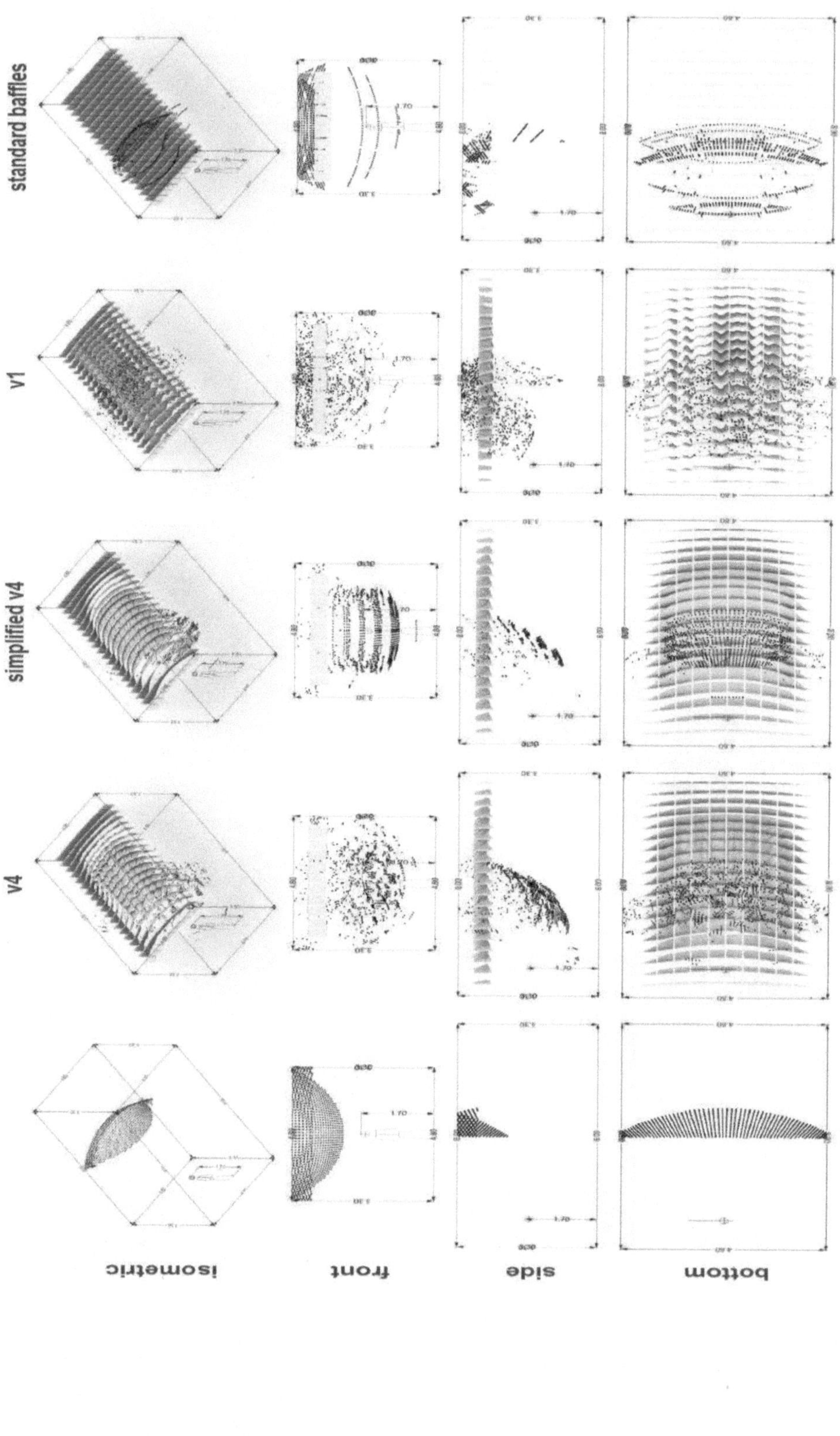

Figure G.48 Continued. The moment when the sound hits the roof and reflect.

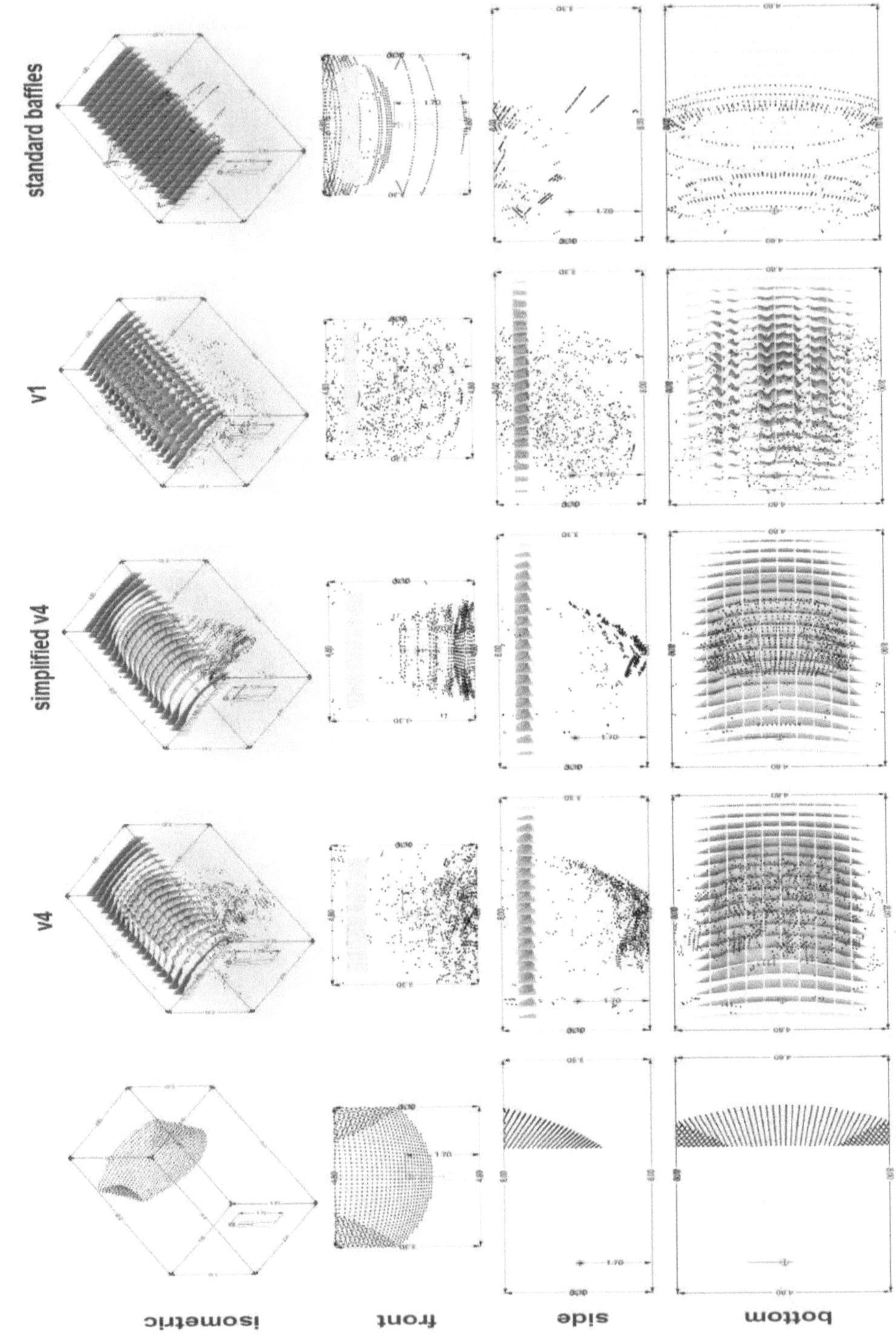

Figure G.48 Continued. The moment when the sound first hits the side walls.

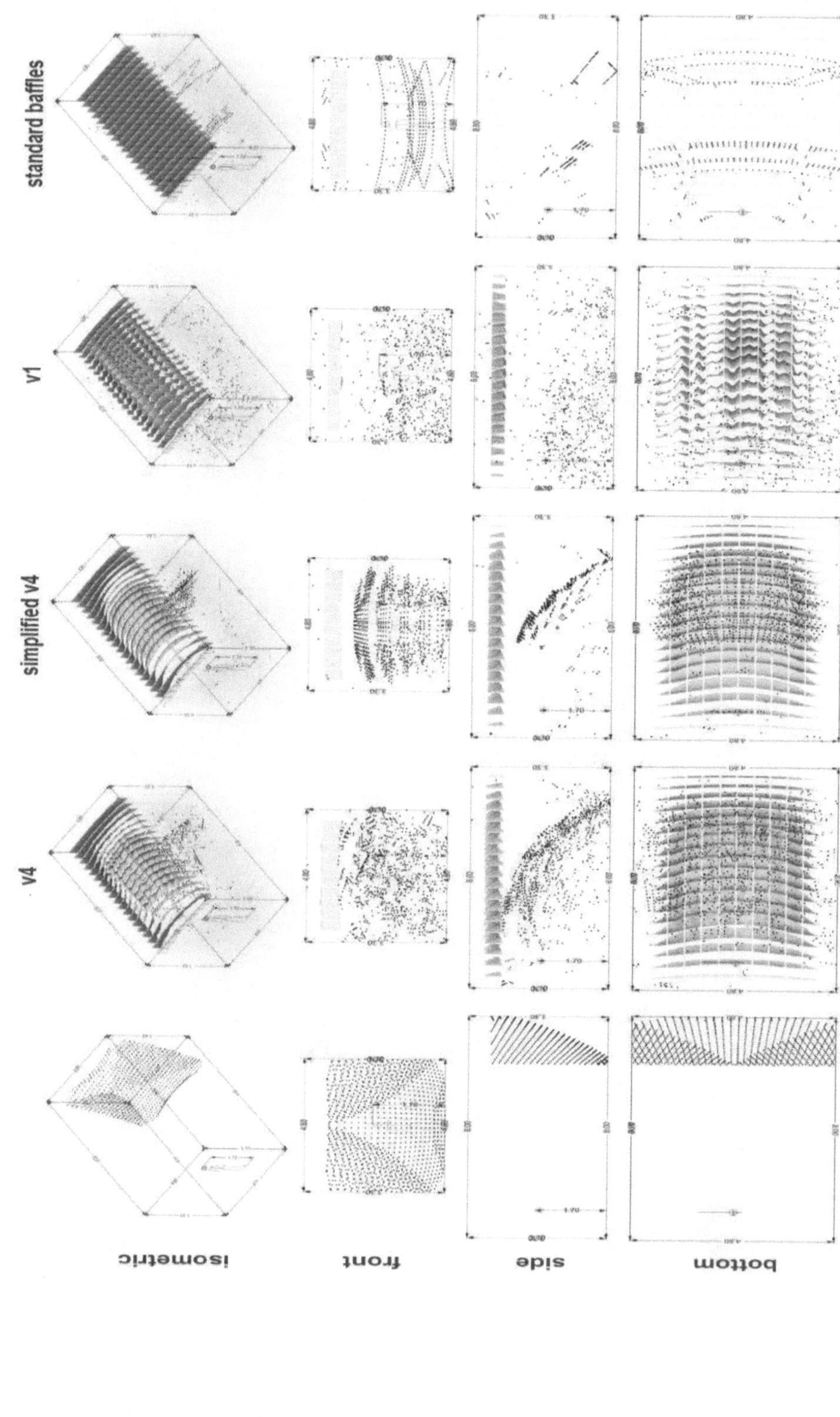

Figure G.48 Continued. The moment when the sound first hits the front wall.

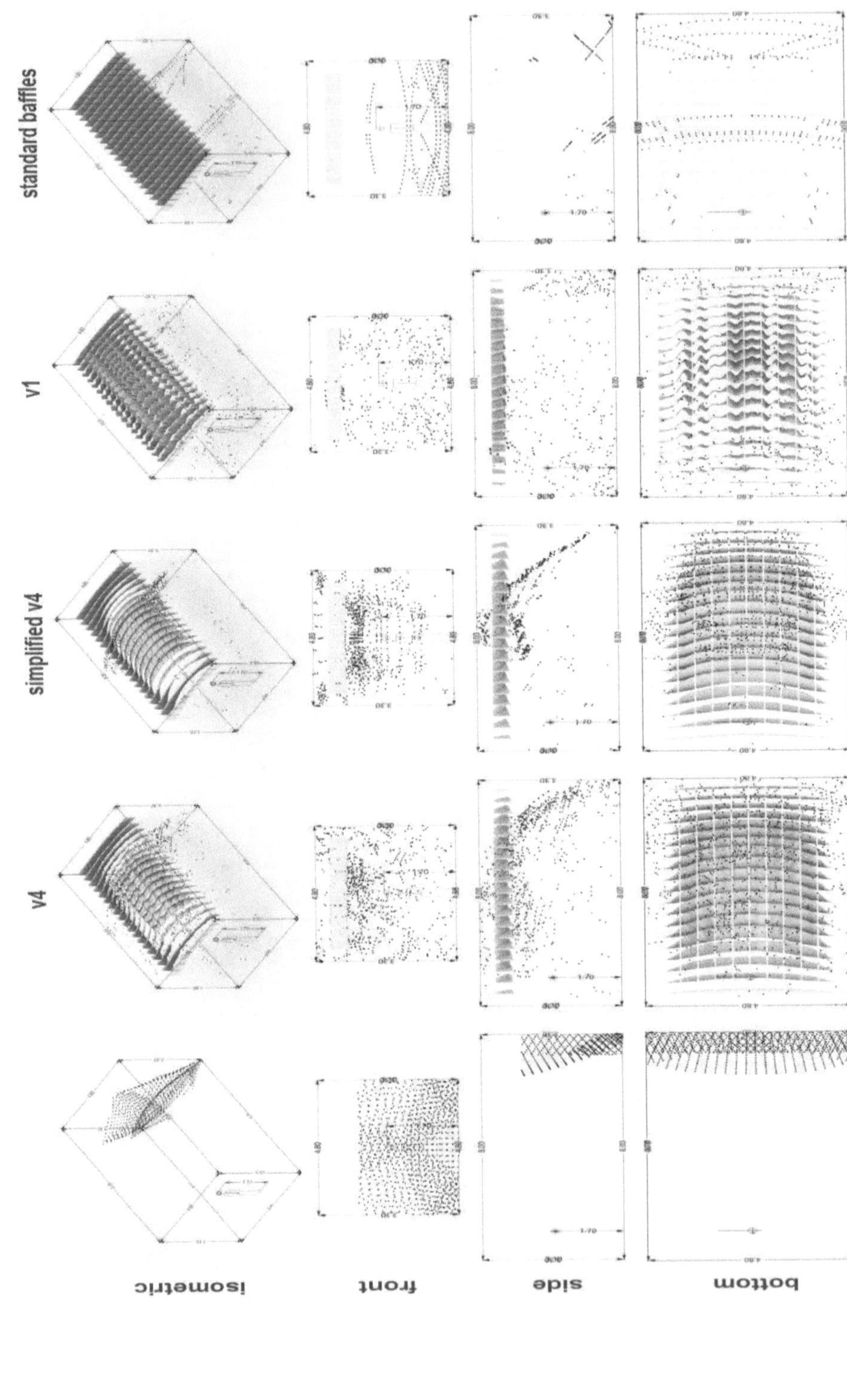

Figure G.48 Continued. The moment when the sound hits the front wall and reflect.

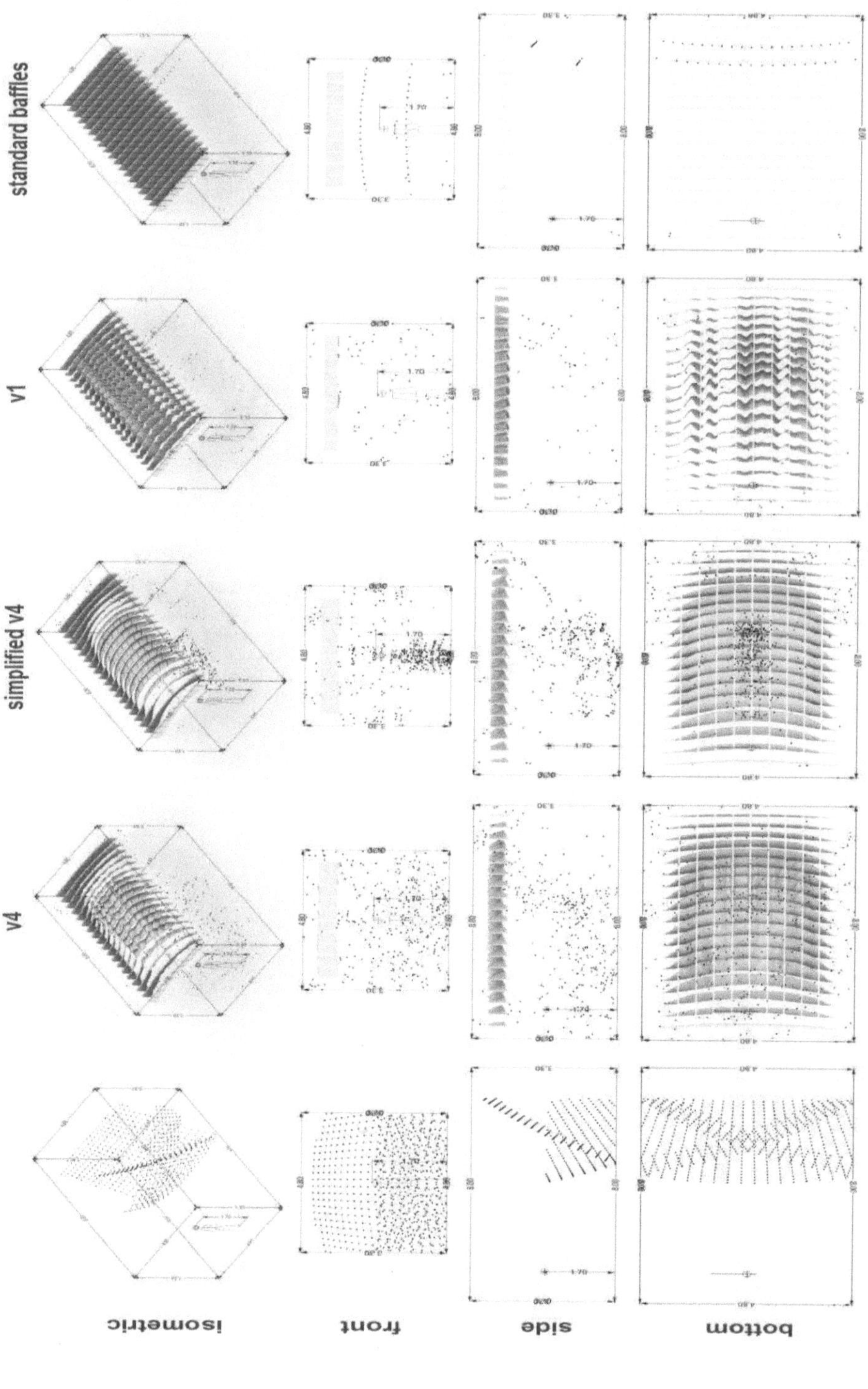

Figure G.48 Continued. The moment when the sound produce echo.

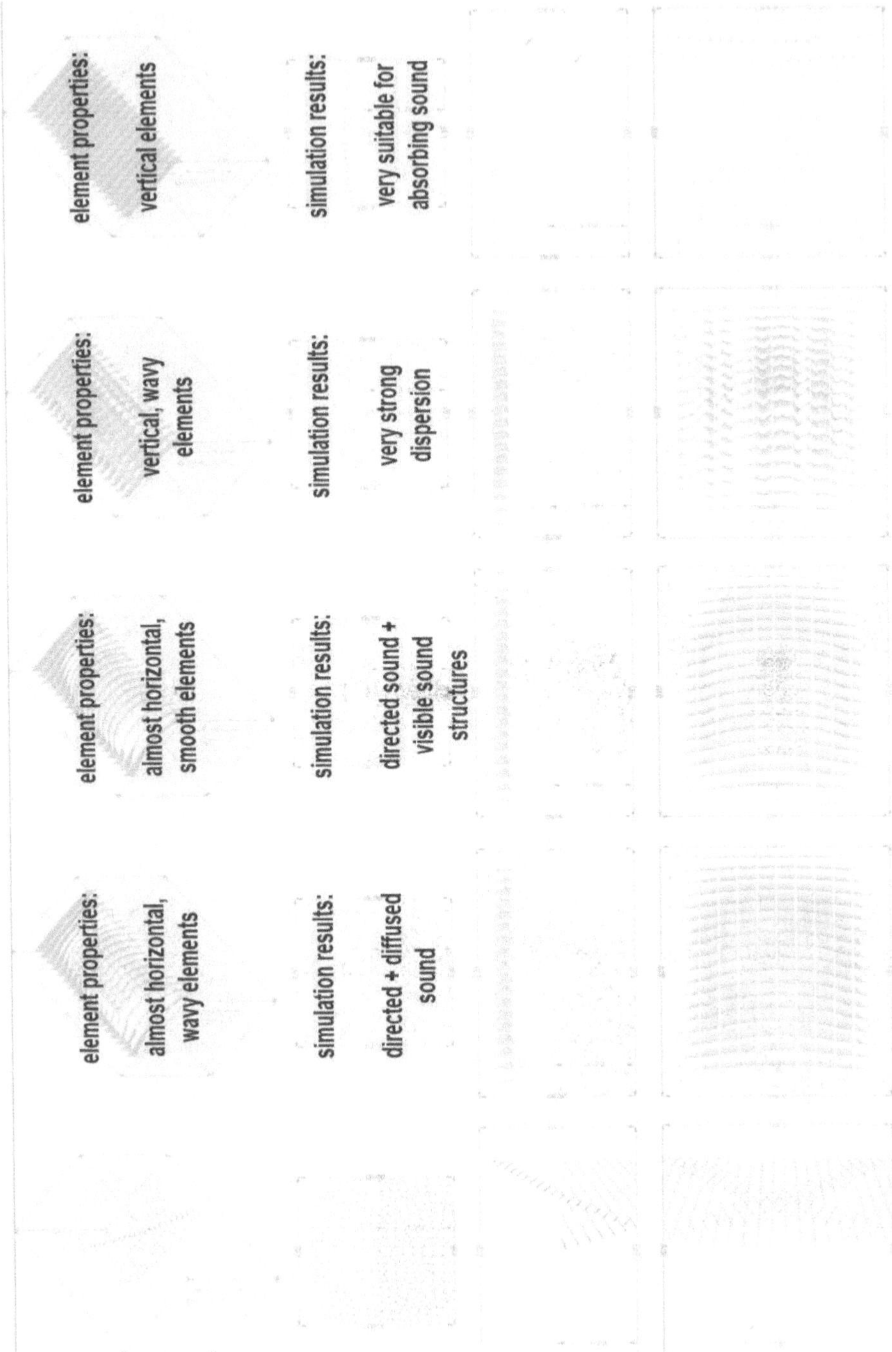

Figure G.48 Continued. An evaluation.

Results

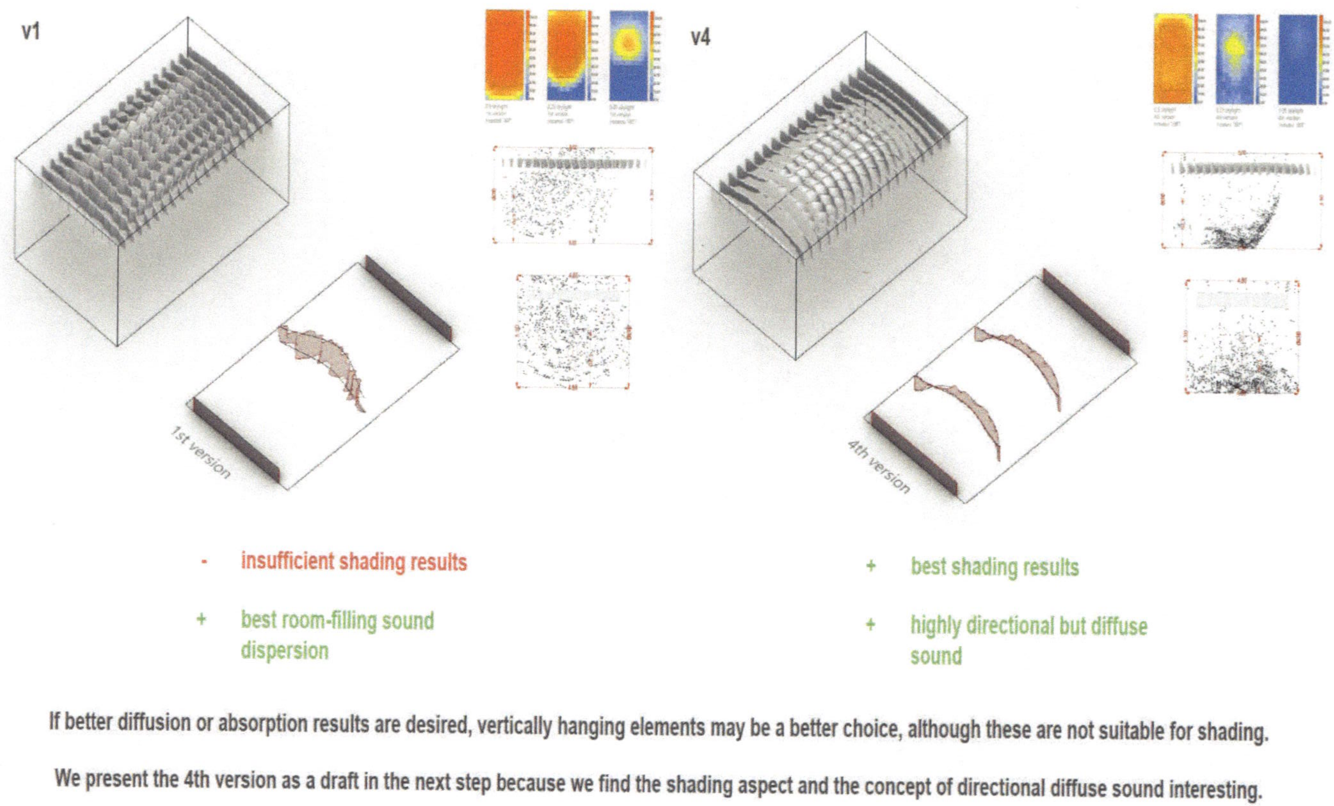

Figure G.48 Continued. Best Results.

Final Design

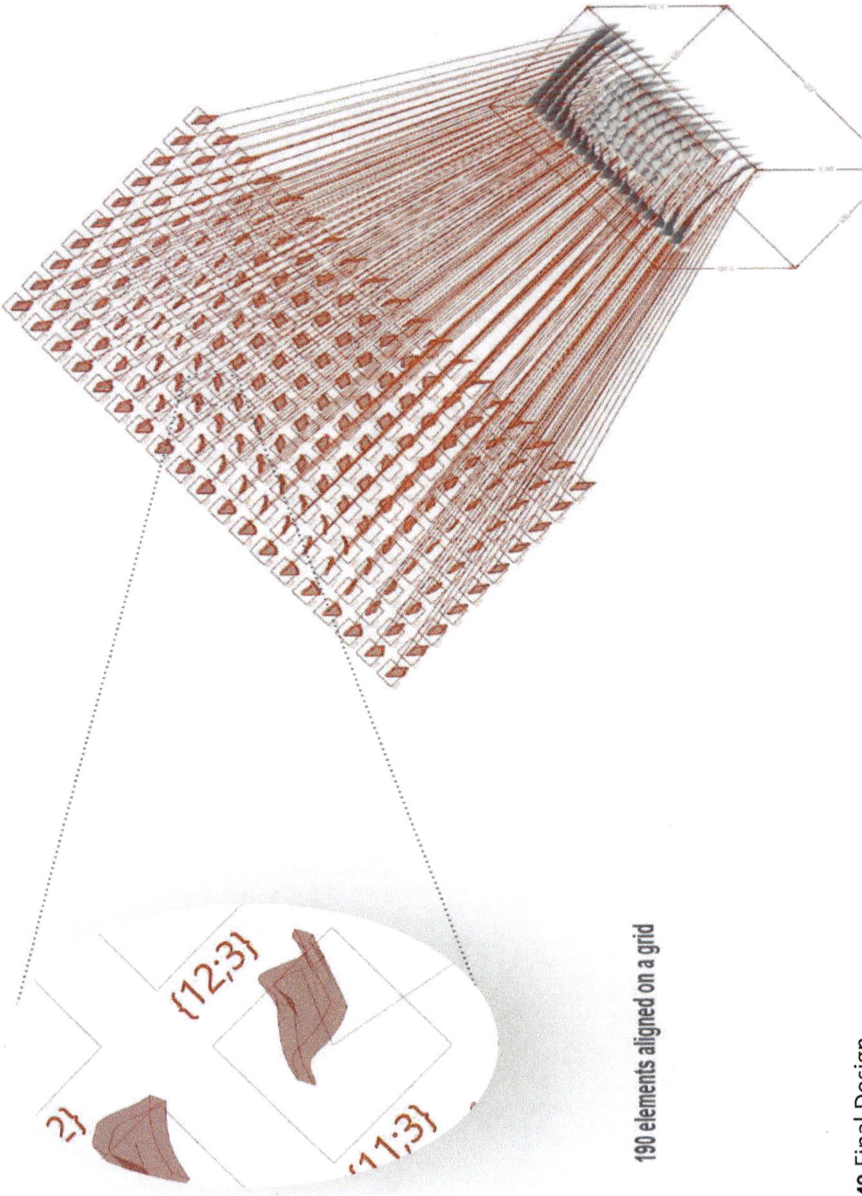

Figure G.49 Final Design.

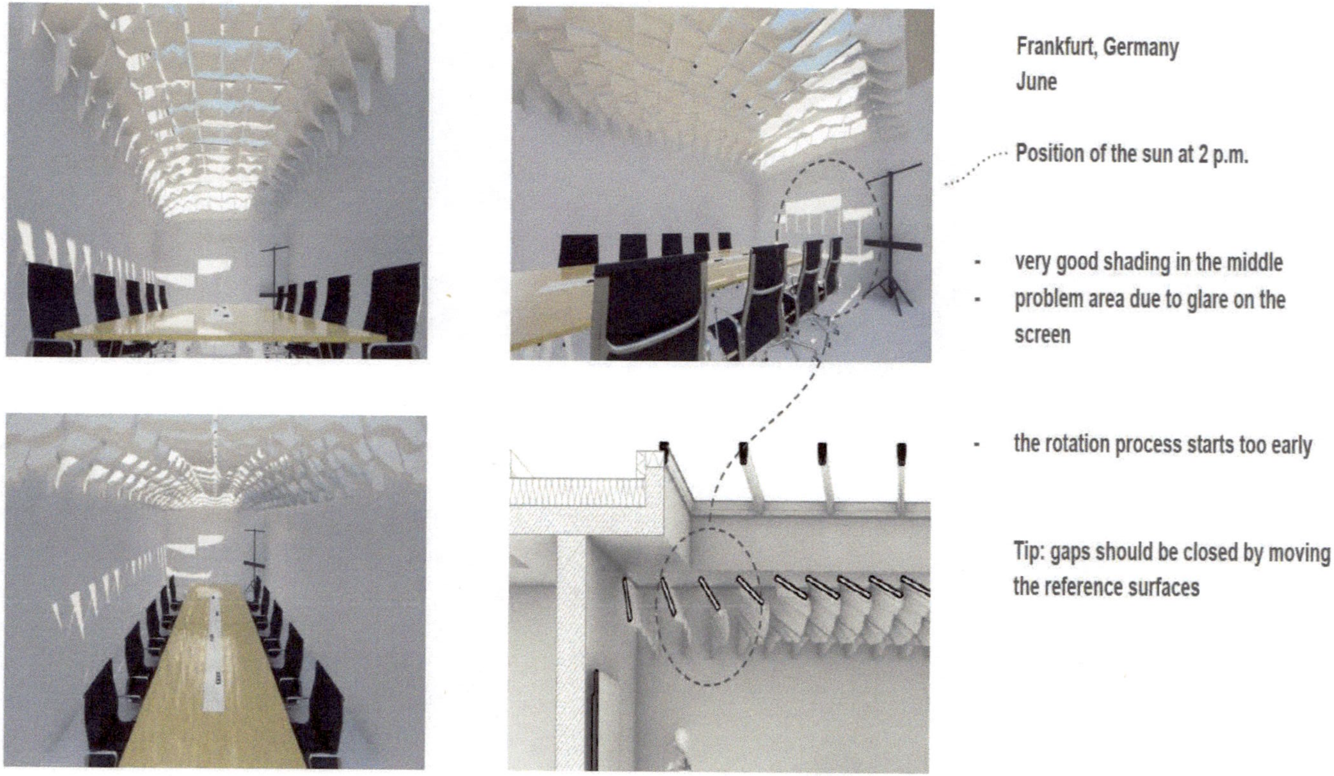

Figure G.50 Some design errors from the Frankfurt simulation time zone, June, sun position at 14:00

Kolibri 360 GmbH
Ferdinand-Jühlke-Straße 7
99095 Erfurt, Deutschland
produktsicherheit@kolibri360.de